Protein Structure by Distance Analysis

edited by

H. Bohr and S. Brunak

Center for Biological Sequence Analysis
The Technical University of Denmark
Lyngby, Denmark

1994

IOS Press
Amsterdam • Oxford • Washington DC

Tokyo • Osaka • Kyoto

ISBN 90 5199 161 4 (IOS Press)
ISBN 4-274-02263-3 (Ohmsha)
Library of Congress Catalog Card Number: 94-075946

Publisher:
IOS Press
Van Diemenstraat 94
1013 CN Amsterdam
Netherlands

Sole distributor in the UK and Ireland:
IOS Press/Lavis Marketing
73 Lime Walk
Headington
Oxford OX3 7AD
England

Distributor in the USA and Canada:
IOS Press, Inc.
P.O. Box 10558
Burke, VA 2209-0558
U.S.A.

Distributor in Japan:
Ohmsha, Ltd.
3-1 Kanda Nishiki-Cho
Chiyoda-Ku
Tokyo 101, Japan

LEGAL NOTICE
The publisher is not responsible for the use which might be made of the following information.

PRINTED IN THE NETHERLANDS

Contents

Overview

Experimental Approaches to Distance–matrix Determination in Proteins and Nucleic Acids

Theoretical Distance–based Approaches to Protein Structure Prediction

Database and Neural Network–based Predictions of Protein Structure

Introduction

Henrik Bohr and Søren Brunak

Center for Biological Sequence Analysis
Department of Physical Chemistry
The Technical University of Denmark, DK-2800 Lyngby, Denmark

The field of protein structure determination contains a vast and ever increasing amount of scientific contributions due to the great importance of protein design and functionality in biology, and, even more owing to the fact that prediction of accurate 3-dimensional structures of proteins from their sequence is still an unsolved problem.

In the light of this vast landscape of scientific information and achievements, aiming ultimately at fulfilling the goal of protein structure prediction from genome sequence data, the present monograph is intended to address the more limited aspect of protein structure determination in the distance geometry approach in order to obtain a clearer picture of the state of the art for a part of the subject while avoiding more general notions of protein folding already described well elsewhere[1, 2]. There has been strong emphasis on presenting experimental as well as theoretical aspects of the subject in order to obtain a balanced presentation of facts and speculation.

The distance geometry approach to protein structure determination is in the following to be understood as protein structure analysis, experimentally as well as theoretically, carried out on the basis of exact distance measures. With respect to experimental techniques this implies that protein structures are described in time or space by means of detailed distance information within the molecule, rather than protein structure formation being described by a phenomenological study of *e.g.* biochemical reactions. The detailed experimental techniques can either be X-ray diffraction crystallography (S. Larsen), nuclear magnetic resonance methods (F. Poulsen), circular dichroism methods (J.M. Hammerstad-Pedersen), infrared spectroscopy, neutron scattering *etc.*, the first technique being the most established, the second dealing with problems of solvents, the third having advantages in particular structural analysis.

As far as theoretical studies are concerned the limitation of distance geometry approaches implies that protein dynamics and protein structure prediction are studied under the constraints of certain given experimental distance information or under the fulfillment of certain distances within the protein in order to limit the degree of uncertainty in structure analysis or prediction.

Although the problem of protein structure prediction from sequence is greatly reduced, given knowledge about certain intermolecular distances, one should still be aware of the complexity in generating a full and detailed 3-dimensional protein structure from often very sparse, and at best, incomplete information about distances within a protein. In fact, many experiments can only give distance inequalities rather than exact real valued distances and often in a 2-dimensional form whereby the mathematical puzzle of generating unambiguously the full 3-dimensional structure is, in principle, rendered unsolvable. However, there are various approximation techniques[3] described in this book that can circumvent these problems mostly with the use of computer simulation techniques. For a very detailed and thorough treatment of the mathematical problems in distance geometry analysis the reader is referred to the book by C.M. Crippen and T.F. Havel: "Distance Geometry and Molecular Conformations" [4, 5].

Apart from generation of 3-dimensional structures of proteins from distance constraints the distance geometry approach to protein structure analysis has also been understood in a wider sense to encompass energy potential methods based on distances and angles in the molecules. One approach[6] (R. Goldstein) is to transform the problem of protein structure prediction into the problem of minimizing an energy function for an analogous spin glass system[18] where the spin states correspond to protein configurations. This method is in line with distance geometry approaches in the sense that such energy function optimization basically implies satisfying a great number of distance constraints and simultaneously comparing sequences corresponding to these protein configurations. Somewhat in the same spirit is comparative protein modelling, performed by satisfying a set of spatial restraints (A. Šali) and aims at making exhaustive enumerations of protein conformations (E. Shakhnovich). Another use of potential function (M. Sippl, G. Crippen) is to identify correctly formed protein structures rather than predicting new structures from sequences. Also constraints from solvent perturbations in NMR spectra can be used to predict surface structures of proteins (A. Lesk) and in similar spirit are glycosylation on proteins analysed (J. Hansen). Moreover a whole new self-consistent molecular field theory is used to predict 3-dimensional structures of globular proteins (A. Finkelstein).

A modern theme recurrent throughout the various contributions has been to discuss general classes of protein folds (C. Sander, W. Taylor, G. Crippen, S. Subramaniam, A. Efimov, A. Aszodi, M. Kanehisa, M. Reczko) rather than describing specific protein structures. It is believed that proteins appearing in organisms are based on a limited repertoire of different core structures or folding motifs. In the past it has been common to classify proteins with respect to sequence similarity for evolutionary purposes or, most commonly, to group proteins with respect to their function so that, for example, proteases go in one group, immunoglobulins in another *etc*. The concept of protein folds[8] is, however, related to topological characteristica so that given folds belong to the same fold class if they share the same topological structure. A fold is a distinct geometrical domain of a protein (*e.g.* a cluster of super secondary structures), either of the whole protein or part of it. Often a necessary requirement, albeit not a sufficient, is that protein folds belong to the same class if they have more than 50% sequence identity. Proteases are for example divided into several fold-classes. A typical example of a fold class is the Tim Barrel class. One of the many questions concerning fold classes, and addressed in this book, is the problem of being able to identify them from sequence studies and from distance geometry analysis (W. Taylor). Another problem is to find an appropriate choice of parameters to link the different classes, such as a parameter for packing of secondary structures. This question arises especially when an entirely new protein, with practically no sequence similarity to any known structure, has to fit into or establish a relationship to one of the known classes. A very relevant question is in this context to ask how the most "extreme" classes could be characterized.

Connected to these protein folds is the recent idea of "threading" [11, 12, 15, 14] (S. Subramaniam, M. Sippl, W. Taylor) meaning that protein sequences are being "threaded" through various different folding motifs in order to identify misfolded structures through an empirical evaluation function that can distinguish incorrect from correct folds. For reasons of simplicity folding motifs have been represented as linear profiles of local environmental properties independent of the type of fold being considered, *e.g.* secondary structures, at each residue in a known protein structure. Specific sequences can be given evaluation scores depending on preferences of the aligned residues for their respective environmental categories. Instead of representing folding motifs as linear profiles they can be represented as 2-dimensional contact matrices or as distance matrices[15, 16, 17, 18, 19] (W. Taylor) in the spirit of this forum.

Predicting which fold-class a given protein belongs to on the basis of its sequence can also be of great help in predicting distance matrices and a whole plan for predicting protein structures in the distance matrix approach could be deviced, perhaps leading to higher accuracy at lower sequence similarity than has yet been achieved. According to this plan neural networks are trained on proteins from each fold-class exclusively, in order to develop an ability to predict distance matrices for new proteins belonging to the fold-class of the training set. There is good reason to believe that distance matrices can be fairly correctly predicted by neural networks for proteins homologous to the ones the network has been trained on[20]. The long term hope is to be able to develop prediction schemes for protein folds and (the inverse folding problem) to understand how many alterations in their sequence are required for transforming a fold from one class into another. In more direct words one could ask how many substitutions are needed to give, for example a lysozyme, the functionality of a cytochrome.

A large section of the book is devoted to the use of computer generated neural networks (B. Rost, M. Reczko, M. Liebman, J. Engelbrecht), genetic algorithms (P. Argos, F. Herrmann) *etc.* for predicting protein structures and folds, and does fairly well represent the state of the art in these methodologies. Special emphasis is devoted to prediction of secondary structure (B. Rost, F. Garnier, J. Engelbrecht) and super secondary structures (A. Efimov). There seems to be an upper limit of around 70% accuracy, and similar for other prediction schemes. One feels tempted to ask if this magic number of 70% is connected to the fact that, on average, 70% of the secondary structures found in the native proteins are formed in the early stages of protein folding, as experiments indicate[21], whether it is because roughly 30% of the side-chains are affected by the surroundings or whether there is a deeper reason originating in the evolution of proteins? The former hypothesis could be verified in the future if it is possible to construct a neural network training set of intermediate folding structures.

The distance geometry approach to protein structure determination is indeed an interesting and versatile forum for scientific discussions of the methodologies of bioinformatics and biotechnology. All considered it is fair to say, concerning the goal of generating new protein structures, that while the experimental efforts focus on still higher accuracy in protein structure determination the theoretical counter part of prediction methodologies is rather, till the present, achieving the gross features of protein structures at low resolution.

This book grew out of the symposium "Distance based approaches to protein structure determination" held at the Center for Biological Sequence Analysis, The Technical University of Denmark, in Lyngby, November 23–26, 1993. The names mentioned in this introduction refer to scientists participating with contributions of high standard and novelty and reproduced here in enlarged and improved versions in this volume. We wish to thank The Danish National Research Foundation for the generous sponsorship supporting the center that made the symposium possible. We also wish to acknowledge Miss Johanne Keiding for providing us with great and indispensable editorial work for this book.

References

[1] D.B. Wetlaufer, "The protein folding problem", AAAS selected Symposium, Vol. 89, Westview Publisher, Boulder, USA (1984).

[2] "Protein folding: Deciphering the second half of the genetic code", eds. L.M. Gierasch and J. King, CIP, Washington, USA, 1989.

[3] J. Bohr, H. Bohr, S. Brunak, R.M.J. Cotterill, H. Fredholm, B. Lautrup and S.B. Petersen, *J. Mol. Biol.*, **231**, 861-869 (1993).

[4] G.M. Crippen and T.F. Havel, Distance Geometry and Molecular Conformation. Wiley, New York (1988).

[5] G.M. Crippen, *J. Math. Chem.*, **6**, 307-324 (1991).

[6] R.A. Goldstein, Z. Luthey-Schulten and P.G. Wolynes, *Proc. Acad. Nat. Sci.*, **89**, 4918-4922 (1992).

[7] M.S. Friedrich and P.G. Wolynes, *Science*, **246**, 371-377 (1989).

[8] T.L. Blundell and M.S. Johnson, *Prot. Sci.*, **2**, 877-883 (1993).

[9] S. Pascarella and P. Argos, *Prot. Eng.*, **5**, 121-137 (1992).

[10] D. Jones and J. Thornton, *L. Comp. Aid. Mol. Design*, **7**, 439-456 (1993).

[11] J. Novotny, A.A. Rashin and R.E. Bruccoleri, *Proteins*, **4**, 19-30 (1988)

[12] D. Eisenberg and A.D. McLachlan, *Nature*, **319**, 199-203 (1986).

[13] J.S. Fetrow and S.H. Bryant, *Biotechnology*, **11**, 479-484 (1993).

[14] L.M. Gregoret and F.E. Cohen, *J. Mol. Biol.*, **211**, 959-974 (1990).

[15] W. Taylor, *Prot. Eng.*, **4**, 853-870 (1991).

[16] J.T. Jones, W.R. Taylor and J.M. Thornton, *Nature*, **358**, 86-89 (1992).

[17] G.M. Crippen, *Biochemistry*, **30**, 4232-4237 (1991).

[18] M.J. Sippl, *J. Mol. Biol.*, **213**, 859-883 (1990).

[19] L. Holm and C. Sander, *J. Mol. Biol.*, **233**, 123-138 (1993).

[20] H. Bohr, J. Bohr, S. Brunak, R.M.J. Cotterill, H. Fredholm, B. Lautrup and S.B. Petersen, *FEBS Lett.*, **261**, 43-46 (1990).

[21] O.B. Ptitsyn, R.H. Pain, G.V. Semisotnov, E. Zerovnik and O.I. Razgulyaev, *FEBS Lett.*, **262**, 20 (1990).

Color Plates

Plate 1

Figure 1 in: Structures from X–Ray Crystallography illustrated by Proteins with prosthetic Groups, page 16 (Sine Larsen *et al.*).

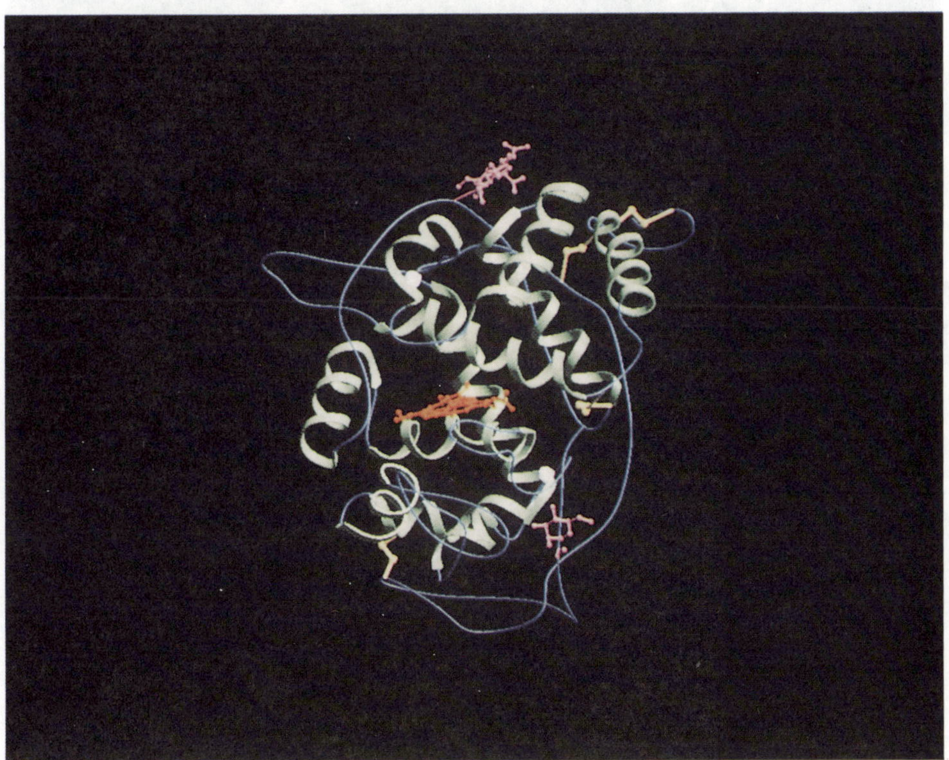

The overall folding of the CiP molecule[5] the heme group (red), Ca^{2+} ion (light green spheres) and bound carbohydrate (pink) are also shown.

Plate 2

Figure 5 in: Function and Three–Dimensional Structure of Proteins Using Nuclear Magnetic Resonance Spectroscopy, page 31 (Flemming M. Poulsen).

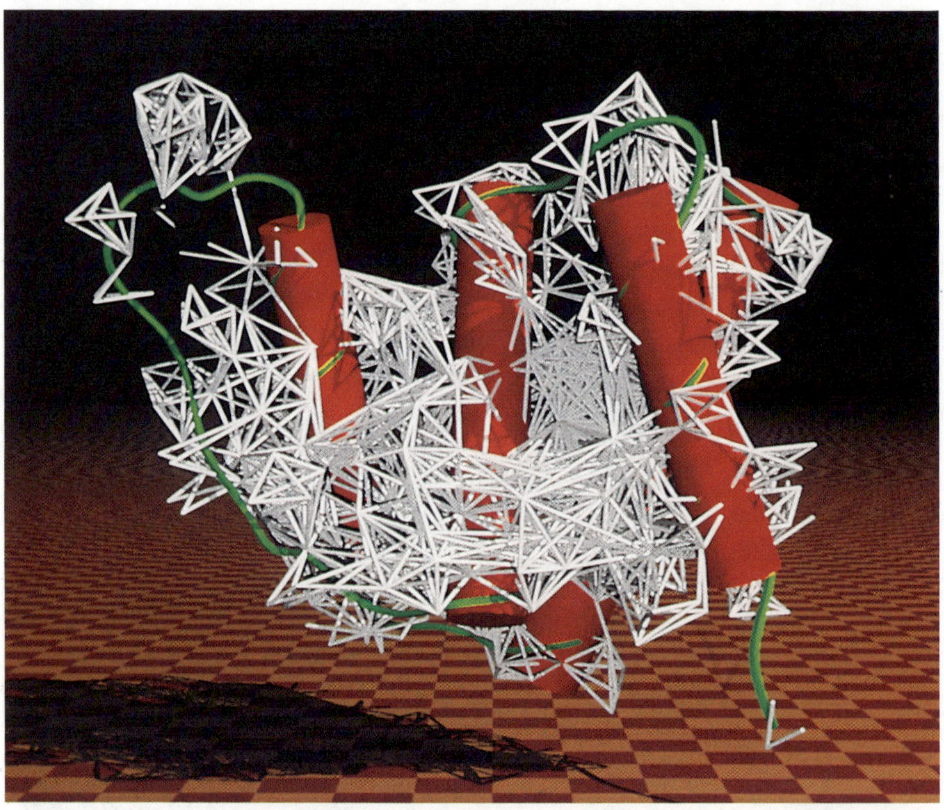

Model of the three-dimensional structure of acyl-Coenzyme A binding protein. The four cylinders present four α-helices in the structure. All the nuclear Overhauser effects used to determine the structure by NMR spectroscopy are indicated as short sticks between the relevant two atoms.

Plate 3

Figure 6 in: Function and Three–Dimensional Structure of Proteins Using Nuclear Magnetic Resonance Spectroscopy, page 31. (Flemming M. Poulsen).

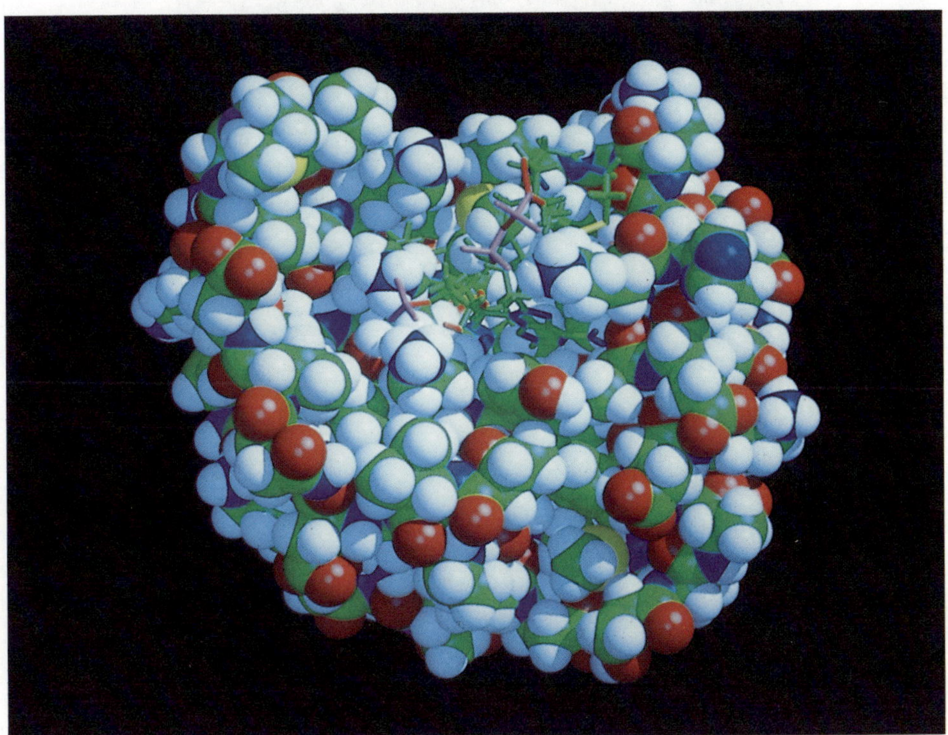

The ligand binding site of acyl coenzyme A binding protein. The atoms of the protein structure are shown using the common color code of green for carbon, blue for nitrogen, red for oxygen and white for hydrogen. The ligand is shown using a stick-model with the same colour code. Phosphor is purple red. The structure is determined by NMR spectroscopy.

Plate 4

Figure 1 in: Experimental Aspects of Ultra-violet and Circular Dichroism Methods for Protein Folding, page 38 (Hans E.M. Christensen *et al.*).

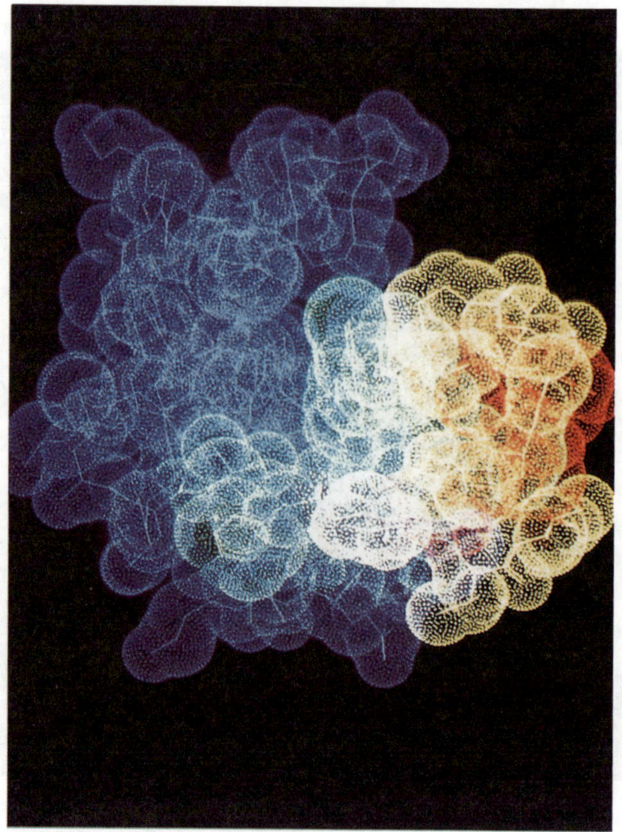

The hot-spot surface of *D. gigas* rubredoxin. Red indicates areas from which electron transfer is facile ("hot" areas), whereas blue indicates "cold" areas where ET to or from the metalcentre is unlikely. Yellow and greenish indicate "luke warm" areas. Since the iron-centre is coordinated to amino acids at the protein surface (top of picture), it is much easier to transfer electrons via these amino acids than via amino acids further away. Therefore, the hottest area is naturally located around the amino acids coordinating the iron. Computed using the program MetProt by Jan M. Hammerstad-Pedersen.

Plate 5

Figure 1 in: Genetic Algorithm Codings used in Protein Structure Prediction by Energy Minimization, page 179 (Frank Herrmann).

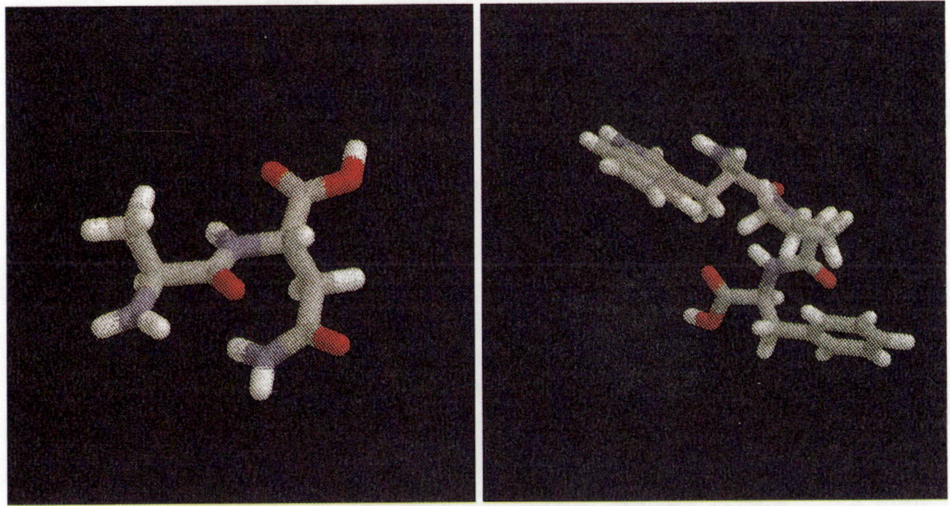

Peptide No. 1: (ALA ASN) Peptide No. 2: (TRP PRO PHE). The conformational space of the peptides is sampled at a resolution of 45 degrees for 8 rotatable bonds. The associated energy values are used to compute average energy values of certain subspaces.

xviii

Plate 6

Figure 1(a) in: Glycosylation and Protein Conformation, page 248 (Jan E. Hansen *et al.*).

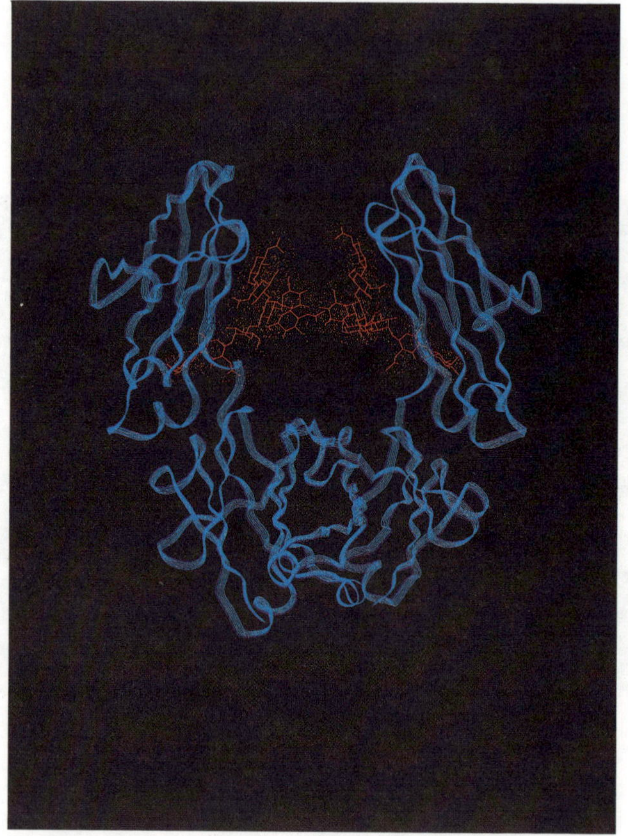

Immunoglobulin Fc from Brookhaven Data Base entry 1FC2 (upper panel) and 1FC1 (lower panel) [25]. Ribbon diagrams of the backbone of the Fc heavy chains of IgG are shown in blue and N-linked carbohydrates are shown in red. See text for details.

Plate 7

Figure 1(b) in: Glycosylation and Protein Conformation, page 248 (Jan E. Hansen *et al.*).

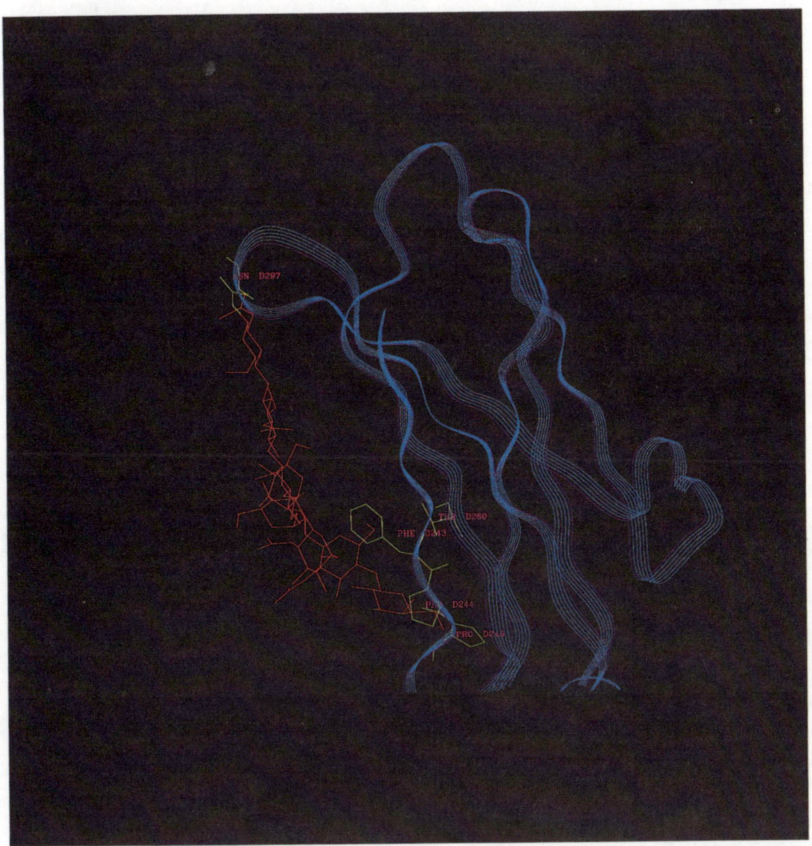

See legend to plate 7, figure 1(a).

xx

Plate 8

Figure 2(a) and 2(b) in: Glycosylation and Protein Conformation, page 248 (Jan E. Hansen *et al.*).

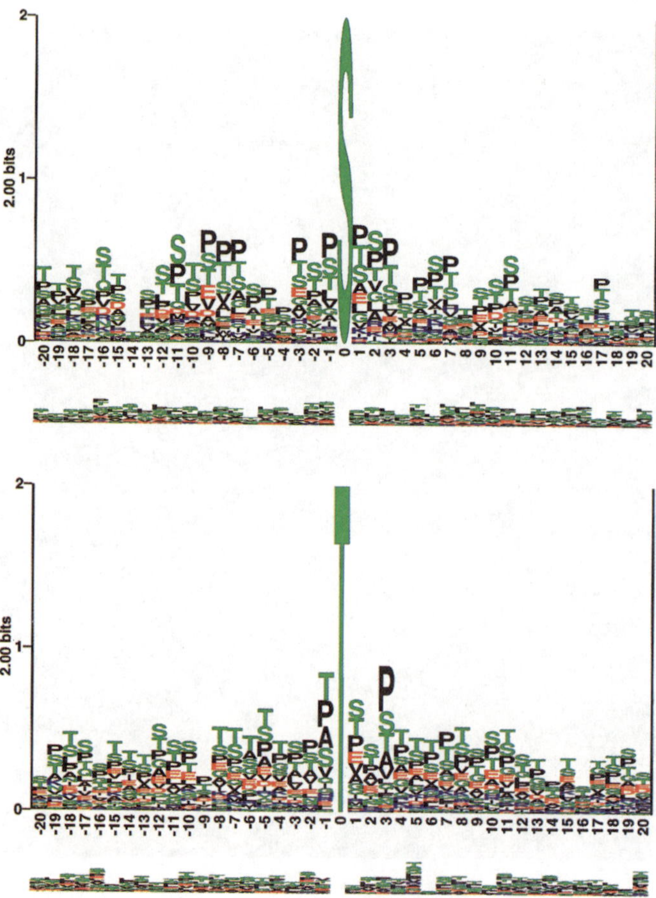

Logos [20] for O-glycosylation sites. Logos for glycosylated (upper logo) and non glycosylated (lower logo) serine sites (a) and threonine sites (b). The height of each column is proportional to the "signal" and the height of a letter within a column reflects the frequency of the corresponding amino-acid on that site. The shown heights of the central S/T have been scaled to elucidate the context and are hence non–informative. The neutral and polar aminoacids are shown in green, the basic in blue, the acidic in red and the neutral and hydrophobic in black. The sequence context around non-glycosylated serine and threonine sites is seen to be more random.

Plate 9

Figure 4 in: Glycosylation and Protein Conformation, page 251 (Jan E. Hansen *et al.*).

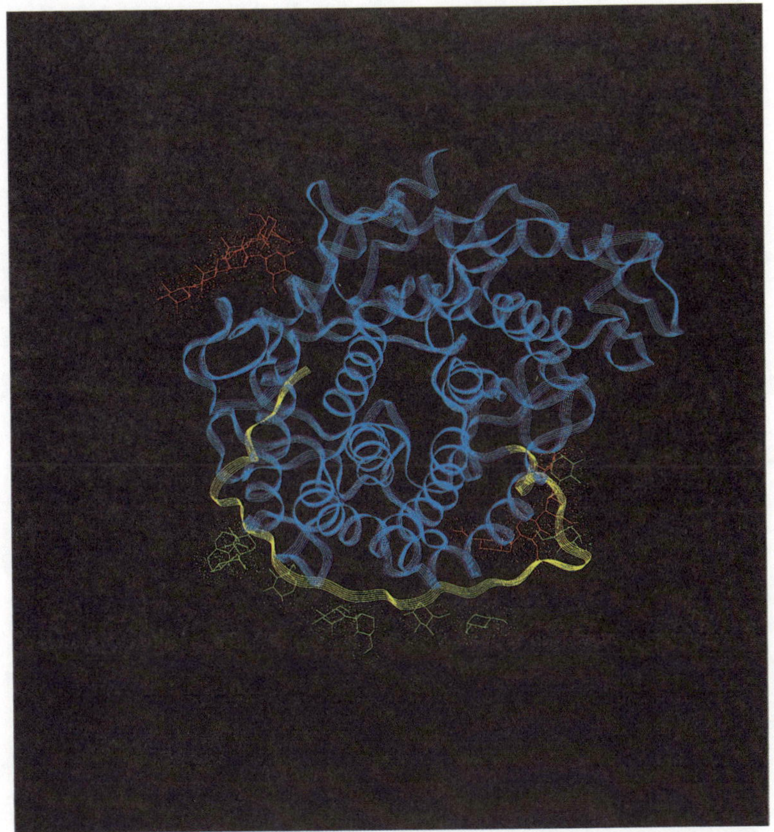

1,4 aspergillus glucanase. (Brookhaven data base entry: P1GLY, Aleshin et. al). N-linked carbohydrates are shown in red and O-linked carbohydrates are shown in light green. The extended peptide strech TSASSVPGTCAATSASGTYSSVTVTSWPSIVA containing the O-glycosylations is shown in yellow.

Plate 10

Figure 2 in: Correlation between protein secondary structure and the mRNA nucleotide sequence, page 329 (Søren Brunak, Jacob Engelbrecht and Can Kesmir).

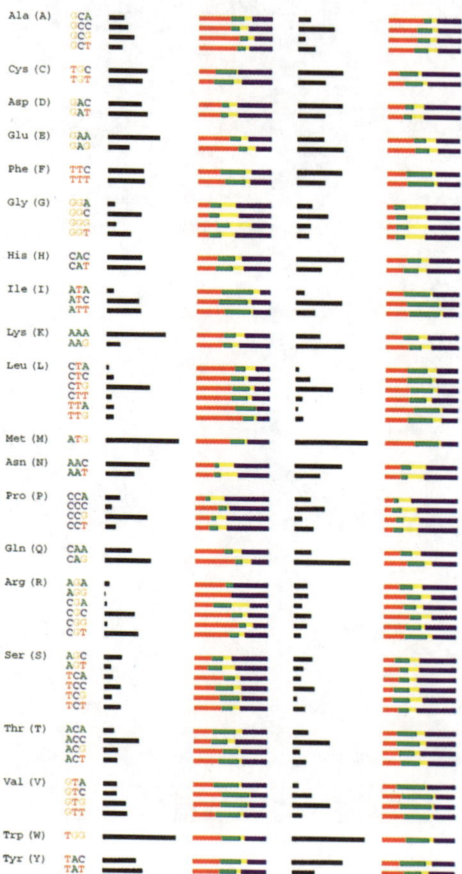

Statistics over preferred secondary structure categories for the the 20 amino acids and their 61 codons in the genetic code. The categories are color coded: helix (red), sheet (green), turn (yellow), and random coil (blue). The histogram in black indicates the codon usage. The left pair of indicators represent data for enterobacteria, the right pair data for mammals.

Plate 11

Figure 4a in: Correlation between protein secondary structure and the mRNA nucleotide sequence, page 331 (Søren Brunak, Jacob Engelbrecht and Can Kesmir).

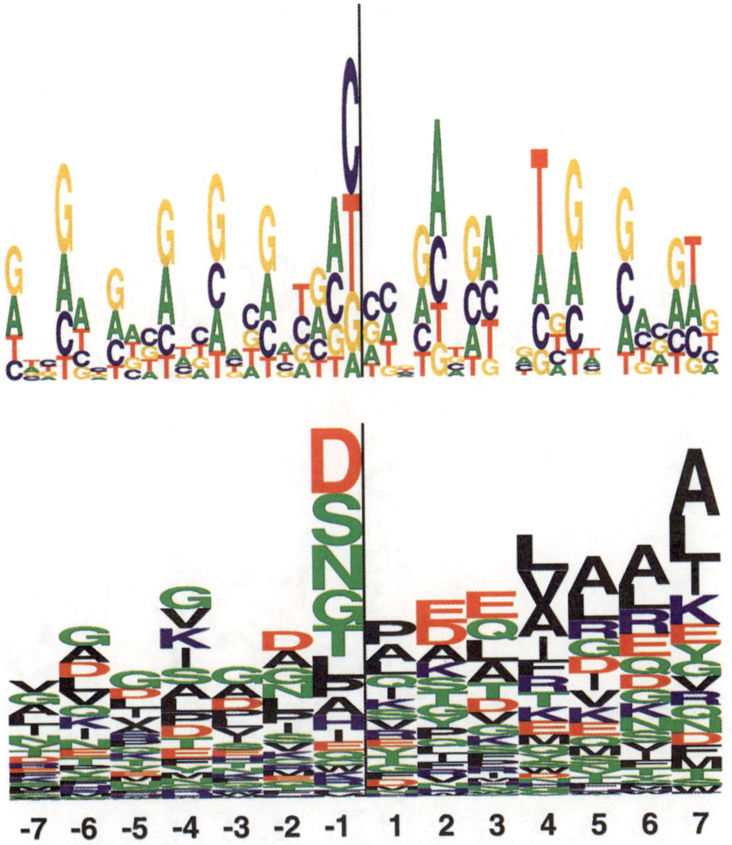

Shannon information content in logo form in the coil–to–helix transition regions in the data set for enterobacteria. Coil has negative positions and helix positive positions. The upper and lower panels in each section correspond to the mRNA and amino acid representations respectively. The height of the letters indicates their frequency at that position; the vertical bar gives the scale, 0.25 bits (mRNA) and 0.50 bits (amino acids).

Plate 12

Figure 4b in: Correlation between protein secondary structure and the mRNA nucleotide sequence, page 331 (Søren Brunak, Jacob Engelbrecht and Can Kesmir).

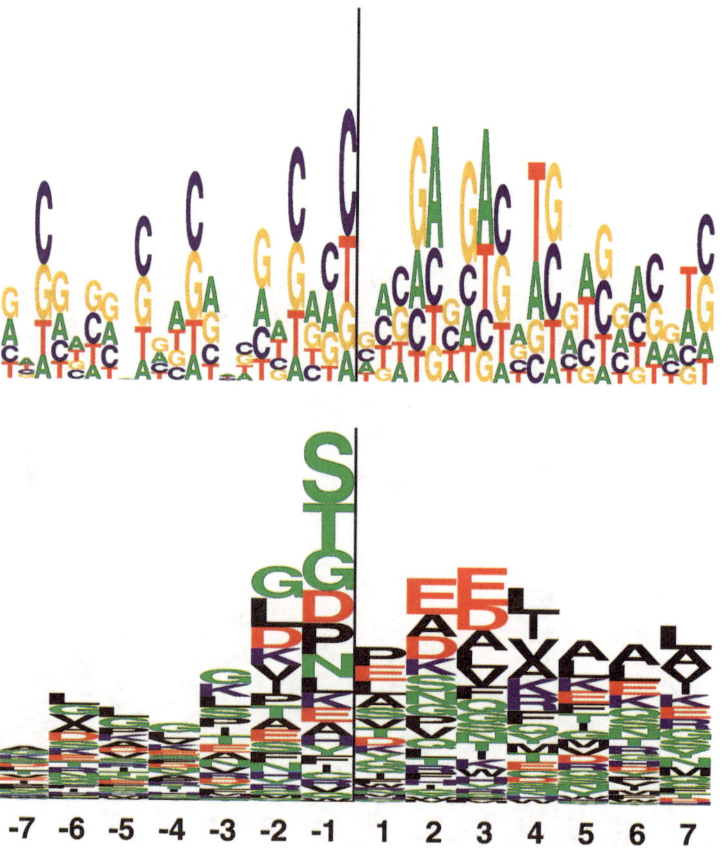

Shannon information content in logo form in the coil–to–helix transition regions in the data set for mammals. Coil has negative positions and helix positive positions. The upper and lower panels in each section correspond to the mRNA and amino acid representations respectively. The height of the letters indicates their frequency at that position; the vertical bar gives the scale, 0.25 bits (mRNA) and 0.50 bits (amino acids).

Overview

Protein Structure Prediction
— Past and Present

Shankar Subramaniam

Department of Physiology and Biophysics, Beckman Institute, University of Illinois,
Urbana, IL 61801

Abstract

There has been an explosion of methods for prediction of protein structures from sequences in recent years. The increasing number of protein structures solved by crystallography and magnetic resonance techniques, the ease with which proteins and their mutants can be expressed and the sophistication in computational hardware has spurred this effort. This paper presents a review of several successful methods.

The methods can broadly be categorized into those that approach the problem from "sequence to structure" perspective and those that address the inverse problem of "structure to sequences". Both efforts have been driven by existing knowledge of protein structures and features. Going beyond sequence identity and homology using sequence profiling methods has proved to be very successful. Building structures from statistically derived mean potentials has paved the way for rapid computation of relative merits of modeled structures. Threading which is arguably the best method for addressing the inverse folding problem uses the knowledge of existing folds to thread sequences and assesses the relative energetic stability by a variety of scoring schemes.

Distance-based methods form a common underlying theme to most structure prediction and evaluation methods. At a first level, the distance matrices constructed from proteins can be used to induce features that in turn can be used to make topological predictions. In principle, a knowledge of all pairwise distances between all atoms in a protein specifies the location of the protein in the energy space. And one can use the existing database of protein structures to derive mean potentials between pairs of residues or subgroups in residues and use this to optimize structures. Such mean potentials or the probability distribution functions themselves can be used to predict mutant structures, assess correct and misfolded proteins, and provide measures of relative stability of structures.

1   Introduction

Protein folding is an extremely complex phenomenon. The question is one of finding out how a polypeptide chain folds into a well–defined three dimensional structure in a given environment. It is established unequivocally that any given protein sequence has a unique native structure, which arguably has the lowest free energy. The assertion of this thermodynamic principle, however, does not provide any clues as to how the folding occurs.

Levinthal posed a very important paradox that seems to establish that it is impossible for a polypeptide chain to fold into a given structure in finite time through a random search of the conformation al space. Simply stated the paradox asks how can proteins fold in the $10^{-2} - 10^2$ second time scales if the unfolded polypeptide chain has to explore the 10^{200} or more energy minima in conformational space [1]. This paradox was even more confounded by the experiments of Anfinsen, which showed that an unfolded polypeptide chain could spontaneously refold to the biologically active form in the absence of all the cellular translation milieu [2]. This implies that sequence, by itself, contains all of the information necessary to define the three dimensional structure of the protein and hence its biological function. Anfinsen and Scheraga stated [3] that

> "It is currently believed that the three dimensional structure of a native protein in a given environment (solvent, pH, ionic strength, presence of other components, temperature, etc.) is the one in which the Gibb's free energy of the whole system is a minimum with respect to all degrees of freedom, i.e. that the native conformation is determined by the various interatomic interactions and hence by the amino acid sequence, in a given environment."

However, the protein folding problem, in principle, is different *in vivo* and *in vitro* and both of these are clearly distinct from the protein structure prediction problem. The last mentioned is the only one that reflects the Anfinsen experiment and enables the use of sequence-structure knowledge-based approaches for structure prediction. Knowledge-based approaches to protein structure prediction [4] can broadly be categorized into those that use sequence profile and homology, structural and topological homology and a combination of the two.

2   Sequence-Based Approaches

At present about 30,000 protein sequences are known and the number is increasing dramatically each year. While the increase in the number of sequences would be expected to enable better understanding of sequence motifs, the absence of significant similarity between the sequences makes it complicated, if not impossible, to derive any significant themes for structure prediction. The earliest efforts at using rigorous statistical techniques to go beyond sequence identity use homology relationships between evolutionary sequences [5]. As pointed out by Dayhoff and coworkers, if in comparing two sequences, if 40% or more of the residues are identical over 25 residues or more without insertions or deletions in either sequence, then the structural relationship is *de facto* established. This however is true only in the most closely related proteins which form a fraction of a percent of all the known sequences. Based on chemical similarity, sub–classes of similarity within the twenty amino acids can be made and within species variants, sequences often have replacements within the sub–class. Using this enables going in a limited way beyond sequence identity to homology. The more rigorous method for obtaining homology is to find out sequence relationships between proteins that occur more frequently than chance and use this to obtain a scoring scheme for replacement of each of the twenty amino acids by another. Such a matrix was first generated by Dayhoff and variations of this are extensively used today for both comparing and aligning sequences and for rational design of point and multiple mutations in proteins. Protein family–specific Dayhoff matrices can also be derived and this is especially useful for exploring structural or sequence diversity within the family of sequences. The quality of alignment of many protein sequences heavily depends on the lengths of the sequences, and the number of mispairings or gaps that occur within the interiors of the sequences. Sequence similarity based on alignment scores establishes a correspondence between the residues in each sequence that preserves order but allows gaps (i.e. the longest common subsequence). The gaps are given costs relative to the value of matches between residues [6]. Algorithms of run-time complexity quadratic in the average length of the two sequences exist to compute an alignment of them that maximizes the overall score, globally balancing the costs of local mismatches with the introduction of gaps [7,8]. Such dynamic programming methods have been extended for multiple sequence alignments and have become a routine tool for molecular biologists [9]. Once multiple sequences are optimally aligned, the protein of unknown structure is said to have a structure similar to the one that has the most sequence homology. This approach has led to predictions of novel protein structures. For instance, use of computer analysis for comparing two protein sequences on concanavalin A and favin (which share the ability to bind to the same type of sugar molecule) showed that the two protein sequences are related by a cyclic permutation [10]. A striking puzzle arising from the apparent absence of a N-terminal signal peptide required for translocation of ovalbumin to the endoplasmic reticulum was solved when it was observed that ovalbumin contains such a sequence in the middle of the protein [11].

3   Sequence Profile–Based Approaches

The main problem with sequence homology based methods is that sequence similarity is not always related to structural similarity and often structurally homologous proteins have little sequence homology. For instance, the muconate lactonizing enzyme and mandelate racemase enzyme that have only 26% sequence homology and are mechanistically distinct in function, yet they display a very high degree of structural homology [12]. While it is true that 50% or better sequence homology is taken to imply structural similarity, smaller peptide fragments in sequences of size smaller than 7 could have different structures. The pentapeptide segment VNTFV, for example, adopts an alpha helical conformation in erythrocruorin while the same segment in ribonuclea se is part of a beta strand conformation [13]. It is obvious that the role of the amino acid residue in the structural context is crucial in determining the conformation it adopts. Methods which exploit the contextual information go under the category of profile-based techniques. Several methods have been developed for profiling residues.

The earliest profiling of the amino acid residue in the structural context was done with respect to frequencies of occurrence of residues in well defined secondary structures[14,15,16]. An additional feature that characterizes residues in a secondary structure is the accessibility to the solvent [17]. Combining secondary structure and a binary classification for solvent accessibility of residues, Eisenberg and coworkers have generated secondary structure-based profiles in a Dayhoff like scheme [18]. Using these profiles for eight–stranded β barrels of the insecticyanin fold, avidin and complement component C8 sequences were identified as belonging to this class [18]. Profile analysis was also built around structural and functional motifs in select classes of proteins in order to probe relatedness of distantly related proteins [19]. An additional profile that characterizes a residue is the three dimensional protein environment. Polar side chains buried inside proteins are more likely to be in proximity to other polar side chains as much as non–polar ones in proximity to non-polar ones. This information is coded neither in solvent accessibility nor in secondary structure-based profiles. Using a combination of secondary structure, fraction of residue polar and fraction of area buried, Eisenberg and coworkers classified each residue as belonging to one of 18 environmental

Table 1: Distance–based potentials for structure prediction.

	Sippl & Coworkers	Maiorov & Crippen	Eisenberg, Luthy, Bowie, Willmans	Goldstein, Schulten, Wolynes	Madej, Mossing	Bryant. Lawrence	Tcheng, Subramaniam
Potential	Main chain/$C\beta$	N, C, O , $C\beta$	Residue	Backbone, $C\beta$	Residue	Residue	All Atoms
Method	Statistical	Statistical	Statistical	Associative Memory	Associative Memory, Profile	Statistical	Statistical
Training Set	187	69	110	42/123	25	161	260
Optimize	Potentials from distance PDFs	Residue Contact Potentials	Profiles	"Spin Glass" like Hamiltonian	"Spin Glass" like Hamiltonian	Residue Contact Potentials	Pairwise all–atom PDFs

classes. Using a library of well determined protein structures, profiles were generated from the frequency of finding a side chain of type i in an environmental class j for each residue. This produces position-specific 3D–1D scores for each structure. Once protein structures are translated into one–dimensional arrays of environmental classes, new sequences are aligned to this 1D templates. This is then used to test if a new sequence is compatible with a given structure [20]. The 3D profiles serve as useful tools to both discriminate correctly and incorrectly folded proteins [21] and recognized the structural relationship between heat-shock cognate protein 70 and actin sequences [20]. As a further extension to this model, the local environment was characterized by the statistical preferences of the profiled residue for specific neighbors of specific residue types, main–chain conformations or secondary structure [22]. This method was used to generate three-dimensional profiles for the β/α barrel fold protein sequences. The enzymes anthranilite phosphoribosyltransferase, glutamine amidotransferase and phosphoriobsylformimino–5–aminoimidazole carboxamide ribotide isomerase were found to belong to the β/α barrel fold [22].

A limitation of the above method is in identifying sequences that do not conform to any existing structural class or in aligning sequences that have large number of insertions and deletions or transpositions as compared to profiled structures. The latter is a computationally intractable problem, while the former rests on the availability of only a limited number of high resolution structures. Only about 400 unique high resolution structures are currently available in the protein data bank [23]. Structure–independent amino acid residue profiles have also been developed for segment structure prediction. In one such method, average physical, chemical and conformational properties [24] are used to select position–specific amino acid probability distribution scores, and generate profiles that are suitable for searching a database of protein sequences. In this method, the dataset of unique proteins are used to generate clusters of N–residue segment length in which members of each cluster are structurally similar. In each position in each cluster scores are assigned for all amino acid attributes to construct property profiles for each cluster. The similarity of an unseen segment to each cluster profile score is measured to find the maximum likelihood cluster. This method has been used to predict structures of seven residue fragments [25]. Blundell and coworkers have also generated topological equivalence in protein structures using a comparison of amino acid properties and relationships using simulated annealing and dynamic programming techniques[26].

Proteins with high resolution structures have been categorized into unique fold families [27,28]. A novel method of comparison of folds using alignment of distance matrices has led to identification of topologically similar proteins even when they are severely distorted or have segments transposed with respect to each other. This method has also turned up unusual topological similarities, for e.g. between the bacterial toxin colicin A and globins and between the eukaryotic POU–specific DNA–binding domain and the bacterial λ repressor[28]. Several fold-based sequence profiles have been generated by researchers. Blundell and coworkers have used structural templates to search for common protein folds [29]. Ioerger et al., have used a constructive induction approach that learns better representation of sequences based on physical and chemical properties. Using this method, they were able to show the similarity of eight pairs of proteins that belong to the same fold, but share little sequence homology [30]. Back propagation neural networks have also been trained to classify and predict tertiary structures from sequences [31]. Bohr et al., [32] and Rezcko et al. [33] have used neural networks to learn distance matrices in order to both predict three–dimensional structures and fold classes.

4   Energy-Based Approaches

Detailed atomic force-fields have been developed by numerous researchers for protein simulations [34]. However the time scales accessible to these simulations are far from those that would result in structural folding. However empirical potentials that intermediate between atomic and residue level descriptions are being developed using knowledge of high resolution structures. Most of them use a combination of statistical and optimization techniques. Table 1 gives a summary of some of the energy and distance-based methods.

Sippl and coworkers obtained residue and sequence specific potentials of mean force from pairwise probability density functions (PDF) of residue $C\alpha$ and $C\beta$ atoms and used this to calculate energies of different conformations of oligopeptide sequences. On rank ordering according to the energies derived from potentials of mean force, they found the lowest energy structures to conform to the actual structures [35,36,37]. Jones et al., have used PDF-derived energy functions and empirical solvation energy functions in conjunction with dynamic programming techniques to thread sequences into different folds and obtain the lowest energy fold for the sequence [38]. All atom PDFs have been derived using kernel density functions from a dataset of 260 unique structures which have been obtained by Tcheng and Subramaniam. These PDFs have been used to classify outliers in the protein data bank. The bacteriophage Gene V Protein in the crystal structure was correctly identified to be an outlier and the PDFs predicted the regions of chain register error [39]. The more recent structure was found to yield optimal PDFs in these regions. A reduced representation model has also been used in conjunction with backbone torsion space genetic algorithm search procedure for predicting the structure of an 18 residue polypeptide, apamin [40]. Mairov and Crippen [41] have used a continuous contact potential to recognize the correct native fold of a sequence as compared to structural fragments of the same size from unnative folds. Miyazawa and Jernigan also obtained effective interresidue contact energies from crystal structures [42]. Bryant and Lawrence [43] have also derived a residue contact potential in terms of mean hydrophobic and pairwise contact energies as a function of residue type and distance interval. They used this method to correctly identify models with the core folding motifs of hemerythrin and immunoglobulin McPC603 VL domain amongst several possible alternatives.

Wolynes and coworkers [44,45] have used concepts from spin glass theory to obtain associative memory hamiltonians to obtain three dimensional structures of proteins. The energy function reflects a collection of sequence/structure patterns obtained from known native structures. The method requires prealignment of the predicted and the memory sequences. Using this method, these researchers have obtained correct folds for proteins using memory set proteins that have very little sequence homology [46]. Madej and Mossing have combined the associative memory methods with the Eisenberg profiles to construct hamiltonians which are then used in molecular mechanics procedures to obtain low energy structures [47]. Skolnick have used molecular mechanics and Monte Carlo methods in conjunction with a topology fingerprint approach, in which they obtain iterative alignment of a sequence to a structure [48].

5   The Protein Design Problem

The problem of protein design is one of the inverse folding problem [49]. Ponder and Richards stated,

> "the problem of protein folding is commonly phrased as the question "what tertiary structure is assumed by a polypeptide chain of a specific sequence?"... it may be useful to express the problem as an inverted question: "what sequences are compatible with a given structure" [50].

Using this as the guiding quest, they designed an algorithm that generates tertiary templates for protein structures. They used all the "common sense" protein themes of preserving covalent geometry, prohibiting atomic overlaps, using the Ramachandran map, avoiding cavities inside proteins, and leaving charge groups on the surface. Using this van der Waals potential model for packing they searched through the rotamer library of side chains to obtain all those sequences that would fit a given tertiary template. Hellinga and Richards further developed this method and applied it to designing a copper binding site in a cupredoxin. These classes of problems required precise knowledge of interactions between residue groups at well-defined distances and the optimization of local geometry using more sophisticated knowledge–based potentials would enable protein engineering and design. Using the PDF methods, the rotamer libraries and a simulated annealing method for optimization, Tcheng and Subramaniam [39] have obtained mutant structures, within an rms deviation of 1 A, of 40 single mutant T4 phage lysozymes starting from the wild-type protein.

6   Perspectives

It is apparent, that there is no *ab initio* procedure for obtaining the three dimensional structure of a prote in from its sequence. All methods used so far have been knowledge–based. Knowledge–based potentials, that appear very promising, incorporate all the distance and contact information, but need further work and improvement and still may require other kinds of knowledge embedded before they can correctly predict three dimensional structures only from the sequence information. The more realistic goal of protein engineering is well-within the grasp of a combination of several of the methods described and it is conceivable that it can be extended to designing simple protein motifs. Modest successes and even failures in this direction in the molecular biology laboratory may shed further light into the question of protein design.

Acknowledgements

The author wishes to thank his coworkers David Tcheng, Laura Walsh, Kara Andosca and George Pappas and Dr. Andrew Wang for providing coordinates of the modified Gene V Protein structure prior to publication. Partial support was provided by the National Science Foundation.

References

[1] C. Levinthal, In: "Mossbauer Spectroscopy in Biological Systems", Ed. P. DeGennes, Urbana, IL: University of Illinois Press, 1969.

[2] C. B. Anfinsen, *Science*, **181**, 223, (1973).

[3] C. B. Anfinsen and H. A. Scheraga, *Adv. Protein Chem.*, **29**, 205(1975).

[4] S. Subramaniam, D. K. Tcheng, K. Hu, H. Raghavan and L. Rendell. In: Proceedings of the Fourth International Conference on Software Engineering and Knowledge Engineering, Capri, Italy, 1992. Los Alamitos, IEEE Computer Society, pp 420-434.

[5] M. O. Dayhoff, R.M. Schwarz and B.C. Orcutt. In "Atlas of Protein Sequence and Structure" Vol 5 suppl. 3, 345-352 (Natl. Biomed. Res. Fnd. Washington DC, 1978).

[6] O. J. Gotoh, *Mol. Biol.*, **162**, 705 (1982).

[7] S. Needleman, and C. Wunsch, *J. Mol. Biol.*, **48**, 443 (1970).

[8] T. F. Smith, and M.S. Waterman, *J. Mol. Biol.*, **147**, 195 (1981).

[9] S. F. Altschul, W. Gish, W. Miller, E.W. Myers and D.J. Lipman, *J. Mol. Biol.*, **215**, 403(1990).

[10] B. A. Cunningham, J. J. Hemperly, T. P. Hopp and G. M. Edelman, *Proc. Natl. Acad. Sci. USA*, **76**, 3218 (1979).

[11] B. W. Erickson, and P. H. Sellers. In: "Time Warps, String Edits, and Macromolecules: The Theory and Practice of Sequence Comparison.", Ed. D. Sankoff and J. B. Kruskal, Reading, MA: Addison Wesley, 1983.

[12] D. J. Neidhart, G. L. Kenyon, J. A. Gerlt and G. A. Petsko, *Nature*, **347**, 692 (1990).

[13] M. J. Sippl, *J. Mol. Biol.*, **213**, 859 (1990).

[14] P. Y. Chou and G. D. Fassman, *Biochemistry*, **13**, 222 (1974).

[15] P. J. Prevelige and G. Fasman. In: "Prediction of Protein Structure and the Principles of Protein Conformation." , Ed. G. Fasman, New York: Plenum Press, 391-416, 1989.

[16] J. Garnier and B. Robson. In: "Prediction of Protein Structure and the Principles of Protein Conformation", Ed. G. Fasman, New York: Plenum Press, 417-465, 1989.

[17] B. K. Lee and F. M. Richards, *J. Mol. Biol.*, **55**, 379 (1971).

[18] R. Luthy, A. D. McLachlan and D. Eisenberg, *Proteins*, **10**, 229 (1991).

[19] M. Gribskov, A. D. McLachlan and D. Eisenberg, *Proc. Natl. Acad. Sci. USA.*, **84**, 4355 (1987).

[20] J. U. Bowie, R. Luthy and D. Eisenberg, *Science*, **253**, 164 (1991).

[21] R. Luthy, J. U. Bowie and D. Eisenberg, *Nature*, **356**, 83 (1992).

[22] M. Wilmanns and D. Eisenberg, *Proc. Natl. Acad. Sci. USA.*, **90**, 1379 (1993).

[23] F. C. Bernstein, T. F. Koetzle, G. J. Williams, E. F. Meyer, M. D. Brice, J. R. Rodgers, O. Kennard, T. Shimanouchi and M. Tasumi, *J. Mol. Biol.*, **112**, 535 (1977).

[24] A. Kidera, Y. Konishi, M. Oka, T. Ooi, and H. A. Scheraga, *J. Protein Chem.*, **4**, 23 (1985).

[25] G. Pappas, D. K. Tcheng, K. Andosca and S. Subramaniam, (unpublished results).

[26] A. Sali and T. L. Blundell, *J. Mol. Biol.*, **212**, 403 (1990).

[27] S. Pascarella and P. Argos, *Prot. Eng.*, **5**, 121 (1992).

[28] L. Holm and C. Sander, *J. Mol. Biol.*, **233**, 123 (1993).

[29] M. S. Johnson, J. P. Overington and T. L. Blundell, *J. Mol. Biol.*, **231**, 735 (1993).

[30] T. Ioerger, L. Rendell and S. Subramaniam, "Constructive induction and the protein structure prediction problem", First International Conference on Intelligent Systems for Molecular Biology, 198-206, 1993.

[31] G. L. Wilcox, M. Poliac and M. N. Liebman, *Tetrahedron Computer Methdology*, **3**, 191 (1990).

[32] J. Bohr, H. Bohr, S. Brunak, M. J. Cotterill, H. Fredholm, B. Lautrup and S. B. Petersen, *J. Mol. Biol.* (in press).

[33] M. Rezcko, *et al.* (in this volume).

[34] J. A. McCammon and S. C. Harvey, Dynamics of Proteins and Nucleic Acids. Cambridge: Cambridge University Press (1987).

[35] M. J. Sippl, *J. Mol. Biol.*, **213**, 859 (1990).

[36] M. J. Sippl and S. Weitckus, *Proteins*, **13**, 258 (1992).

[37] M. J. Sippl, M. Hendlich and P. Lackner, *Prot. Sci.*, **1**, 625 (1992).

[38] D. T. Jones, W. R. Taylor and J. M. Thornton, *Nature*, **358**, 86 (1992).

[39] D. K. Tcheng and S. Subramaniam, (to be published).

[40] S. Sun, *Prot. Sci.*, **2**, 762 (1993).

[41] V. N. Maiorov and G. M. Crippen, *J. Mol. Biol.*, **227**, 876 (1992).

[42] S. Miyazawa and R. L. Jernigan, *Macromolecules*, **18**, 534 (1985).

[43] S. H. Bryant and C. E. Lawrence, *Proteins*, **16**, 92 (1993).

[44] M. S. Frederichs and P. G. Wolynes, *Science*, **246**, 371 (1989).

[45] M. S. Frederichs, R. A. Goldstein and P. G. Wolynes, *J. Mol. Biol.*, **222**, 1013 (1991).

[46] R. A. Goldstein, Z. A. Luthey-Schulten and P. G. Wolynes, *Proc. Natl. Acad. Sci. USA*, **89**, 9029 (1992).

[47] T. Madej and M. C. Mossing, *J. Mol. Biol.* , **233**, 480 (1993).

[48] A. Godzik, A. Kolinski and J. Skolnick, *J. Mol. Biol.*, **227**, 227 (1993).

[49] C. Pabo, *Nature*, **301**, 200 (1993).

[50] J. Ponder and F. M. Richards, *J. Mol. Biol.*, **193**, 775 (1987).

Experimental Approaches to Distance–matrix Determination in Proteins and Nucleic Acids

Structures from X-Ray Crystallography illustrated by Proteins with prosthetic Groups

Sine Larsen, Anders Kadziola and Jens F.W. Petersen

Center for Crystallographic Studies, University of Copenhagen, Universitetsparken 5,
DK-2100 Copenhagen, Denmark

Abstract

X-ray crystallographic methods are widely used for the determination of the three dimensional structure of proteins. It is the quality of the protein crystals that ultimately determines how accurately a protein structure can be determined by X-ray diffraction. Some of the indicators used in making the accuracy of a protein structure are described, using the structure of a peroxidase from *Coprinus cinereus* as an example. This protein consists of a single amino acid chain with 343 residues, a heme group and two Ca^{2+} ions. Its structure illustrates the significant role divalent cations (Ca^{2+}) and prosthetic groups like heme can play in protein folding. The two Ca^{2+} ions in the protein structure are not replaced by Mg^{2+} ions though the crystals were obtained with Mg^{2+} present in large excess. The heme group is interacting with helices from the two domains that constitute the Coprinus peroxidase structure.

1   Introduction

The present knowledge of the three dimensional structures of proteins relies to a large degree on results obtained by X-ray crystallographic methods. Recently protein crystallography has experienced almost a renaissance due to the development of DNA techniques, new detection devices used for measurements of X-ray diffraction intensities and computer graphics.

The determination of protein structure by X-ray diffraction methods is a complicated process consisting of the steps listed in Table 1. The number of protein structures in the Protein Data Bank is growing so rapidly that it could lead to the assumption, that determination of protein structures has become routine work. However, this is still not the case, as serious difficulties may be encountered in the structure determination process.

A necessary prerequisite for determining the three dimensional structure of a protein by X-ray diffraction methods is the presence of protein crystals. Crystallization of the protein can constitute one of the more tricky and time consuming steps in the process. The crystals obtained should be of an appreciable size, preferably larger than 0.2 mm and be well ordered, otherwise they will not diffract the X-rays strongly to give an interpretable diffraction pattern.

The structure (phase) determination step may also present serious difficulties. Knowledge of the protein folding from sequence homology to proteins with known structure

Table 1: Steps in Structure Determination by X-ray Crystallographic Methods.

Crystallization

Characterization of crystals

Data collection

Structure (Phase) determination

Structure refinement

Figure 1: See Color Plate 1, page xiii

is valuable information that can be employed in the structure determination. Otherwise heavy atoms derivatives must be made of the protein in order to accomplish the phase determination. If good derivatives cannot be obtained; it may ultimately be necessary to study a mutant of the protein, where selected residues have been replaced with other amino acids like Cys which are known to be good candidates for heavy atom sites.

The development of the new refinement methods based on simulated annealing [1] has made it possible to refine protein structures to a level of accuracy that was difficult to achieve 10 years ago. However, it should not be forgotten that it is the quality of the crystals indicated by the resolution limit in the diffraction pattern, that sets the limit to the accuracy of structure determination by X-ray crystallography. Fortunately, there are important indicators that show how reliable the results from a protein structure determination are, some of these indicators will be decribed in the following.

The three dimensional structures of proteins are known to be closely linked to their amino acid sequence. For some proteins the presence of divalent cations and prosthetic groups plays a significant role for the structure of the folded protein. We will illustrate these aspects of protein structures using, as an examples the structure of a heme group containing protein, that has recently been determined by us.

2   Heme Proteins, Coprinus Cinereus Peroxidase

A facinating aspect of proteins that contain heme groups is their diversity in biological function e.g. oxygen carrier, electron transfer, O2 activator. The diversity in their function is intimately connected with the electronic structure of iron surrounded by a porphyrin ring system. Iron is usually found in the oxidation states (II) and (III), where it can exist in high spin/low spin or intermediate spin states, likewise there may be variations in the coordination number. The results from a structure determination of a peroxidase isolated from the ink cap *Coprinus cinereus* is reported here. It is the only peroxidase isolated from this basidiomycete. It has been cloned and expressed in *Aspergillus oryzae* [2]. It consists of a single polypeptide chain of 343 amino acid residues with four S-S bridges, Mr = 38000. Besides the heme group, it contains two structurally bound Ca^{2+} ions and from protein chemical investigations three possible glycosylation sites have been suggested [3].

Like the related plant peroxidases, *Coprinus cinereus* peroxidase (CiP) is an important enzyme that catalyse the reduction of hydrogen peroxide to water and the simultaneous oxidation of a large variety of substrates. The broad substrate specificity of CiP makes it an attractive candidate for the study of structure/function relationships for peroxidases.

Table 2:

Recombinant Coprinus cinereus peroxidase (CiP) crystals grown from 18% w/v PEG 6 K, 0.35 M $MgCl_2$ buffered at pH=0.7 with 0.M HEPES [4].

CiP crystallizes in the orthorhombic space group $P2_1 2_1 2_1$ with unit cell dimensions a = 127.4 Å, b = 75.4 Å, c = 76.7 Å. Two independent CiP molecules in the asymmetric unit related by a non-crystallographic two-fold axis parallel to the c-axis.

Structure solution was achieved by molecular replacement [5] and structure refinement by simulated annealing [1]. The present model is based on data to 2.2 Å, each CiP molecule consists of residues 8-43, two calcium ions, two covalently bound sugar moieties (two N-glucosamines and one mannose), in addition the asymmetric unit contains 350 water molecules.

Based on 36729 independent reflections with $I/\sigma(I) > 0$, the present R-value is 18.9%.

A summary of the X-ray crystallographic investigations of CiP is presented in Table 2. The overall folding of the protein is shown in Fig. 1. As shown in this diagram CiP is largely helical like many other heme containing proteins. The helical topology of CiP is similar to the folding observed in the two other known peroxidase structures Cytochrome c peroxidase [6] and a ligninase from Phanerochaete chrysosporium [7]. The structure can be described as consisting of two domains with the heme group tucked in between them. The domains are traditionally named after the two His residues adjacent to the heme group. The domain that contains the coordinating His[181] is called the proximal domain, whereas the non-coordinating His[55] is part of the distal domain.

2.1 The Accuracy of the Structure

The R-value listed in Table 3 indicates that the structure is fairly accurate but, as has been shown [8], this may not always mean that the structure is correct. It is important to relate this number to the deviation from ideal geometry of the structural model obtained by the refinement. In the refined model of the CiP structure the bond lengths and bond angles on the average deviate 0.015 Å and 1.9^0, respectively, from ideal geometry. This shows that the low R-value is not a result of an unrealistic model.

One of the best indicators of accuracy is a direct inspection of the electron density map. This enables an evaluation of the quality of the structure as well as the fit to the density. For structures solved by the molecular replacement methods it is important to check that the final model is not biased by the search structure.

Fig. 2 shows a socalled omit electron density map of a region in the CiP structure where it differs significantly from the related LiP structure, which was used for the molecular replacement solution. The structure factors were calculated without the contribution from the amino acids 303 to 306. The resulting $(2F_0\text{-}F_c)$ electron density map in Fig. 2 is in excellent agreement with the model used for the omitted residues.

Researchers that use the results obtained by X-ray crystallography may ask about the

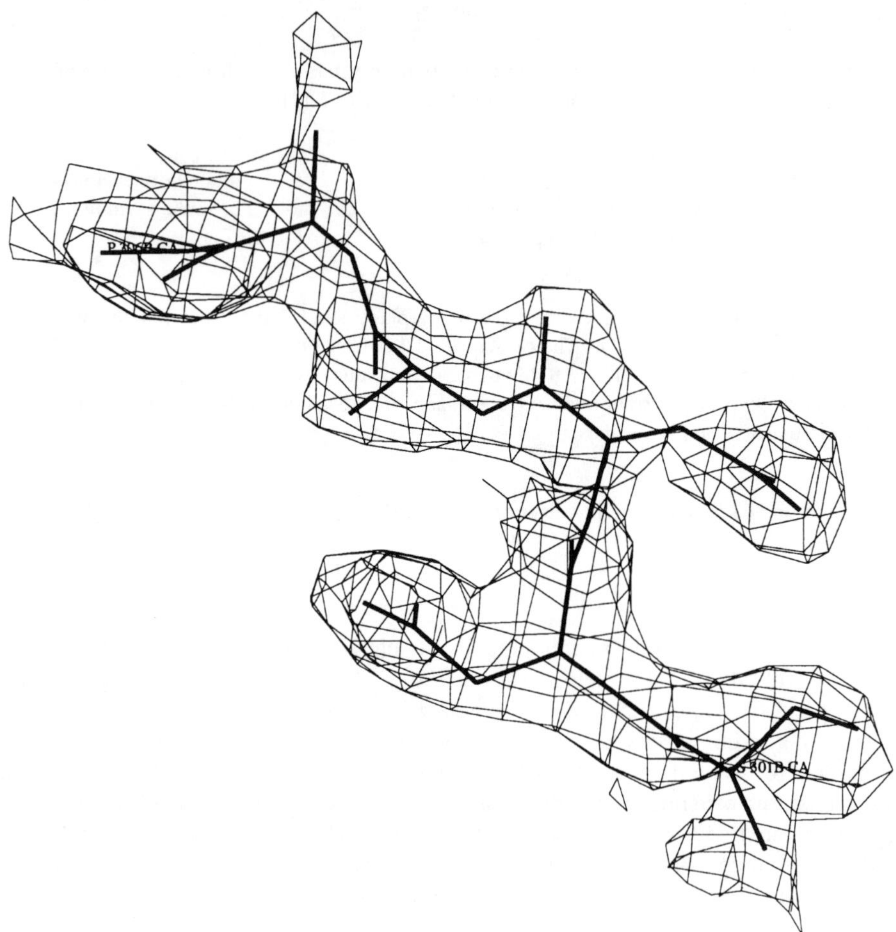

Figure 2: 2 F_0-F_c omit electron density map of the region containing the amino acid residues 303 to 306. F_c was calculated without the contribution from these amino acids. TURBO-Frodo [10] was used for the illustrations here and in Fig. 4, 5 and 6.

accuracy of the atomic positions within the structure. Luzzati [9] has shown how an analysis of R-values as a function of resolution and the magnitude of structure amplitudes can provide estimates of the error in the atomic coordinates. The CiP structure contains two crystallographically independent molecules. They were treated independently during the refinement procedure. A comparison of the two molecules should provide a realistic estimate of the errors in atomic positions in the structure. The root mean square difference between the positions of the C_α atoms is 0.25 Å. The difference between the two molecules measured as deviations in the positions of the C_α atoms as a function of amino acid residue are shown in Fig. 3. The thermal parameters used for the atoms are not only indicators of the thermal displacements of the atoms, but a large thermal parameter may also indicate poorly defined electron density and disorder of the protein structure. This disorder can be dynamic all due to large thermal movements or static corresponding to the existence of different conformations of the protein in a given region. The thermal parameter as a

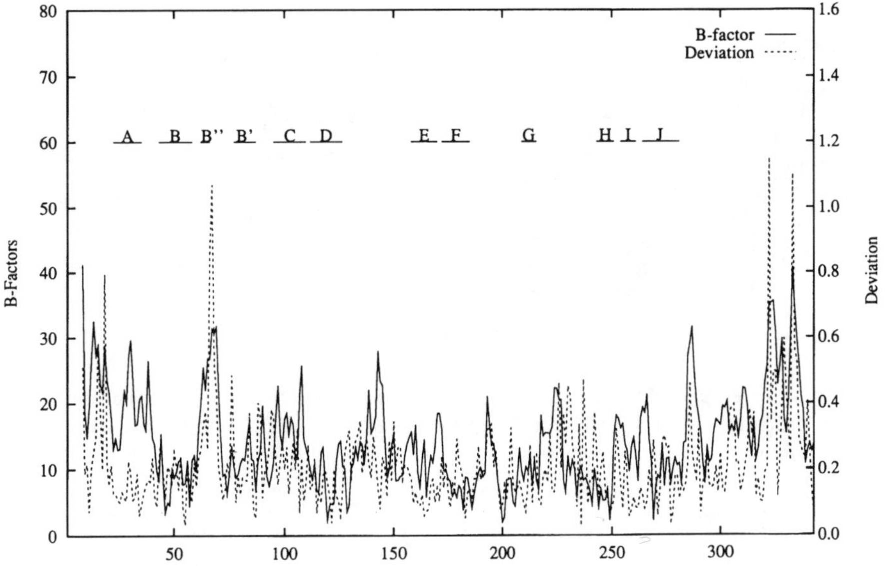

Figure 3: The absolute difference between the C_α positions of the two independent molecules after super position. The average B-factor for C_α atoms in units of $\overset{\circ}{A}^2$ as a function of residual number is also shown. The letters indicate the helical regions in the structure.

function of amino acid residue number is also listed on Fig. 3. It is obvious that differences in the structures of the two independent molecules are confined to loop and the N-terminal regions. In the helical regions both the temperature factors and the differences between the positions of the C_α atoms are generally smaller.

3   The Role of divalent Cations

The two crystallographically independent molecules in the *Coprinus cinereus* peroxidase structure are related by a non-crystallographic twofold axis parallel to the crystallographic c-axis. This arrangement gives rise to pseudo translational symmetry that was apparent in the diffraction pattern [4]. The presence of divalent cations in large concentrations of 0.35 M turned out to be essential for the crystal growth. The results presented here are based on data from crystals precipitated with 0.35 M $MgCl_2$ present, but crystals could also be obtained with $CaCl_2$ in similar concentrations. This need for divalent cations in order to get crystals can be understood from an analysis of the crystal packing. A Mg^{2+} ion is found on the pseudo two-fold axis that relates the two independent molecules, Asp^{316} from each molecule coordinates to the Mg^{2+} ion, as shown in Fig. 4. The distances from Mg^{2+} to the carboxylate oxygen atoms of the Asp residues correspond to outer sphere coordination distances to a $Mg(H_2O)_{2+}^6$ ion. The density around Mg^{2+} supports this interpretation, as it contains peaks that can be interpreted as water molecules.

Based on the sequence homology to the ligninase structure, it was predicted that each CiP molecule contains two structurally bound Ca^{2+} ions. The Ca^{2+} ion found in the proximal

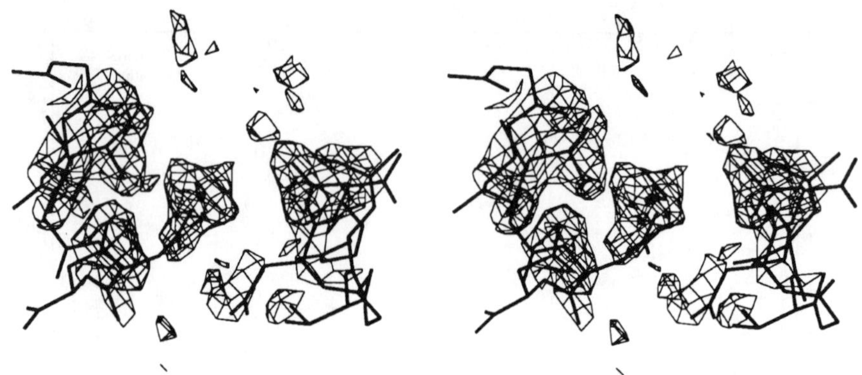

Figure 4: A stereo pair of the density viewed along the pseudo two-fold axes. The density in the center corresponds to $Mg(H_2O)_6^{2+}$ ion with two carboxylate groups from Asp^{316} in each independent molecule pointing towards the ion.

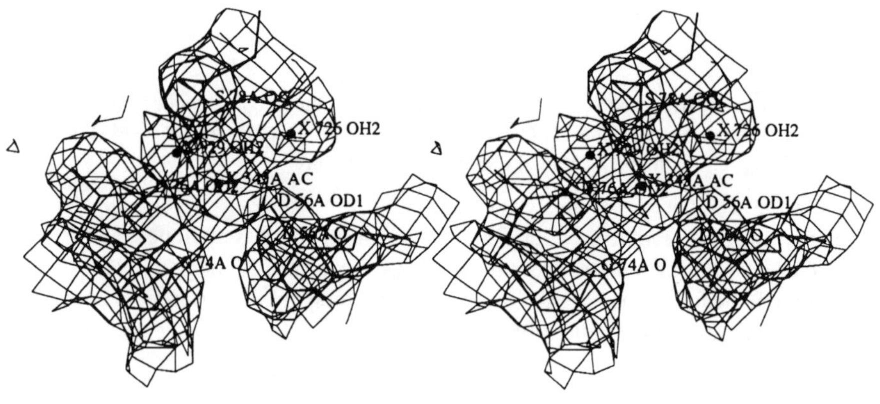

Figure 5: A stereo pair showing the electron around the Ca^{2+} ion found in the distal (top) domain of Fig. 1.

domain is eight coordinated. All the ligating oxygen atoms are from the main and side chains of the following residues Ser^{184}(O+OD1), Asp^{201}(OD1+OD2), Thr^{203}(O+O-G1), Val^{206} (O) and Asp^{208} (OD1), ligands equivalent to those observed in the LiP structure [7]. The distal domain of the structure contains the other Ca^{2+} ion. This ion is heptacoordinated with five ligands from the main and side chains of the residues: Asp^{56}(O+OD1),Gly^{74}(O), Asp^{76}(OD2) and Ser^{78}(O6), two water molecules constitute the remaining ligands. Fig. 5 shows the electron density around this Ca^{2+} ion.

As the crystals are prepared with large concentrations of $MgCl_2$ (0.35 M) it could be expected that Mg^{2+} had replaced some of the Ca^{2+} ions in the structure. To check this, refinements were performed with either Mg^{2+} or Ca^{2+} ions on the sites. The ligand distances to the central metal ion do not differ significantly if the Ca^{2+} ions are replaced with Mg^{2+}, but significant differences are observed in the thermal parameters of the cations. Table 3

Table 3: Thermal parameters ($\mathring{A}^2$) including either Ca or Mg at the proximal and distal metal binding sites during the refinement. The numbers in parentheses are average B-factors for the ligands.

Mg–refinement		Ca–refinement	
distal site			
Mg345a	B$\leq$ 2.0 (11.6)	Ca345a	B = 15.1 (11.1)
Mg345b	B$\leq$ 2.0 (16.8	Ca345b	B = 14.9 (11.3)
proximal site			
Mg 346a	B$\leq$ 2.0 (11.9)	Ca346a	B = 11.1 (11.6)
Mg346b	B$\leq$ 2.0 (8.9)	Ca346b	B = 10.0 (7.8)

summarizes the results from these refinements, which give equivalent values for the two independent molecules. If Mg^{2+} is introduced at both sites the thermal parameters for Mg^{2+} are less than 2 $\mathring{A}^2$. This corresponds to physically unrealistic values as they are significantly smaller than those of the ligating atoms. The refinement with Ca^{2+} ions gave thermal parameters for the distal Ca^{2+} sites similar in magnitude and comparable to the thermal parameters of the ligand. The agreement between the thermal parameters of the ligands from the two independent molecules is also better. However, it should be noted that the thermal parameters of the Ca^{2+} ion in the distal domain have parameters, that are larger than those of ligating atoms. This could indicate that a very small fraction of the Ca^{2+} ions in this site has been replaced with Mg^{2+}. However, these results are very sensitive to the model employed in the refinement. Refinements based on a data set to 2.6 $\mathring{A}$ resolution gave thermal parameters for the distal Ca^{2+} site that indicated a much larger and unrealistic degree of Mg^{2+} substitution. This illustrates that one should be cautious when using the results derived from data to a medium resolution and also bear in mind that the results may depend on the model used in the description of the structure.

The present structure determination has demonstrated that despite the large concentrations of Mg^{2+} ions only an insignificant amount of the Ca^{2+} ions have been replaced with Mg^{2+} ions. This shows that CiP has a large affinity for Ca^{2+} ions and their presence in the structure is essential for the folding of the protein.

4　The Heme Groups

The heme group appears to have a stabilizing effect on the folding of CiP interacting with the two domains of the structure. The catalytic function is intimately connected with the heme group and its surroundings. Through the reaction with hydrogen peroxide iron in the heme group is oxidised from Fe(III) to Fe(IV) and a free radical associated with the porphyrine ring is formed. The environment of the heme group is shown in Fig. 6. In the native enzyme the formal oxidation state of iron is III and besides the four nitrogen atoms of the porphyrine ring system His[185] form the proximal domain coordinates to iron. The ring nitrogen of the His residue on the opposite side of the porphyrine ring system in the distal domain has a distance about 3 $\mathring{A}$ to the iron atom, which is so long that this His cannot be considered as a ligand. Electron density that can be attributed to a water molecule is found between iron and the distal histidine. It has been discussed whether this water molecule is a ligand to iron, thus making iron hexacoordinated. In the related ligninase structure Poulos[7] employed different models in the refinement and found that the Fe-O(water) distance displayed large

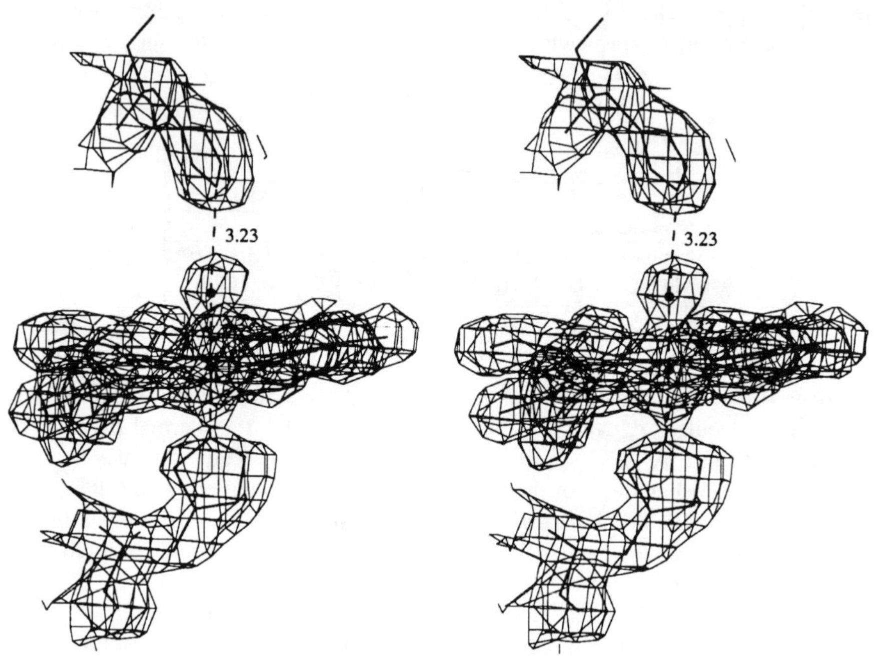

Figure 6: A stereo pair of the electron density of the heme group environment in the peroxidase from *Coprinus cinereus*.

variations with the model used. If virtually all restraints are released, this distance was 2.7 Å and conversely a distance of 2.4 Å was found, if conventional restraints were used, showing that X-ray crystallography is not the most sensitive method for the study of metal ligand interactions.

Acknowledgements

The Centre for Crystallographic Studies is supported by The Danish National Research Foundation. This work was also supported by the Danish Technology Council, Grant B3/001-90.0090.

References

[1] A.T. Brünger. X-PLOR Version 3.0. Yale University, New Haven, 1992.

[2] L. Baunsgaard, H. Dalbøge, G. Houen, E.M. Rasmussen and K.G. Welinder, Amino acid sequence of Coprinus macrorhizius peroxidase: and cDNA sequence encoding Coprinus cinereus peroxidase: a new family of fungal peroxidases. *Eur. J. Biochem.*, **213**, 605-611 (1993).

[3] M. Kjalke, M.B. Andersen, P. Schneider, B. Christensen, M. Schülein and K.G. Welinder Comparison of structure and activities of peroxidases from Coprinus cinereus, Coprinus macrorhizus and Arthromyces ramosus, *Biochim. et Biophys. Acta*, **1120**, 248 (1992).

[4] J.F.W. Petersen, J.W. Tams, J. Vind, A. Svensson, H. Dalbøge, K.G. Welinder and S. Larsen Crystallization and X-ray Diffraction Analysis of Recombinant Coprinus cinereus Peroxidase, *J. Mol. Biol.*, **33**, 491-497 (1993).

[5] J.F.W. Petersen, A. Kadziola and S. Larsen Three-dimensional structure of a recombinant peroxidase from Coprinus cinereus at 2.6 Å resolution, *FEBS Letters*, **314**, (1994), to appear.

[6] B.C. Finzel, T.L. Poulos and J. Kraut Crystal structure of Yeast Cytochrome c peroxidase Refined at 1.7 Å resolution, *J. Biol. Chem.*, **259**, 13027-13036 (1984).

[7] T.L. Poulos, S.L. Edwards, H. Wanishi and M.H. Gold, Crystallographic Refinement of Lignin Peroxidase at 2 Å, *J. Biol. Chem.*, **268**, 4429-4440 (1993).

[8] C.E. Bräden and T.A. Jones, Between objectivity and subjectivity *Nature*, **283**, 687 (1991).

[9] P.V. Luzzati, Traitement Statistique des Erreurs dans la Determination des Structures Cristallines, *Acta Cryst.*, **5**, 802-810 (1952).

[10] A. Roussel and C. Cambillau, Turbo-Frodo, Biographics, LCCMB-CNRS, Bvd, Pierre Dramard, F-13916 Marseille Cedex 20, France 1989.

Function and Three-Dimensional Structure of Proteins using Nuclear Magnetic Resonance Spectroscopy

Flemming M. Poulsen

Carlsberg Laboratorium, Kemisk Afdeling, Gamle Carlsberg Vej 10, DK-2500 Valby, Copenhagen, Denmark

Abstract

Although the three-dimensional structure of a protein can provide valuable information and stimulate rational investigation of other important features of the protein it is important to stress that a structure *per se* is rarely a revelation of the biological function of the protein. This paper emphasizes the importance of acquiring results that measure the fundamental physical chemical parameters in protein function events and the importance of getting quantitative information to support our understanding of the link between physical parameters that describe function and the biological relevance of a protein molecule. It is emphasized that nuclear magnetic resonance spectroscopy, because it combines the ability of measuring three–dimensional structure and the ability of measuring many physical parameters related to both structure and function, is one of the key techniques in structural biology.

1   Introduction

Studies of function and structure are intimately associated in the aims to understand the biological roles of proteins. In recent years the emphasis in structure and function studies of proteins has been strongly directed towards structure determination. A very large number of protein structures have been solved and in many cases three-dimensional structures have immediately provided important information about protein function. The three-dimensional structure of a protein, though, does not directly provide a quantitative description of the functional aspects of the protein as for example association constants, rate constants and thermodynamic properties. One of the aims in structural biology is to compile the experience required to derive this quantitative description of protein function from the geometrical organization of the functional groups in the protein structure alone. In order to achieve this, further advance of research a very strong emphasis on protein function is required. Nuclear magnetic resonance spectroscopy combines the capability of quantitative studies of function with a potential of three-dimensional structure determination for a large group of proteins in the size range up to 30 kDa. This is one of the major reasons for the success of the method. This paper therefore advocates that more effort is addressed towards understanding protein function, and describes NMR spectroscopy as one of the most useful techniques for the advancement of this aspect of protein science.

NMR spectroscopy and x-ray crystallography are the only two methods available today for three-dimensional structure determination at the atomic level of proteins [1, 2, 3]. This paper is especially concerned with the application of nuclear magnetic resonance spectroscopy in structural biology. As an introduction to this it is relevant to contemplate some of the questions that are typically being addressed in this field. Subsequently, it may be of interest to examine why, in particular, NMR spectroscopy is so excellently suited to assist the structural biologist. The consideration is that understanding the mechanisms of function is the overall aim; solving structures only provides one prerequisite of many others necessary to unravel the complexity of protein function. The advantages of NMR spectroscopy in this particular field will be emphasized.

1.1   Function versus Structure in Structural Biology

One aim of structural biology is to link information about biological function and structural organization of biological molecules. Proteins play numerous important roles in the living cells, and understanding these roles, therefore, is closely associated with insight into the characteristics of the individual protein molecules at the atomic level. This has led to an enormous interest in three-dimensional structures of protein molecules and the methods that can determine such structures have received great attention. Vast amounts of research have been devoted to develop the techniques that can provide us with accurate three-dimensional outlines of protein structures. The result has been an impressive increase in the number of protein structures determined with an equally impressive and still increasing rate of one structure a day. The techniques may still see breakthroughs that will improve both speed, efficiency, and accuracy of the structures. So in the future we may envisage protein structure determination methods becoming fully or at least partly automated routine facilities in the structural biology research groups. However, with these developments in technology the scientific challenge in structure determinations of proteins *per se* will disappear and we may envisage the front lines moving towards more complex and important structures with the challenges to solve structures such as the ribosome, the membrane bound receptors, and the cell skeleton and many other complex biostructures that we know only little about at present as well as systems that can not easily be determined on a routine basis.

With the wealth of structures available and an even larger number in the future, an increasing interest is envisaged in advancing the methods for understanding how these structures are effectual for the living cell. But, even with the assistance of a three-dimensional structure many aspects of a structure-function correlation are not always so easily understood. To unravel the function of a protein many sets of additional information are needed that may be so complex that computer technology will be required to correlate structure and function. It may be suggested that we consider the problems of understanding proteins and their functions as a multi-dimensional problem where the geometrical structure is only one set of dimensions. Therefore, solving the three-dimensional structure is only one prerequisite of solving the structure-function relationship.

With more than a thousand protein structures determined already we are facing the enormous task of measuring the parameters that describe the other dimensions of protein structure function relationships and then correlate these with structure. This is to address questions like: How are the physical characteristics of ligand binding and enzyme catalysis related to the structural geometry, for example the binding constants, the thermodynamics, the "on" and "off" rates, the rates of catalysis? Can we rationalize protein properties like stability from the details in the geometry of the structure, and the effects on structure and functional parameters of pH, ionic strength, temperature? Does the geometry of the ligand/substrate binding site add on to the understanding of the biological function of the protein?

Protein structures are also characterized by a variety of dynamic processes with large and small amplitudes and time constants ranging from nanoseconds to thousands of seconds. A relevant question is therefore: Does the static three-dimensional structure of proteins provide a realistic description of a protein? And: Are the dynamic properties of a protein structure relevant for the function? And if so, how?

One of the most challenging problems in molecular and structural biology is the unravelling of the folding mechanism of proteins. The emphasis here is to understand the pathways of the folding process as well as the driving forces, and when these processes are understood, questions regarding the correlation between amino acid sequence, secondary structure, tertiary structure and folding pathways are to be solved. In this context one of the most exciting observations has been that protein structures with completely different amino acid sequences can fold up into very closely related three-dimensional structures as well as the finding that proteins with very similar structures may have very different functions. These observations may seem to be a blow to the traditional structure prediction attempts as well as to certain efforts to correlate structure and function. However, larger samples of non homologous proteins that are topologically similar and fold into similar three-dimensional structures may provide an excellent source for understanding the relations between primary and three-dimensional structure.

To most of these questions there is no straight positive answer, but it is encouraging that work required to advance research of these aspects in molecular and structural biology is in progress and significant results are being achieved. The task is immense requiring considerable and combined efforts from experiments, computing, and theoretical chemistry. Given that resources are not unlimited the rational solution to this problem will probably be to subject a selected set of proteins to the entire ensemble of examinations essential for a quantitative understanding of all the facets of a protein structure. Eventually, the knowledge will become so sophisticated that all the functional aspects of a protein can be obtained from the precise details of its structure and its dynamics.

2   Proteins and NMR Spectroscopy

NMR spectroscopy as compared to x-ray crystallography is the younger of the two techniques. The first protein structure determined by NMR spectroscopy was described in 1985 [4]. At the time, this was a major event in protein science because it was the first time a protein structure had been determined from measurements of a protein dissolved in water. Till then, the only source to structures of proteins had been x-ray diffraction of proteins in the crystal state. For the first time it became possible to compare structures determined in the two phases and subsequently several examples have shown that protein structures are normally folded in a very similar fashion in crystal and in solution [1]. This is most reassuring, and there have been only minor differences reported for structures determined in both phases. In a few cases though, significant differences between structures determined by the two techniques have been observed that did not originate from erroneous structure determination [2, 5].

NMR spectroscopy is the only method available for determination of three-dimensional structures at the atomic level in solution. In addition to this, NMR spectroscopy has a potential for the measuring of important physical and chemical properties at the atomic level that often cannot be achieved by other techniques. This is the advantage of NMR spectroscopy over other methods, that it from direct measurements has the potential to provide the multi-dimensional description at the atomic level of the protein structure that is required for a comprehensive understanding of protein function. At the same time the method measures proteins in solution which is a condition closely related to the natural environment of most proteins.

2.1 Application of NMR Spectroscopy to Protein Studies

This paragraph will describe in outline a set of examples of the application of NMR spectroscopy to studies of proteins. These examples are in some cases illustrated by examples from the author's work. Otherwise, references to recent work of other research groups are provided. This presentation of proteins and NMR spectroscopy is not meant to be comprehensive. NMR like any other technique certainly has its limitations [6]. These have not been dealt with in details here, primarily to keep clear the scope of this presentation, which has been primarily to list the possibilities of NMR spectroscopy to studies of the function of proteins and such factors that may be important for comprehension of the fundamental mechanisms.

2.1.1 Protein NMR Parameters

The reason NMR is so well suited to studies of molecules is its ability to record directly the influence of physical and chemical properties at the individual atom in a molecular structure. This is because the individual NMR active atom is characterized by one signal, the nuclear magnetic resonance, and very often this individual signal can be resolved. The proton, the ^{1}H nucleus, is the most abundant NMR active nucleus present in organic materials. Similarly in proteins, hydrogen is the most important NMR reporter atom because every single proton in a protein has an NMR signal that in principle can be measured. This signal is characterized by the resonating frequency, the chemical shift. Identification of the individual signals by their chemical shift is the prerequisite for any application of NMR spectroscopy to protein studies. A large number of similar hydrogen atoms in a protein may give difficulties because the chemical shifts of resonances of similar protons often differ only slightly and give rise to spectral overlaps that may hamper the assignment of individual resonances and the direct inspection of them. Therefore, assignment procedures have been developed to facilitate assignment of as many signals in the protein NMR spectrum as possible. These procedures are described in great detail elsewhere [7]. The NMR techniques that have been developed for the assignment of otherwise severely overlapping NMR signals apply principles that increase the dispersion of the overlapping signals [8]. This is accomplished by identification of the signals by the coupling of one nuclear spin to another (see Fig. 1).

A coupling of two nuclei with NMR signals at different frequencies can be recorded in a two-dimensional data matrix and identified in the matrix at positions given by the coordinates of the two chemical shift values. A coupling of three spins involves three chemical shifts and can be recorded in a three-dimensional matrix and observed in positions of the coordinates of the three chemical shifts. By dispersing the signals into several dimensions the resolution and subsequently the identification of the individual signal has become possible. Further improvements both in identification and resolution have been accomplished by the introduction of the NMR active nuclei ^{13}C and ^{15}N. These nuclei have a very low natural abundance, however, if the proteins of interest can be produced in cell cultures or in microorganisms, they can be grown on isotope enriched media, whereby the isotopes are incorporated into the proteins. This provides further possibilities for signal

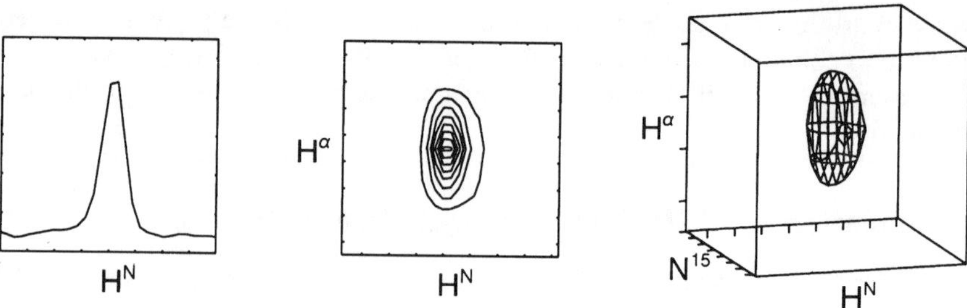

Figure 1: From one– to multidimensional NMR spectroscopy. In a) what could have been the amide resonance H^N in a one dimensional spectrum (one chemical shift axis); in b) a contour diagram of the coupling of H^N and H^α in the chemical structure $\underline{H}^N$-N-C$^\alpha$-$\underline{H}^\alpha$ in a two-dimensional spectrum (two chemical shift axis); in c) the same coupling as in b) as observed from the ^{15}N chemical shift in a three-dimensional spectrum (three chemical shift axis).

recognition and dispersion. These methods are now being widely used and a whole range of hetero– and homonuclear multidimensional techniques are available for the detection of the coupling of these nuclei. The result is that for a protein molecule it is possible to obtain signal assignments for virtually all NMR active nuclei. It thereby becomes possible to use signals of these to monitor essentially any perturbation of the molecular environment of each individual nucleus in the protein.

2.1.2   Protein stability, pK Values

In the folded protein the chemical shifts of the individual nuclei depend on the environment of the nuclei in the structure. In particular carbonyl groups, aromatic side chains and charged groups can influence the chemical shift. Any modification in the structure, therefore, can cause a change of a chemical shift of a nucleus in the vicinity of the structural reorganization. Therefore, any perturbation of the structure that is large enough to induce a change in chemical shift can be monitored by NMR spectroscopy. At one extreme the folding and unfolding of a protein can often be monitored directly by NMR spectroscopy because the structure dependent chemical shift dispersion collapses as the structure does (see Fig. 2). Folding and unfolding of a protein induced by temperature, pH, denaturants like urea or guanidinium chloride can be monitored directly by NMR spectroscopy and provide information about the stability, the pH dependence of unfolding, temperatures of denaturation, folding and unfolding rates, and the thermodynamics of the equilibrium process.

In the pH range where the protein is stable NMR can provide detailed information about the pK values of individual ionizable groups in the structure, and monitoring the atoms affected by the protonation or deprotonation of a given group can provide information about conformational changes resulting from this. In particular, it is important to know pK values of groups in the catalytic sites of enzymes as these can be important guidelines for our understanding of the reaction mechanisms of the enzyme. However, measuring the pK values of other groups in a protein can be of great importance for understanding the stability of the protein.

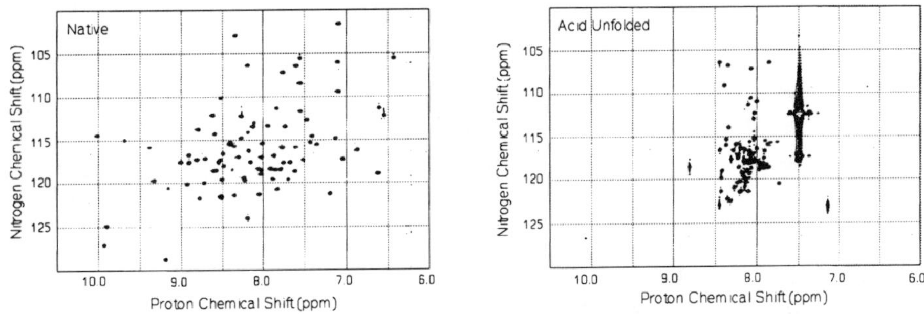

Figure 2: Two-dimensional ^{15}N single quantum coherence spectrum of the protein acyl coenzyme A binding protein. ^{1}H chemical shifts are marked on the horizontal axis, and ^{15}N chemical shift on the vertical axis. a) at pH 7.0 where the protein is folded; and b) at pH 1.0 where the protein is unfolded. The chemical shift dispersion is dramatically reduced for the unfolded protein particular for the amide hydrogen shifts.

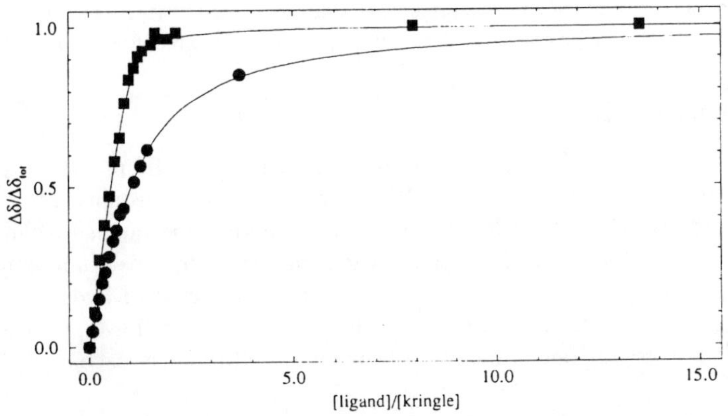

Figure 3: Binding curves for trans-aminomethyl-cyclohexane carboxylic acid, a lysine analog, to plasminogen kringle-4 and to a protein engineered derivative, where Arg71 has been substituted with Gln. The binding of the ligand to the modified protein is reduced considerably suggesting that the arginyl residue is relevant for the binding of the ligand. Squares represent measurements of the kringle binding; circles measurements of the derivative.

2.1.3 Ligand Binding

The binding of small molecules to proteins can give rise either to small conformational changes or influence charges in groups near the binding site. These reorganizations often result in changes in the chemical shifts of resonances of the involved nuclei and by monitoring these relevant information regarding binding constants, "on" and "off" rates, conformation of the ligand[9] as well as perturbations in the protein can be obtained from the NMR data (see Fig. 3[10]).

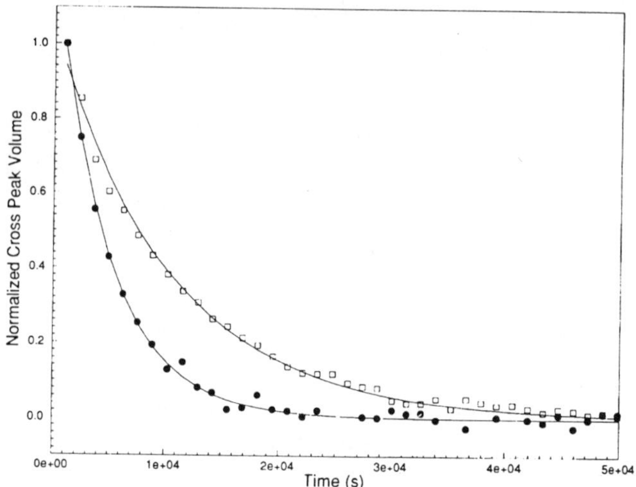

Figure 4: Amide hydrogen exchange of one individual amide group in acyl coenzyme A binding protein The results have been obtained from a series of ^{15}N nuclear single quantum coherence spectra (see Fig. 2). The intensity of the amide resonance is reduced as the H^N is replaced by deuterium in a hydrogen deuterium exchange reaction. The experiments have been recorded at pH 7.4 and 8.1.

2.1.4   Reaction Kinetics

In cases where a protein is involved in a chemical reaction NMR may be instrumental in measurements of the reaction kinetics[9] and measure rate constants in the range of milliseconds and upwards either directly or by rapid mixing and quench techniques. In the simplest case the disappearance of an NMR signal from the spectrum as a result of the chemical reaction can be followed (see Fig. 4). In the case of hydrogen exchange of the amide groups in a protein all exchanging amides can be followed in one series of experiments demonstrating one of the real advantages of NMR spectroscopy.

2.1.5   Protein Engineering

The modification typically introduced in a protein structure by protein engineering[11] is the substitution of one amino acid residue for another. Such changes are often introduced in order to change a property of the protein, the stability, enzyme specificity, or they may be made in order to prove or disprove the relevance of potentially important groups in the active site of the protein. In order to understand the effects of such modifications it is important to have methods that can measure the impact on the structure. NMR spectroscopy offers an easy method to examine the effects of such modifications. Subsequently the methods mentioned above can be used to further investigate the effects of the protein engineering (see Fig. 3[10]).

2.1.6   Protein Folding

An important task in structural biology is understanding the mechanisms of protein folding [12, 13, 14, 15, 16, 17, 18, 19, 20, 21, 22, 23, 24, 25, 26, 27, 28]. NMR spectroscopy is a key methodology in protein folding research because no other techniques offer methods that can study protein folding intermediates at the atomic level. Intermediates in protein folding

Figure 5: See Color Plate 2, page xiv.

Figure 6: See Color Plate 3, page xv.

as well as several partly unfolded forms of proteins are attracting much interest. The molten globule is just one of these forms. If these are intermediates of the protein folding pathway it is of considerable interest to study the structure of these forms in a context of the known and folded form. Also for the folding mechanism it is relevant to measure which parts of the secondary structure is being established first. Here the techniques of pulse labelling using quench flow techniques have provided important information for several proteins. These techniques measure the time at which the hydrogen bond formation is established in the folding process by recording the degree of amide hydrogen exchange at different times in the folding sequence. These methods are only informative when the assignment of each of the individual amide hydrogen resonances is known for the protein being studied.

2.1.7 Protein Structure Determination

NMR spectroscopy[8] can measure the distance between two nuclei by recording the size of the dipolar coupling between the two. This coupling is measured by the strength of the so called nuclear Overhauser effect. In a small globular protein it is often possible to determine the distance between a thousand or more pairs of nuclei, typically ^{1}H atoms (see Fig. 5[29, 30, 31]).

NMR spectroscopy also offers methods to determine dihedral angles both in the peptide backbone and in the amino acid side chains. By a combination of this information and the covalent geometry of the amino acid residues it is possible by the assistance of a number of structure calculation programs to derive a set of three-dimensional structures that are in agreement with the experimentally determined structural constraints. A prerequisite for such a structure determination is a set of fully assigned NMR spectra.

Protein structure determination may also be applied to determination of the structure of a protein-ligand complex. In this case the additional task is to measure nuclear Overhauser effects between nuclei of ^{1}H atoms in the protein and the ligand (see Fig. 6[29, 30, 31]).

2.1.8 Water Molecules in Protein Structures

Water molecules in protein structures[32, 33, 34, 35] often play a structural or a functional role. They are important parts of the structure and it is of interest to determine their position in the structures. NMR spectroscopy can determine water molecules by the nuclear Overhauser effects between the protons of the water and close hydrogen atoms of the protein, like for any other ligand. In addition to these NMR spectroscopy has methods by which the exchange rate of the water molecules in proteins are interchanged with bulk solvent molecules.

2.1.9 Protein Dynamics

Protein molecules are not static. Are, however, the dynamic properties of proteins relevant for their function and stability? NMR can measure the dynamics of proteins[35, 36, 37, 38, 39, 40] at a whole range of time constants from the picosecond time scale to seconds. By measuring the dynamic properties of the individual groups either in the backbone or

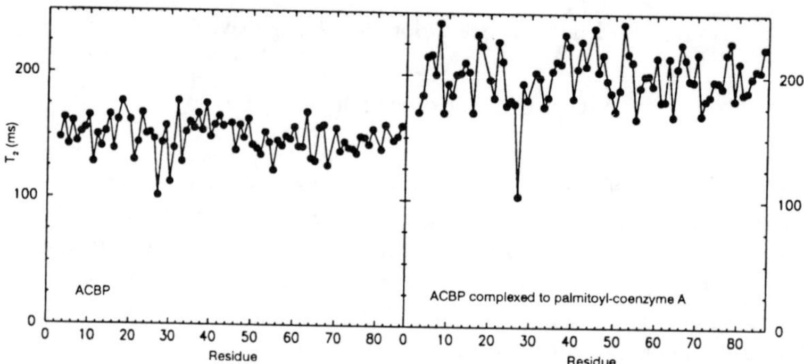

Figure 7: Measurements of the transverse relaxation time, T_2, for the peptide backbone ^{15}N nuclei of fully ^{15}N enriched acyl coenzyme A binding protein. a) of the free protein; and b) of the protein ligand complex. The horizontal axis marks the residue number of the ^{15}N nuclei, the vertical is marking T_2. Apart from local variations a global shift in the average T_2 relaxation from 150 ms in (a) to 200 ms in (b) suggests that the free protein has additional dynamic properties that influence the transverse relaxation. A contribution that apparently disappears upon ligand binding.

in the sidechains of a protein information is obtained that may provide an answer to many of the questions related to protein dynamics. Amide hydrogen exchange was one of the first indications of protein dynamics. With NMR spectroscopy it is possible to measure the kinetics for each individual peptide group in the protein. The kinetic information of amide hydrogen exchange is being employed in protein folding studies, both in the pulse labelling experiments and in determination of protection of the individual amides in the unfolded or partly folded forms of the protein. Other NMR measuring techniques can provide information about protein dynamics. Protein relaxation measurements are providing information about the overall tumbling time of the protein molecule as well as sub– and super-nanosecond dynamics at individual sites in the structure. These methods have revealed sites in the protein structure where the dynamics are correlated either to structural or to functional properties of the protein (see Fig. 7).

3   Summary

There is strong emphasis on structure determination of proteins. This is because knowledge of the three-dimensional structure is considered a key to understanding protein function and also to understanding the sequence and structure relationship to the protein folding problem. A geometrical outline of a structure is indeed most useful for our understanding of proteins, however, the structure on its own is not a sufficient information. Additional information is required, and perturbation of the structures is often necessary for improving our knowledge and understanding. NMR spectroscopy is a method excellently suited to provide not only the structural information but also the additional information. In contrast to any other method NMR can provide useful results in the characterization of protein stability, electrostatic properties (pKs), ligand binding, enzymatic reactions, derivatives of protein engineering, the three-dimensional structure of proteins, the structures of protein-ligand complexes, the presence of structural and functional water molecules, protein dynamics and protein mobility. NMR spectroscopy is also a most useful technique when applied in combination with other techniques, such as fluorescence, circular dichroism, and x-ray crystallography.

Acknowledgements

I wish to acknowledge the contributions to this work of my colleague members of the Protein Structure and NMR Group at Carlsberg Laboratory, Kim Vilbour Andersen, Birthe Brandt Kragelund, Mogens Kjær, Jens Chr. Madsen, Peter Reinholt Nielsen, Peter Osmark, Christian Rischel, Pia Skovgaard, Poul Sørensen and Niels Kirk Thomsen. Also thanks to Professors Jens Knudsen and Karsten Kristiansen of Odense University for their collaboration on acyl Coenzyme A binding protein, and to Professor H. Chr. Thøgersen for the collaboration on Plasminogen Kringle-4 Domains. FMP is a member of the Danish Protein Engineering Research Centre, PERC.

References

[1] M. Billeter, Comparison of Protein Structures Determined by NMR in Solution and by X-Ray Diffraction in Single Crystals. *Q. Rev. Biophys.*, **3**, 325-377 (1992).

[2] W. Braun, M. Vasak, A. H. Robbins, C. D. Stout, G. Wagner, J. H. R. Kägi and K. Wüthrich, Comparison of the NMR Solution Structure and the X-Ray Crystal Structure of Rat Metallothionein-2, *Proc. Natl. Acad. Sci. USA*, **89(21)**, 10124-10128 (1992).

[3] B. Shaanan, A. M. Gronenborn, G. H. Cohen, G. L. Gilliland, B. Veerapandian, D. R. Davies and G. M. Clore, Combining Experimental Information from Crystal and Solution Studies - Joint X-Ray and NMR Refinement, *Science*, **257**, 961-964 (1992).

[4] M. P. Williamson, T. F. Havel and K. Wüthrich, Solution Conformation of Proteinase Inhibitor IIA from Bull Seminal Plasma by 1H Nuclear Magnetic Resonance and Distance Geometry, *J. Mol. Biol.*, **182**, 295-315 (1985).

[5] M. Ikura, G. M. Clore, A. M. Gronenborn, G. Zhu, C. B. Klee, and A. Bax. Solution Structure of a Calmodulin-Target Peptide Complex by Multidimensional NMR. *Science* **256**, 632-638 (1992).

[6] G. Wagner, Prospects for NMR of Large Proteins, *J. Biomol. NMR*, **3**, 375-385 (1993).

[7] K. Wüthrich, NMR of proteins and Nucleic Acids, Wiley, New York, U.S.A., 1986.

[8] A. Bax, and S. Grzesiek, Methodological Advances in Protein NMR, *Account. Chem. Res.*, **26**, 131-138 (1993).

[9] G. Otting, Experimental NMR Techniques for Studies of Protein Ligand Interactions, *Curr. Opin. Struct. Biol.*, **3**, 760-768 (1993).

[10] P. R. Nielsen, K. Ejnar-Jensen, T. Holtet, B. D. Andersen, F. M. Poulsen and H. C. Thøgersen, The Protein-Ligand Interactions in The Binding Site of Plasminogen Kringle 4 in Solution and Crystals are Different. Electrostatic Interactions Studied by Site Directed Mutagenesis Exclude Lys35 as an Important Acceptor in Solution, *Biochemistry*, **32**, 13019-13025 (1993).

[11] A. Fersht and G. Winter, Protein Engineering, *Trends. Biochem. Sci.*, **17**, 292-294 (1992).

[12] D. Barrick and R. L. Baldwin, The Molten Globule Intermediate of Apomyoglobin and the Process of Protein Folding, *Protein Sci.*, 869-876 (1993).

[13] D. Neri, M. Billeter, G. Wider and K. Wüthrich, NMR Determination of Residual Structure in a Urea-Denatured Protein, the 434-Repressor, *Science*, **257**, 1559-1563 (1992).

[14] M. Buck, S. E. Radford and C. M. Dobson, A Partially Folded State of Hen Egg White Lysozyme in Trifluoroethanol - Structural Characterization and Implications for Protein Folding, *Biochemistry*, **32**, 669-678 (1993).

[15] H. J. Dyson, G. Merutka, J. P. Waltho, R. A. Lerner and P. E. Wright, Folding of Peptide Fragments Comprising the Complete Sequence of Proteins - Models for Initiation of Protein Folding. 1. Myohemerythrin, *J. Mol. Biol.*, **226**, 795-817 (1992).

[16] H. J. Dyson, J. R. Sayre, G. Merutka, H. C. Shin, R. A. Lerner and P. E. Wright, Folding of Peptide Fragments Comprising the Complete Sequence of Proteins - Models for Initiation of Protein Folding. 2. Plastocyanin, *J. Mol. Biol.*, **226**, 819-835 (1992).

[17] P. A. Jennings and P. E. Wright, Formation of a Molten Globule Intermediate Early in the Kinetic Folding Pathway of Apomyoglobin, *Science*, **262**, 892-896 (1993).

[18] S. E. Radford, C. M. Dobson and P. A. Evans, The Folding of Hen Lysozyme Involves Partially Structured Intermediates and Multiple Pathways, *Nature (London)*, **358**, 302-307 (1992).

[19] P. Sørensen, J. R. Winther, N. C. Kaarsholm and F. M. Poulsen, The Pro-Region required for Folding of Carboxypeptidase Y is a Stable Molten Globule-like Structure, *Biochemistry*, **32**, 12160-12166 (1993).

[20] C. L. Chyan, C. Wormald, C. M. Dobson, P. A. Evans and J. Baum, Structure and Stability of the Molten Globule State of Guinea-Pig alpha-Lactalbumin - A Hydrogen Exchange Study, *Biochemistry*, **32**, 5681-5691 (1993).

[21] A. Miranker, C. V. Robinson, S. E. Radford, R. T. Aplin and C. M. Dobson, Detection of Transient Protein Folding Populations by Mass Spectrometry, *Science*, **262**, 896-900 (1993).

[22] H. Vandael, P. Haezebrouck, L. Morozova, C. Aricomuendel and C. M. Dobson, Partially Folded States of Equine Lysozyme - Structural Characterization and Significance for Protein Folding, *Biochemistry*, **32**, 11886-11894 (1993).

[23] J. P. Waltho, V. A. Feher. G. Merutka, , H. J. Dyson and P. E. Wright, Peptide Models of Protein Folding Initiation Sites. 1. Secondary Structure Formation by Peptides Corresponding to the G-Helice and H-Helice of Myoglobin, *Biochemistry*, **32**, 6337-6347 (1993).

[24] H. C. Shin, G. Merutka, J. P. Waltho, P. E. Wright and H. J. Dyson, Peptide Models of Protein Folding Initiation Sites. 2. The G-H Turn Region of Myoglobin Acts as a Helix Stop Signal, *Biochemistry*, **32**, 6348-6355 (1993).

[25] H. C. Shin, G. Merutka, J. P. Waltho, L. L. Tennant, H. J. Dyson and P. E. Wright, Peptide Models of Protein Folding Initiation Sites. 3. The G-H Helical Hairpin of Myoglobin, *Biochemistry*, **32**, 6356-6364 (1993).

[26] P. Osmark, P. Sørensen and F. M. Poulsen, Context dependence of protein secondary structure formation. The three-dimensional structure and stability of a hybrid between Chymotrypsin inhibitor 2 and Helix E from Subtilisin Carlsberg, *Biochemistry*, **32**, 11007-11014 (1993).

[27] A. R. Fersht, Protein Folding and Stability - The Pathway of Folding of Barnase, *FEBS Lett.*, **325**, 5-16 (1993).

[28] J. Eder, N. Rheinnecker and A. R. Fersht, Folding of Subtilisin BPN' — Role of the Pro-Sequence, *J. Mol. Biol.*, **233**, 293-304 (1993).

[29] K. V. Andersen and F. M. Poulsen, Three-dimensional structure of acyl-coenzyme A binding protein from bovine liver. Structural refinement using heteronuclear and multi–dimensional NMR spectroscopy, *J. Biomol. NMR*, **3**, 271-284 (1993).

[30] J. Knudsen, S. Mandrup, J. T. Rasmussen, P. H. Andreasen, F. M. Poulsen and K. Kristiansen, The function of acyl-CoA-binding protein (ACBP)/Diazepam binding inhibitor, *Molecular and Cellular Biochemistry*, **123**, 129-138 (1993).

[31] B. B. Kragelund, K. V. Andersen, J. C. Madsen, J. Knudsen and F. M. Poulsen, Three-dimensional structure of the complex between acyl-coenzyme A binding protein and palmitoyl-coenzyme A, *J. Mol. Biol.*, **230**, 1260-1277 (1993).

[32] R. M. Brunne, E. Liepinsh, G. Otting, K. Wüthrich and W. F. van Gunsteren, Hydration of Proteins - A Comparison of Experimental Residence Times of Water Molecules Solvating the Bovine Pancreatic Trypsin Inhibitor with Theoretical Model Calculations, *J. Mol. Biol.*, **231**, 1040-1048 (1993).

[33] G. M. Clore, M. A. Robien and A. M. Gronenborn, Exploring the Limits of Precision and Accuracy of Protein Structures Determined by Nuclear Magnetic Resonance Spectroscopy, *J. Mol. Biol.*, **231**, 82-102 (1993).

[34] J. D. Forman-Kay, A. M. Gronenborn, P. T. Wingfield, and G. M. Clore, Determination of the Positions of Bound Water Molecules in the Solution Structure of Reduced Human Thioredoxin by Heteronuclear Three-Dimensional Nuclear Magnetic Resonance Spectroscopy, *J. Mol. Biol.*, **220**, 209-216 (1991).

[35] E. Liepinsh, G. Otting, and K. Wüthrich, NMR Observation of Individual Molecules of Hydration Water Bound to DNA Duplexes - Direct Evidence for a Spine of Hydration Water Present in Aqueous Solution, *Nucleic. Acids. Res.*, **20**, 6549-6553 (1992).

[36] G. Wagner, NMR Relaxation and Protein Mobility, *Curr. Opin. Struct. Biol.*, **3**, 748-754 (1993).

[37] J. Clarke, A. M. Hounslow, M. Bycroft and A. R. Fersht, Local Breathing and Global Unfolding in Hydrogen Exchange of Barnase and Its Relationship to Protein Folding Pathways, *Proc. Natl. Acad. Sci. USA*, **90**, 9837-9841 (1993).

[38] T. G. Pedersen, N. K. Thomsen, K. V. Andersen, J. C. Madsen, and F. M Poulsen, Determination of the rate constants k1 and k2 of the Linderstr$\dot{Z}$m-Lang model for protein amide hydrogen exchange. A study of the individual amides in hen egg white lysozyme, *J. Mol. Biol.*, **230**, 651-660 (1993).

[39] N. K. Thomsen and F. M. Poulsen, Low energy of activation for amide hydrogen exchange reactions supports a local unfolding model, *J. Mol. Biol.*, **234**, 234-241 (1993).

[40] B. L. Grasberger, A. M. Gronenborn and G. M. Clore, Analysis of the Backbone Dynamics of Interleukin-8 by N-15 Relaxation Measurements, *J. Mol. Biol.*, **230**, 364-372 (1993).

Experimental Aspects of Ultra-violet and Circular Dichroism Methods for Protein Folding

Hans E.M. Christensen[*], Jan M. Hammerstad-Pedersen[†§], Arne Holm[‡], Gitte Iversen[†] and Jens Ulstrup[†]

[*] Institute of Molecular and Cell Biology, National University of Singapore, Singapore
[†] Bioinorganic Group, Chemistry Department A, The Technical University of Denmark, Lyngby, Denmark
[‡] The Royal Veterinary and Agricultural University, Frederiksberg, Denmark

[§] To whom correspondence should be addressed

Abstract

Synthetically made peptide chains can be folded around metals to produce either exact copies of natural metalloproteins, smaller analogues or 'mutated' analogues. This offers an important alternative to microbiological mutagenesis. Furthermore, with chemical synthesis it is possible to insert unnatural segments into the peptide chain, thereby allowing us to probe the influence of specific structural changes on the protein function. The folding of the peptide chain around metal ions can be followed by UV/VIS spectroscopy, and the folding geometry of both the metal-centre and the peptide chain can be described by a range of spectroscopic methods, eg. UV/VIS, CD and resonance Raman spectroscopy.

We have successfully synthesized $D.\ gigas$ rubredoxin, which contains iron coordinated to four cysteines at the redox site. Also two different 25-residue analogues have been prepared after careful design of the peptide to mimic the properties of the native rubredoxin. Charge-transfer from sulphur to iron gives rise to absorption bands in the UV/VIS spectrum suitable for following the folding of the peptide chain around the iron. UV/VIS, CD, resonance Raman and mass spectra are available for the folded metalloproteins.

1 Introduction

It is well established that the three-dimensional structure of proteins is determined by their amino acid sequence. This was first elegantly demonstrated by Anfinsen $et\ al.$[1], who denatured and then refolded bovine pancreatic ribonuclease A (RNaseA). He demonstrated that the natural folding could be regained by adding catalytic amounts of a reducing agent (e.g. mercaptoethanol) after denaturation of RNaseA and rearrangement of the S-S bridges. He concluded that the free energy of folding provides the driving force not only for the correct folding but also for the correct pairing of the S-S bridges, which he regarded as fasteners reinforcing the correctly folded structure.

This implies that folding does not occur by a random search of the possible conformations but that the information needed for correct folding of the protein is contained entirely in the primary amino acid sequence. According to this information the polypeptides fold into secondary structures, helices and β strands, separated by turns and peptide segments without any apparent organized structure (random coil). The secondary structures are packed together, and form the three-dimensional structure of the protein. Packing is governed by kinetic and thermodynamic forces[2, 3], including hydrophobic and electrostatic interactions[4, 5], hydrogen bonding and the need for hydration of charged groups. Proteins found in plasmas, i.e. surrounded by water, fold to expose as many of the charged amino acids and other hydrogen bonding side chains as possible to water solvent. The hydrophobic amino acids are placed inside the protein structure with the peptide amide and carbonyl groups paired within the backbone. The regions of random coil are often on the surface of the protein where amide and carbonyl groups can form hydrogen bonds with water. For membrane bound proteins the forces that keep the hydrophobic groups on the surface of water soluble proteins, now exclude them from the interior of the lipid bilayer of the membrane[4]. The part of the protein bound inside the bilayer should therefore be arranged in either α- or β-structures formed primarily by hydrophobic amino acids.

The folding of the peptide chain into the three-dimensional structure is possible because of the ability of the peptide chain to rotate around the single bonds in the backbone. The peptide bond itself is rigid with a partly double-bond nature, but the bonds on each side of the peptide bond are normal single bonds with free rotation. A comparison between the appearance of the different amino acid residues in different secondary structures can be made[6]. From such work it is evident that some of the amino acids are found far more often in certain helices and sheets, and therefore could be promoting that specific structure element. For example, Ala, Leu and Met are often found in helices, Val, Ile and the aromatic amino acid residues in sheets, and Gly and of course Pro mainly in turns. Since Gly is able to accommodate many different geometries, it is a very versatile turn-residue.

2   Designing Peptides and Proteins

Early attempts to mimic the behaviour of metalloproteins were predominately based on inorganic iron sulphur complexes. This evolved into the use of small peptide fragments as ligands to the metal-ion, but there were still major problems trying to mimic especially the redox properties of the proteins[7]. However, during the past decade both gene cloning and chemical peptide synthesis have developed into powerful techniques. These new techniques have made it possible to approach synthetic metalloproteins from new angles. From gene cloning or mutagenesis it is possible to make small alterations to the protein structure. The protein is made by the naturally occurring enzyme systems within the host cell and the folding and incorporation of metals is, therefore, very reliable. Chemical peptide synthesis requires knowledge of the peptide sequence, and only peptides of up to about 100 amino acid residues can be synthesized reliably. Furthermore, the folding and coordination of metals must be given special attention. But chemical synthesis makes it possible to incorporate unnatural amino acids or other species into the peptide chain. A chemical approach which incorporates solid–state peptide synthesis and suitable folding procedures offers similar perspectives as genetically engineered proteins in relation to molecular details of protein structure-function relations. This would relate for example to details of long–range, directional electron transport of redox metalloproteins, specific dependence on the intermediate protein matter, structural surface sites etc. In addition, in chemical metalloprotein synthesis not only

Figure 1: See Color Plate 4, page xvi.

the sequence but also the size of the protein can be controlled. This offers interesting perspectives not found in the microbiological approach in relation to design of artificial proteins with specific, pre–determined properties.

2.1   Synthetic Metalloproteins

Apart from small peptide segments used as ligands in iron-sulphur protein models[7, 8, 9], there have recently also been reports of synthesis of other proteins. Smith *et al.* have synthesized *C. pasteurianum* 2[4Fe–4S]–ferredoxin[10]. UV/VIS, CD and EPR spectra were in good agreement with those of the native protein. They have also demonstrated its redox capability by reducing it using both *C. pasteurianum* hydrogenase and dithionite. Zawadzky and Berg[11] have synthesized both the D– and L–form of *D. desulfuricans* (strain 27774) rubredoxin. The UV/VIS and CD spectra of the 45–residue synthetic proteins are also found to agree well with those of natural rubredoxin. Furthermore, the Fe(II)–form of the proteins can be oxidized by air. A 104–residue horse heart cytochrome c has been synthesized by Di Bello *et al.*[12]. Mitochondria were used to insert the heme–group into the protein.

2.2   Investigation of Electron transfer Routes in Proteins

One very intriguing aspect of redox metalloproteins such as cytochromes, blue copper proteins and iron–sulphur proteins is their ability for long–range electron transfer (ET)[13]. Such ET patterns are frequently described by means of specific surface sites and ET routes between the surface site and the metal ion inside the protein. Areas on the protein surface from which it is easy to transfer electrons to the metal centre are denoted as 'hot–spots'. Using computer models these can be used to identify the most suitable surface–areas for electron transfer, i.e. for docking of redox partners (fig. 1).

Mutants and protein analogues of redox proteins are important for the studies of ET routes because they can help describe the routes on a molecular level. By changing or deleting one amino acid residue in the ET route, details about the ET can be elucidated. Mutagenesis has, therefore, become a very important tool in today's ET research[14, 15, 16]. With some knowledge about protein folding, chemical synthesis of metalloproteins and their 'mutants' is also feasible. However, much is still to be learned about metal containing proteins, and especially about how folding around the metal–ion is accomplished. This paper deals with such issues, and especially with the chemical synthesis of the iron containing redox protein rubredoxin.

3   Rubredoxin

Rubredoxins are a group of small iron containing metalloproteins found in a wide variety of organisms, spanning aerobic and anaerobic bacteria, algae, fungi, higher plants and even mammals. Despite this wide spread and the detailed knowledge of their structure, very little is known about the biological function of rubredoxins. It is generally believed though that they play an important role in electron transfer (ET) reactions. This is substantiated by the

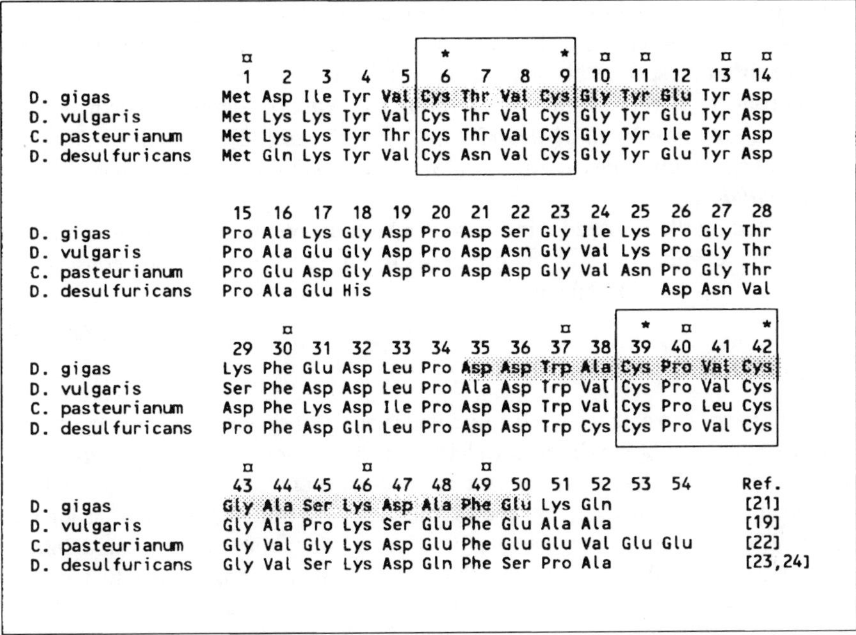

Figure 2: The amino acids of some anaerobic bacterial rubredoxins. The variations in the sequence around the two iron binding sites are small, while the deletion of several residues in the middle of D. desulfuricans rubredoxin could indicate that this part of the protein is less important to its function. The synthetic rubredoxin analogue discussed later in this chapter is composed of the residues in the shaded areas. The active site segments are shown in the boxes. * indicates the iron ligands, and ◻other conserved residues in all of the 10 known rubredoxins from anaerobic bacteria. References are given in the figure.

fact that they can undergo reversible oxidation and reduction, and that they can function as an electron carrier in certain ferredoxin-requiring reactions[17].

Rubredoxin was first discovered in the early sixties as a red fraction during isolation of ferredoxin, which is dark brown. This red colour of the oxidized protein caused the name rubredoxin. It was first purified from *Clostridium pasteurianum* and analyzed by Lovenberg and Sobel[18] in 1965. Within the iron–sulphur proteins rubredoxins are defined by the content of one iron and the absence of acid–labile sulphur.

Rubredoxins from anaerobic bacteria generally contain 50–60 amino acid residues and have a molecular mass around 6000; they contain a single iron directly chelated to the polypeptide chain by four cysteinyl sulphur groups. The iron is bound in a distorted tetrahedral geometry. The chelating site always consists of two Cys-X-Y-Cys segments separated by a peptide chain of varying length. In *Desulfovibrio vulgaris*[19, 20] and *D. gigas*[21] rubredoxin these two segments are identical: Cys6-Thr7-Val8-Cys9 and Cys39-Pro40-Val41-Cys42. In *Clostridium pasteurianum*[22] rubredoxin, however, Val41 of the second segment has been replaced by Leu41, and in *D. desulfuricans*[23, 24] rubredoxin Thr7 of the first segment is replaced by Asn7. Fig. 2 gives the total amino acid sequence of a number of rubredoxins from anaerobic bacteria. A high degree of homology is seen.

From the deletions in the middle of the peptide chain connecting the two chelating segments in *D. desulfuricans* it might be expected that this rather large part of the protein is not important to its function. This part of the protein is also where the least homology between the different species occurs.

Apart from these rubredoxins, the amino acid sequence is also known for six other species: *C. thermosaccharolyticum*[25], *C. perfringens*[26], *Butyribacterium methylotrophicum*[27], *Peptococcus aerogenes*[28], *Megasphera elsdenii*[29] and *Chlorobium thiosulfatophilum*[30].

Even including these rubredoxins there is a high degree of conservation of amino acid residues in the terminal parts of the proteins. In fig. 2 the amino acids conserved in all known rubredoxins are marked with ¤.

3.1 Spectroscopic Properties

Rubredoxins have been characterized with a wide variety of spectroscopic methods. UV/VIS spectra of oxidized rubredoxin are available for 20 species, including *D. gigas*[31] and *C. pasteurianum*[18]. Only five spectra of reduced rubredoxin are available. Oxidized rubredoxin absorbs around 280, 380 and 490 nm with absorption coefficients around 20000, 10000 and 8000 $M^{-1}cm^{-1}$, respectively. Reduced rubredoxin absorbs around 280 (25000 $M^{-1}cm^{-1}$), 310 (10000) and 330 nm (6000). Oxidized rubredoxin has CD bands around 340(-), 400(+) 440(+), 505(-), 560(+) and 630(-), whereas reduced rubredoxin has bands around 310(-) and 330(+)[32]. The bands at 380 nm and 490 nm in the oxidized form, and the bands at 310 nm and 330 nm in the reduced form are assigned to Cys-S $\rightarrow$ Fe charge–transfer in the active site.

4 Synthetic Rubredoxin

D. gigas rubredoxin was chosen because it is a very well characterized protein[21], and because its three-dimensional structure is available through Brookhaven Protein Databank[33]. Furthermore, the active site segments of *D. gigas* rubredoxin are identical to those of *D. vulgaris*, and differ only by one amino acid substitution from those of *C. pasteurianum* and *D. desulfuricans*.

4.1 Designing and synthesizing a 25–residue Rubredoxin Analogue

First we aimed at synthesizing a *D. gigas* rubredoxin analogue consisting of 25 amino acid residues[34]. The residues were selected so that details of the structure–function relations could be expected to be carried over from the native protein. The idea was to make the peptide chain, and to fold the chain correctly around the centre iron, thereby making the spectroscopic and redox properties of the analogue resemble those of the native rubredoxin as closely as possible. By comparing computer models of *D. gigas* rubredoxin and amino acid sequences of the different anaerobic rubredoxins the two segments consisting of residue 5-12 and 35-50 were found to contain the most important structural elements. These two segments are shown in the shaded areas of fig. 2. The loop connecting the two segments is the area where most differences between the sequences are found. As noted, *D. desulfuricans* rubredoxin even has a large deletion in this region. This loop might therefore be envisaged not to be a central element in the general function of rubredoxin. Due to the same lack of conservation in the C–terminal, residue 51–52 were also omitted. The N-terminal residues

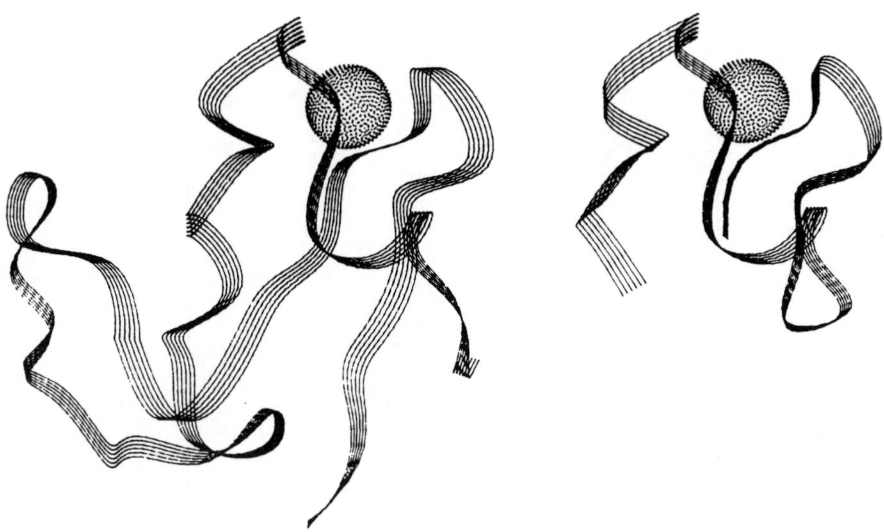

Figure 3: Left: The three-dimensional structure of D. gigas rubredoxin [21]. Right: The calculated three-dimensional structure of the 25-residue rubredoxin analogue. Both structures are drawn as 'ribbon-structures' indicating the backbone folding. The spherical dotted surface shows the position of the iron [34].

(1– 4), on the other hand, have a higher degree of conserved residues, and it cannot be ruled out that its deletion would have more significant implications on the function of the protein.

From computer graphics the shortest gap is 5.4 Å between Val5 and Glu50 ($C\alpha$–$C\alpha$) when rubredoxin is cut into these two fragments. Since it would require some sort of turn to connect the two fragments at this point, Gly seemed a natural choice as a link. This combined 25-residue peptide was energy minimized using the Kollman minimization algorithm provided by Sybyl. The resulting structure showed only minor differences from the original native rubredoxin, primarily in the region around the link. The three–dimensional structure of native *D. gigas* rubredoxin and the predicted structure of the 25–residue peptide including the iron site, is shown in fig. 3. The 25-residue rubredoxin analogue has a calculated molecular mass of 2626 g/mole. Based upon an average of pK_A–values for the individual amino acids, the peptide has a pI of 5.3, and the overall charge is estimated at -6 at pH 7 compared to -8 for the native protein.

The 25-residue peptide chain (H-DDWACPVCGASKDAFEGVCTVGCYE-OH) was synthesized by solid phase peptide synthesis using the Fmoc α-amino protection strategy, and purified by reverse phase chromatography. By dissolving the peptide chain in 8 M urea solution containing dithiothreitol and Fe^{2+}–ions, and slowly lowering the urea concentration, the peptide chain was allowed to fold around Fe^{2+}. The folding process can be followed spectroscopically by rapidly scanning the range 200-700 nm on a diode–array spectrophotometer throughout the reaction. The product revealed UV/VIS absorption bands in good agreement with native rubredoxin[34] (fig. 4). Matrix assisted laser desorption ionization mass spectroscopy (MALDI-MS) spectra of the peptide revealed two peaks, one corresponding to the unfolded peptide and one to the peptide/iron complex. Since non–covalent associations between metal and protein most frequently are found to be labile in

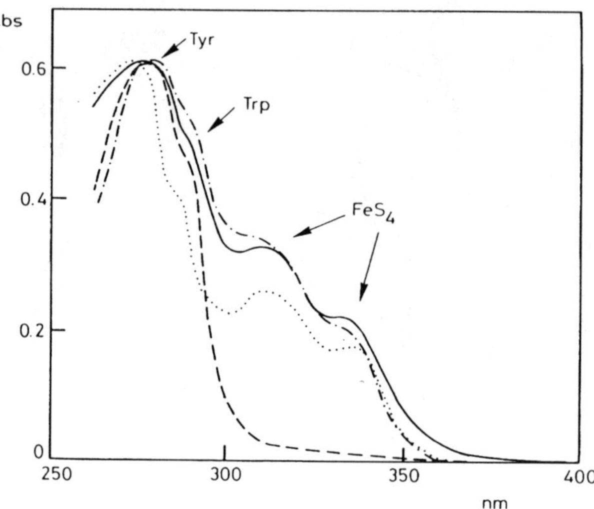

Figure 4: The absorption spectra of the synthesized 25-residue *D. gigas* rubredoxin(II) analogue (——), its Val41Leu 'mutant' (· − ·−), the unfolded peptide (− − −) and of *C. pasteurianum* rubredoxin(II), (· · ·) see [18]. The spectra have been normalized with respect to the 278 nm Tyr absorption maximum. Contributions to the spectrum from different components of the iron- sulphur proteins are indicated.

MALDI-MS, this strongly indicates that the peptide has folded around the iron. However, the 25-residue rubredoxin analogue was very unstable when exposed to oxygen, and all attempts to purify it were unsuccessful. Kept under argon atmosphere, however, the metallopeptide was stable for at least 80 hours.

4.2 Designing Mutants

As discussed earlier the iron binding site in *C. pasteurianum* rubredoxin differs from those of *D. gigas* and *D. vulgaris* by only one amino acid. In *C. pasteurianum* the Val41 has been exchanged with a Leu41. When comparing ET rates and reaction patterns in the 25–residue 'native' analogue and a Val41 → Leu41 'mutant' of the analogue, details about the influence of these amino acids on ET can be elucidated. This appears therefore to be a natural first choice for a synthetic 'mutant'. The Val41 → Leu41 mutant was synthesized in the same manner as the "native" 25-residue metalloprotein. Its spectra have some small shifts compared to natural rubredoxin (fig. 4) possibly due to changes in the geometry of the active site. The mutant's stability properties corresponded to those of the "native" 25–residue peptide.

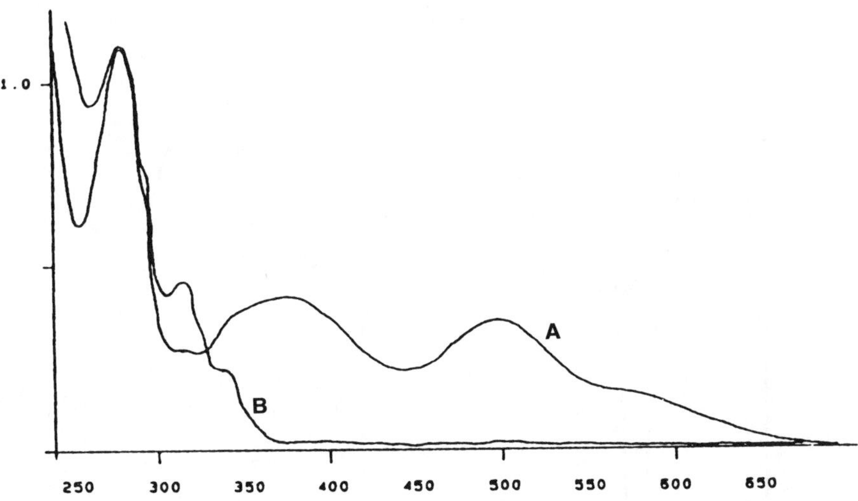

Figure 5: UV/VIS spectrum of oxidized (A) and reduced (B) totally synthetic rubredoxin. The concentration of rubredoxin is approximately 48 μM.

4.3   Full Rubredoxin

We have also successfully synthesized the full 52–residue peptide chain from *D. gigas* rubredoxin[35] (fig. 2). Using the same folding method as with the 25–residue analogue, this chain was folded around iron(II), and the UV/VIS spectrum obtained is almost identical to the spectrum of reduced *C. pasteurianum* rubredoxin. In contrast to the 25–residue analogue the synthetic full rubredoxin was stable and could withstand redox–cycles consisting of O_2-oxidation and KBH_4-reduction. The position of the bands in the UV/VIS spectra of both oxidized and reduced forms of the synthetic rubredoxin correspond very nicely to the spectra of native rubredoxin (fig. 5). However, the absorption ratio A_{280}/A_{490} is larger. This has also been noticed for rubredoxin obtained from synthetic genes incorporated into *E. coli*[36], and was ascribed to iron substitution by ubiquitously occurring trace amounts of zinc. Zinc-rubredoxin absorbs only around 280 nm, and zinc incorporated into rubredoxin instead of iron would therefore disturb the absorption ratio. This would agree with the fact that the absorption ratio does not change significantly, when the protein is run through HPLC, if iron- and zinc-rubredoxins are chromatographically too similar for separation.

The secondary structure of the folded protein is substantiated by CD spectra. The bands around 230 nm are identical to those seen in native rubredoxin which have been assigned to the β-structure in rubredoxin[37]. The circular dichroism is very sensitive towards changes in the coordination geometry, and the bands corresponding to S→Fe charge transfer for both oxidized and reduced forms are in very good agreement. This indicates that the coordination geometry around iron(II) is essentially identical to native rubredoxin, which in turn means that the protein is folded correctly around iron(II). Both UV/VIS and resonance Raman spectroscopy supports this conclusion. Resonance Raman reveals a dominating breathing mode, ν_1, at 312 cm^{-1} and a weakly resolved asymmetric stretch peak in the range 360-370 cm^{-1}, which is in agreement with a distorted tetrahedral geometry[35].

The MALDI-MS spectra obtained also indicate a tight binding between the protein chain and iron(II). The main peak is observed at 5735 D $\pm$ 10 D which is very close to the calculated molecular mass 5729 D, especially considering that eight protons are required to

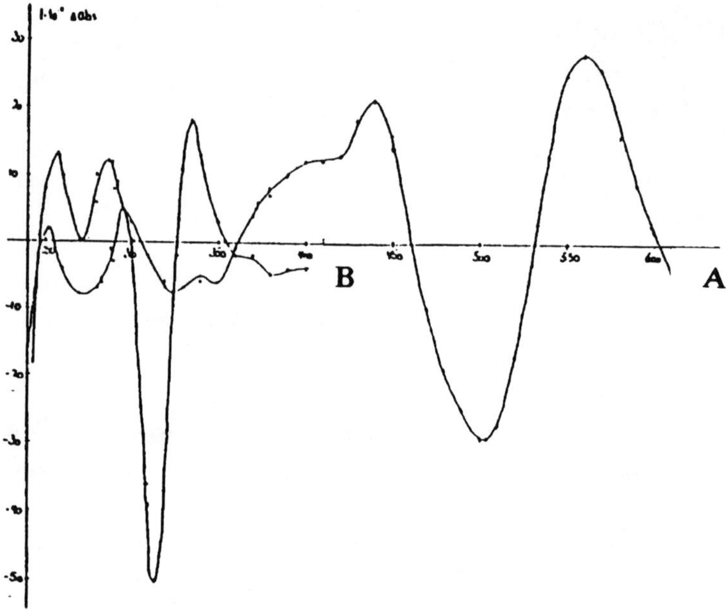

Figure 6: CD spectrum of oxidized (A) and reduced (B) synthetic rubredoxin. The concentration of rubredoxin is approximately 50 μM.

compensate for the negative charge of the protein. This demonstrates that the iron–protein bonds are strong and substantiates that iron has been correctly incorporated into the protein structure[35].

5 Concluding Remarks

As shown by recent reports and in the present work real metal-containing proteins can be prepared synthetically. Chemically synthesized proteins can be used to elucidate for example the ET properties of redox proteins, and more generally the relations between structure and function of proteins and enzymes. Chemical synthesis of proteins is a prospective alternative to mutagenesis. Although the scale of peptide synthesis is presently much smaller than what can be achieved by growing mutated *E. coli*, chemical synthesis would offer a powerful alternative to microbiological protein modification once the procedure has been improved to provide adequate sample amounts. This extends to the possibility of implementing unnatural changes in the proteins, for example by insertion of atypical amino acids. Folding of synthetic proteins can be followed and characterized by UV/VIS, CD and other spectroscopic techniques. Especially metal containing proteins are suited for VIS absorption and CD spectroscopy because of the chromophores associated with most transition metals incorporated into protein. When specific charge-transfer bands can be assigned to particular metal-ligand interactions, these bands are highly sensitive to the geometry and ultimately to the complete three-dimensional protein folding around the metal.

Acknowledgements

We would like to thank Prof. P. Roepstorff and O. Worm, Odense University, Denmark, for doing the MALDI mass spectroscopy, and Dr. O. Faurschou Nielsen, Institute of Chemistry at The University of Copenhagen, for access to the Raman instrument and for supervising the acquisition of the resonance Raman spectra. This work was supported financially by The Danish Natural Science Research Council.

References

[1] C. B. Anfinsen, E. Haber, M. Sala and F. H. White *Proc. Natl. Acad. Sci. USA*, **47**, 1309 (1961).

[2] P. E. Leopold, M. Montal and J. N. Onuchic, Protein folding funnels: A kinetic approach to the sequence-structure relationship, *Proc. Natl. Acad. Sci. USA*, **89**, 8721-8725 (1992).

[3] R. L. Baldwin and D. Eisenberg, Protein Stability, In: Protein Engineering. (Eds. D. L. Oxender and C. F. Fox) 127-148, Alan R. Liss Inc., New York 1987.

[4] G. J. Lesser, R. H. Lee, M. H. Zehfus and G. D. Rose, Hydrophobic Interactions in Proteins, In: Protein Engineering. (Eds. D. L. Oxender and C. F. Fox), 175-179, Alan R. Liss Inc., New York 1987.

[5] J. T. Kellis, Jr., K. Nyberg, D. Sali and A. Fersht, Contribution of hydrophobic interactions to protein stability, *Nature*, **333**, 784-786 (1988).

[6] T. E. Creighton, Proteins - Structure and Molecular Properties, W.H. Freeman & Co, New York 1984.

[7] A. Nakamura and N. Ueyama, Cysteine-containing oligopeptide model complexes of iron-sulfur proteins, *Adv. Inorg. Chem.*, **33**, 39-67 (1989).

[8] J. R. Anglin and A. Davison, Iron(II) and Cobalt(II) Complexes of Boc-(Gly-L-Cys-Gly)$_4$–NH$_2$ as Analogs for the Active Site of the Iron-Sulfur Protein Rubredoxin. *Inorg. Chem.*, **14**, 234-237 (1975).

[9] A. Nakamura and N. Ueyama, Chemistry of Iron-Sulfur Complexes of Cysteine-Containing Peptides, In: Iron-Sulfur Protein Research, (Ed. H. Matsubara) pp. 302-314, Japan Sci. Soc. Press, Tokyo 1986.

[10] E. T. Smith, B. A. Feinberg, J. H. Richards and J. M. Tomich, Physical Characterization of a Totally Synthetic 2[4Fe-4S] Clostridial Ferredoxin, *J. Am. Chem. Soc.*, **113**, 688-689 (1991).

[11] L. E. Zawadzke and J. M. Berg, A Racemic Protein, *J. Am. Chem. Soc.*, **114**, 4002-4003 (1992).

[12] C. Di Bello, C. Vita and L. Gozzini, Total Synthesis of Horse Heart Cytochrome C, *Biochem. Biophys. Res. Commun.*, **183**, 258-264 (1992).

[13] Several contributions in Metal Ions in Biological Systems, Vol. 27. (Eds. H. Sigel and A. Sigel) Marcel Dekker, New York 1991.

[14] A. Kuki, Electronic tunneling paths in proteins, *Structure and Bonding*, **75**, 49-83 (1991).

[15] H. E. M. Christensen, L. S. Conrad, J. M. Hammerstad-Pedersen and J. Ulstrup, Resonance effects in strongly exothermic long-range electron transfer and their possible implications for the behaviour of site-directed mutant proteins, *FEBS Letters*, **296**, 141-144 (1992).

[16] M. Nordling, K. Sigfridsson, S. Young, L. G. Lundberg and Ö. Hansson, Flash-photolysis studies of the electron transfer from genetically modified spinach plastocyanin to photosystem I, *FEBS Letters*, **291**, 327-330 (1991).

[17] W. A. Eaton and W. Lovenberg, The Iron-Sulfur Complex in Rubredoxin, In: Iron Sulfur Proteins, Vol. II. (Ed. W. Lovenberg) pp. 131-162, Academic Press, New York 1973.

[18] W. Lovenberg and B. E. Sobel, Rubredoxin: A New Electron Transfer Protein from Clostridium pasteurianum, *Proc. Natl. Acad. Sci. USA*, **54**, 193-199 (1965).

[19] E. T. Adman, L. C. Sieker and L. H. Jensen, Structure of Rubredoxin from Desulfovibrio vulgaris at 1.5 Å Resolution, *J. Mol. Biol.*, **217**, 337-352 (1991).

[20] Z. Dauter, L. C. Sieker and K. S. Wilson, Refinement of Rubredoxin from Desulfovibrio vulgaris at 1.0Å with and without restraints, *Acta Cryst.*, **B48**, 42-59 (1992).

[21] M. Frey, L. Sieker, F. Payan, R. Haser, M. Bruschi, G. Pepe and J. LeGall, Rubredoxin from Desulfovibrio gigas. A Molecular Model of the Oxidized Form at 1.4Å Resolution, *J. Mol. Biol.*, **197**, 525-541 (1987).

[22] K. D. Watenpaugh, L. C. Sieker and L. H. Jensen, The Structure of Rubredoxin at 1.2Å Resolution, *J. Mol. Biol.*, **131**, 509-522 (1979).

[23] R. E. Stenkamp, L. Sieker and L. H. Jensen, The Structure of Rubredoxin From Desulfovibrio desulfuricans Strain 27774 at 1.5Å Resolution, *Proteins: Struct. Funct. Genet.*, **8**, 352-364 (1990).

[24] S. Hormel, K. A. Walsh, B. C. Prickril, K. Titani, J. LeGall and L. C. Sieker, Amino acid sequence of rubredoxin from Desulfovibrio desulfuricans strain 27774, *FEBS Letters*, **201**, 147-150 (1986).

[25] J. Meyer, J. Gagnon, L. C. Sieker, A. van Dorsselaer and J.-M. Moulis, Rubredoxin from Clostridium thermosaccharolyticum, *Biochem. J.*, **271**, 839-841 (1990).

[26] Y. Seki, S. Seki, M. Satoh, A. Ikeda and M. Ishimoto, Rubredoxin from Clostridium perfringens: Complete Amino Acid Sequence and Participation in Nitrate Reduction, *J. Biochem.*, **106**, 336-341 (1989).

[27] K. Saeki, Y. Yao, S. Wakabayashi, G.-J. Shen, J. G. Zeikus and H. Matsubara, Ferredoxin and Rubredoxin from Butyribacterium methylothrophium: Complete Primary Structures and Construction of Phylogenetic trees, *J. Biochem.*, **106**, 656 (1989).

[28] H. Bachmayer, A. M. Benson, K. T. Yasunobu, W. T. Garrard and H. R. Whiteley, Nonheme Iron Proteins. IV. Structural Studies of Micrococcus aerogenes Rubredoxin, *Biochemistry*, **7**, 986-996 (1968).

[29] H. Bachmayer, K. T. Yasunobu, J. L. Peel and S. Mayhew, None-heme Iron Proteins. V. The Amino Acid Sequence of Rubredoxin from Peptostreptococcus elsdenii, *J. Biol. Chem.*, **243**, 1022-1030 (1968).

[30] K. J. Woolley and T. E. Meyer, The complete amino acid sequence of rubredoxin from the green phototrophic bacterium Chlorobium thiosulphatophilum strain PM, *Eur. J. Biochem.*, **163**, 161-166 (1987).

[31] I. Moura, A. V. Xavier, R. Cammack, M. Bruschi and J. LeGall, A Comparative Spectroscopic Study of Two Non-heme Iron Proteins Lacking Labile Sulphide from Desulphovibrio gigas, *Biochim. Biophys. Acta*, **533**, 156-162 (1978).

[32] N. M. Atherton, K. Garbett, R. D. Gillard, R. Mason, S. Mayhew, J. L. Peel and J. E. Stangroom, Spectroscopic Investigation of Rubredoxin and Ferredoxin, *Nature*, **212**, 590-593 (1966).

[33] E. E. Abola, F. C. Bernstein, S. H. Bryan, T. F. Koetzle and J. Weng, In Crystallographic Databases - Information Content, Software Systems, Scientific Applications. (Eds. F. H. Allen, G. Bergerhoff and R. Sievers) pp. 107-132, Data Commission of the International Union of Crystallography, Born/Cambridge/Chester 1987.

[34] H. E. M. Christensen, J. M. Hammerstad-Pedersen, A. Holm, P. Roepstorff, J. Ulstrup, O. Vorm and S. Østergård, Synthesis and characterization of a 25-residue D. gigas rubredoxin(II)-like metalloprotein and its Leu41 Mutant, *FEBS Letters*, **312**, 219-222 (1992).

[35] H. E. M. Christensen, A. Holm, J. M. Hammerstad-Pedersen, G. Iversen, M. H. Jensen and J. Ulstrup, Synthesis and Characterization of Desulfovibrio gigas Rubredoxin and Rubredoxin fragments, Submitted (1994).

[36] M. K. Eidsness, S. E. O'Dell, D. M. Kurtz,Jr., R. L. Robson and R. A. Scott, Expression of a synthetic gene coding for the amino acid sequence of Clostridium pasteurianum rubredoxin, *Protein Eng.*, **5**, 367-371 (1992).

[37] I. Mus-Veteau, D. Diaz, J. Gracia-Mora, B. Guigliarelli, G. Chottard and M. Bruschi, Spectroscopic studies of the nickel-substituted Desulfovibrio vulgaris Hildenborough rubredoxin: implication for the nickel site in hydrogenases, *Biochim. Biophys. Acta*, **1060**, 159-165 (1991).

Theoretical Distance–based Approaches
to
Protein Structure Prediction

Probing Protein Structure by solvent Perturbation of NMR Spectra: III. Combination of Experiment and Theory

Gennaro Esposito[*], Arthur M. Lesk[†], Henriette Molinari[‡] Andrea Motta[§], Neri Niccolai[¶] and Annalisa Pastore[‖]

[*] Istituto di Biologia, Facoltà di Medicina, Università di Udine, Viale Gervasutta, 33100 Udine, Italy

[†] Department of Haematology, University of Cambridge Clinical School, MRC Centre, Cambridge, CB2 2QH, U.K.

[‡] Dipartimento di Chimica Organica e Industriale, Via Golgi 19, 20133 Milano, Italy

[§] Istituto per la Chimica di Molecole, di Interesse Biologico, CNR, Via Toiano 6, 80072 Arco Felice, Napoli, Italy

[¶] Dipartimento di Biologia Molecolare, Centro Didattico dell'Università ≪Le Scotte≫, 53100 Siena, Italy

[‖] European Molecular Biology Laboratory, Meyerhofstrasse 1, Postfach 1022.09, 6900 Heidelberg, Germany

Abstract

The experimental assigment of most residues in a protein to the surface or interior is in principle possible without prior solution of a complete three-dimensional structure. The method described is based on NMR measurements which determine the amino acid composition of the surface of a protein (Petros, Mueller & Kopple, 1990; Esposito, Lesk, Molinari, Motta, Niccolai & Pastore, 1992). We describe the experimental controls that establish the validity of this approach, and show how to derive from these data partial information about which positions in the sequence contain exposed residues. Next we examine a method for *a priori* prediction of the residues on the surface of a protein (Holbrook, Muskal & Kim, 1990). We find that a combination of experiment and theory can result in a more correct and complete prediction than either alone.

1 Introduction

The knowledge of which residues in a protein are exposed to the solvent or accessible to it, and which others are buried, is important for structure determination, studies of intermolecular interactions with other molecules, and protein dynamics. It has been shown that mapping of protein surfaces can be obtained by NMR studies using soluble nitroxides. In these experiments, dipolar interactions between the unpaired electron spin of the label and proximal protein hydrogens increase the relaxation rate of the protons. Since the resonance linewidth is an inverse function of the transverse relaxation time (T2), these resonances are consequently broadened or disappear. They are effectively "bleached" out of the spectrum. Difference spectra computed from observations in the presence and in the absence of the bound spin label identify those resonances which are affected by the proximity of the paramagnetic probe.

This method is not new, and has found several applications in the past. Nitroxide stable spin labels covalently bound to proteins or nucleic acids (Wien *et al.*, 1972; Dwek *et al.*, 1975; Schmidt & Kuntz, 1984; Moonen *et al.*, 1984; Kosen *et al.*, 1986; Anglister *et al.*, 1984ab; De Jong *et al.*, 1988) or extrinsic probes, such as paramagnetic metal ions (Campbell *et al.*, 1975; Arean *et al.*, 1988; Williams, 1989) have been widely used in biological applications.

As a development of the chemical perturbation approach, soluble radicals such as TEMPOL have recently been used for mapping protein surfaces (Esposito *et al.*, 1989; Petros *et al.*, 1990). This approach, an extension of a 1D relaxation method previously proposed (Niccolai *et al.*, 1982, Niccolai *et al.*, 1984), holds the promise that nitroxide-protein interactions might be developed into a general approach for the study of protein folding and dynamics. Once it is demonstrated that nitroxide molecules approach the protein randomly and that no specific interactions occur, the extent of the observed paramagnetic effects may be applied to reflect the native folding pattern of the protein.

2 Experimental measurements using hen egg-white lysozyme: establishment of the technique

To be able to rely on this method, extensive preliminary studies were necessary. The goals of our previous work were: (a) to identify possible specific interactions by both ESR data and analysis of the NMR chemical shifts, (b) to ascertain that no conformational changes are induced by the spin-label, (c) to suggest a mechanism compatible with the experimental data, and (d) to check whether a correlation is possible between our results and the secondary and tertiary structure.

Hen egg-white lysozyme was used as a model because of the availability of detailed NMR data, including the complete assignment of the spectrum, and a high-resolution crystal structure (Wien, Morrisett & McConnell, 1972; Redfield & Dobson, 1988; Imoto *et al.*, 1972; Artymiuk & Blake, 1981). The spin label TEMPOL (4-hydroxy-2,2,6,6-tetramethylpiperidine-1-oxyl) was employed because of its solubility in water.

A summary of our results follows:

(a) ESR spectra exclude specific TEMPOL-protein site binding

ESR is a sensitive method which should show the presence of specific interactions. Spectra in aqueous solution of lysozyme with different protein/TEMPOL ratios were shown to be identical to the spectrum of free TEMPOL. This indicates a freely tumbling nitroxide species both in the absence and in the presence of the protein, excluding strong interactions between the free label and the protein, and preferential binding sites.

(b) TEMPOL does not alter the conformation of the protein

Possible perturbing effects of TEMPOL on the conformation of lysozyme were assessed from chemical shift variations in the NMR spectrum of lysozyme in the presence of TEMPOL and following TEMPOL titrations. Close analysis of the 1D spectra of the two less crowded regions of the lysozyme spectrum (at high and low frequencies) shows no differential shifts which would occur if there were specific interactions with protons having resonances around this region.

The effect on the chemical shift was analysed by TEMPOL titration. For concentrations of the free radical up to 1.0 M, the experimental points can be fitted by a linear function of the concentration. This is an important point. It implies that TEMPOL does not perturb the native structure of the protein under the conditions of these measurements and confirms the ESR data.

(c) Nature of the interactions

The understanding of the mechanism of interaction between TEMPOL and the protein is essential for confident application of our results to structural or dynamical problems.

Short collisions or very weak transient complexes between a protein and a radical are sufficient for dipolar interactions between the unpaired electron of the radical and the hydrogen spins of the protein, which are modulated by random relative translational motion of the molecules.

Evidence for weak, random interactions between TEMPOL and lysozyme is contained in the linear dependence of chemical shifts on TEMPOL concentration. If intermolecular interactions involve only weak van der Waals forces, then to a good approximation, ideal solution behaviour may be assumed and the binding constants will be small (Draney & Kingsbury, 1981).

(d) Mapping the protein surface

To show that it is possible to correlate our results with the protein surface, we considered one representative class of protons, the amides, and correlated the spectral perturbation of TEMPOL with elements of secondary and tertiary structure. We considered the behavior of amide protons only, because 1) they are backbone protons and therefore they should have a more mutually similar correlation time; 2) the results may be correlated with exposure to the solvent as measured by hydrogen/deuterium exchange. A good correlation was observed between amide exchange rates, exposed surface areas and TEMPOL accessibility (Esposito, *et al.*, 1992). Fourteen exceptions to this correlation were observed, eleven of which are in fast exchange although inaccessible to TEMPOL. This may easily be explained by considering that sites exposed to water may be completely sheltered from the bulkier spin-label. For all but two of the amide protons in lysozyme (those of residues 63 and 86) the exposure to nitroxide parallels exposure to the solvent and/or the presence of hydrogen bonds.

Lysozyme is a typical globular protein (Artymiuk & Blake, 1981). The secondary structure includes four helices, formed by residues 5–15, 25–35, 80–84, and 89–101. The first two pack together, with the second mostly buried in the core of the protein. The third and fourth helices are partially buried. A three-stranded antiparallel β-sheet and a very short two-stranded β-sheet are also present, formed by residues 42–46, 50–54, and 57–60; and 1–3 and 38–40, respectively. The general exposure of a secondary structure element does not necessarily imply, of course, that all the amide groups of the region are exposed to the solvent; indeed many helices in proteins have a hydrophilic face exposed to solvent and a hydrophobic face buried in the protein interior.

This feature shows up quite clearly in our results, as the amide protons located in helical regions exhibit an alternating pattern of TEMPOL exposure, in good agreement with the alternating pattern of the corresponding computed surface areas or observed exchange rates. In addition, the overall decreasing trend of H2 amide exposure nicely parallels the progressive burying of this region in the structure. The TEMPOL exposure of the β-sheet regions reflects both the hydrogen bond network and the location of these elements with respect to the overall protein surface. Within the three-stranded β-sheet region 42–60, for instance, the H-bonded amides (44, 46, 51, 52, 53, 57 and 60) invariably exhibit lower TEMPOL exposures and the overall exposure level appears higher for the segment 42–46, consistent with its outer location.

Most of the tight turns appear quite accessible to the spin label, as expected from the crystal structure. The only exceptions are the 74–77 and 115–118 turns (type I and type II respectively). The relative inaccessibility to TEMPOL of the former can be ascribed to steric hindrance from the covering loop 61–64 from one side, and the sidechains of Leu75 and Cys76 from the other.

Our results can therefore be fairly well correlated with the structural and dynamical features of the protein surface. The analysis presented so far, however, was possible in this application only because the structure of lysozyme is known and the proton NMR spectrum has been fully assigned. Dealing with a protein of unknown structure, we could obtain the amino acid map of the TEMPOL-exposed residues. How to apply this information is the subject of the next section.

3   Derivation of exposed residues from experiments on homologous proteins

There has been considerable recent interest in methods for determining the residues in a protein of unknown structure that are on the surface, and for deriving inferences about the secondary structure and folding pattern from this information. Several investigators have developed entirely theoretical methods based on aligned sequences for distinguishing surface from buried positions (Nishikawa & Ooi, 1986; Benner & Gerloff, 1989, 1991; Holbrook, Muskal & Kim, S.-H., 1990).

NMR spectral measurements using TEMPOL provide a direct *experimental* assignment of surface and buried residues in a protein of unknown structure. We have derived a method based on experimental data from paramagnetic perturbation of protein NMR spectra of several homologous proteins of known sequence, to determine which positions in the sequence contain exposed residues. In principle the method is general and the data reflecting surface exposure could be obtained by fluorescence or CIDNP or any other experimental source. It requires only the measurement of the degree of exposure for each residue *type*. If measurements of degree of exposure are carried out on a series of related proteins for which the amino acid sequences are known and aligned, it is possible to correlate the surface amino acid compositions of the individual molecules with the variations in the sequence at different positions, to identify positions containing buried or surface residues.

We make the following assumptions: We are given a set of M homologous proteins. The sequences are all known and can be aligned — allowing for insertions and deletions — with the alignment table containing a total of N residue positions. Suppose that for each

Table 1: Alignment of twelve acylphosphatase sequences from mammals and birds; the lines underneath the blocks of residues show the absolutely conserved residues.

```
                                    10        20        30
                                     .         .         .
DUCK            STLGKAPGALKSVDYEVFGRVQGVCFRMYTEEEAR
CHICKEN         SALTKASGSLKSVDYEVFGRVQGVCFRMYTEEEAR
TURKEY          SALTKASGALKSVDYEVFGRVQGVCFRMYTEEEAR
CHICKEN2          AGSEGLMSVDYEVSGRVQGVFFRKYTQSEAK
HUMAN            AEGNTLISVDYEIFGKVQGVFFRKHTQAEGK
BOVINE          STGRPLKSVDYEVFGRVQGVCFRMYTEDEAR
HORSE           STARPLKSVDYEVFGRVQGVCFRMYAEDEAR
HUMAN2          STAQSLKSVDYEVFGRVQGVCFRMYTEDEAR
PIG             STARPLKSVDYEVFGRVQGVCFRMYTEDEAR
RABBIT          STAGPLKSVDYEVFGRVQGVCFRMYTEGEAK
GUINEA PIG      SAAAQLKSVDYEVFGRVQGVCFRMYTEGEAK
RAT                 ?KSVDYEVFGTVQGVCFRMYTEGEAK

                S L         SVDYE   G VQGV FR       E

                                    45        55        65
                                     .         .         .
DUCK            KLGVVGWVKNTSQGTVTGQVQGPEDKVNAMKSWLT
CHICKEN         KLGVVGWVKNTSQGTVTGQVQGPEDKVNAMKSWLS
TURKEY          KLGVVGWVKNTRQGTVTGQVQGPEDKVNAMKSWLS
CHICKEN2        RLGLVGWVRNTSHGTVQGQAQGPAARVRELQEWLR
HUMAN           KLGLVGWVQNTDRGTVQGQLQGPISKVRHMQEWLE
BOVINE          KIGVVGWVKNTSKGTVTGQVQGPEEKVNSMKSWLS
HORSE           KIGVVGWVKNTSKGTVTGQVQGPEEKVNSMKSWLS
HUMAN2          KIGVVGWVKNTSKGTVTGQVQGPEDKVNSMKSWLS
PIG             KIGVVGWVKNTSKGTVTGQVQGPEEKVNSMKSWLS
RABBIT          KIGVVGWVKNTSKGTVTGQVQGPEDKVNSMKSWLS
GUINEA PIG      KIGVVGWVKNTSKGTVTGQVQGPEEKVNSMKSWLS
RAT             KRGLVGWVKNTSKGTVTGQVQGPEEKVNSMKSWLS

                G VGWV NT  GTV GQ QGP    V      WL

                                    80        90        100
                                     .         .         .
DUCK            KVGSPSSRIDRTNFSNEKEISKLDFSGFSTRY
CHICKEN         KVGSPSSRIDRTKFSNEKEISKLDFSGFSTRY
TURKEY          KVGSPSSRIDRTNFSNEKEISKLDFSGFSTRY
CHICKEN2        KIGSPQSRISRAEFTNEKEIAALEHTDFQIRK
HUMAN           TRGSPKSHIDKANFNNEKVILKLDYSDFQIVK
BOVINE          KVGSPSSRIDRTNFSNEKTISKLEYSSFNIRY
HORSE           KVGSPSSRIDRTNFSNEKTISKLEYSNFSVRY
HUMAN2          KVGSPSSRIDRTNFSNEKTISKLEYSNFSIRY
PIG             KIGSPSSRIDRTNFSNEKTISKLEYSNFSIRY
RABBIT          KVGSPSSRIDRTNFSNEKTISKLEYSNFSIRY
GUINEA PIG      KVGSPSSRIDRTNFSNEKSISKLEYSNFSIRY
RAT             KVGSPSSRIDRADFSNEKTISKLEYSNFSIRY

                GSP S I   F NEK I   L      F
```

position the corresponding residues are either buried in every protein in the set or on the surface of every protein in the set, or deleted from one or more of the sequences; that is, we assume that the proteins share a common fold.

Our goal is to determine the positions of the residues that lie on the surface of the protein from measurable quantities — including but not limited to the results of NMR experiments using soluble spin labels as described in the preceding section. Initially we assume that for each single protein, an experiment determines the number of amino acids of each type accessible to the probe. Later we shall discuss the complications that arise if all twenty amino acids cannot be individually distinguished.

On the basis of measured amino acid compositions of the surfaces of several homologous molecules, we wish to assign the individual positions in the sequences to the interior or to the surface of the structure. The basic idea is this: If we knew the correct assignment of positions in the table of aligned sequences to the surface, we could calculate directly the amino acid composition of the surface of each molecule. We need to be able to invert this relationship to infer the assignments from the measured surface amino acid compositions. The relationship between the assignments of positions to the surface and the experimental data may be expressed mathematically as a matrix equation of the form $Ax = b$, in which the matrix A contains the table of aligned sequences, the vector b contains the experimentally-determined surface amino-acid compositions, and the vector x contains the assignments of positions in the sequences to the surface and to the interior. The elements of A and b are known, and we wish to determine the vector x. This system differs from the more common type of linear matrix equation because of the constraint that the elements of x be either 0 or 1. However, techniques exist to solve such a system of equations under these constraints (Hammer & Rudeanu, 1968; Lesk, 1973). (For further details, see Esposito, *et al.,* 1993).

A general characteristic of such a set of equations is that, in various cases, no solution, a unique solution, or multiple solutions exist. In the case of the problem which we have formulated in these terms, the existence of multiple solutions would imply that the experimental data do not uniquely determine the surface positions.

There is in fact an intrinsic limitation to this and related approaches: if the same amino acid is absolutely conserved at several positions in the sequences, and if the residues at these positions are not all exposed or all buried, the method can determine how many of these positions contain exposed residues but cannot distinguish which subset of these positions contains exposed residues. (This limitation might be overcome by using additional sequences, perhaps generated by site-directed mutagenesis.) Thus suppose that positions i, j and k are all contain the same absolutely conserved amino acid. If the data are consistent with one of the three being on the surface and two buried, then there must be three solutions corresponding to assigning each of the three to the surface. An example may clarify this: Consider the following region in the acylphosphatase sequences shown in Table 1 (positions 38-41).

```
DUCK          L  G  V  V  G
CHICKEN       L  G  V  V  G
TURKEY        L  G  V  V  G
CHICKEN2      L  G  L  V  G
HUMAN         L  G  L  V  G
BOVINE        I  G  V  V  G
HORSE         I  G  V  V  G
HUMAN2        I  G  V  V  G
PIG           I  G  V  V  G
RABBIT        I  G  V  V  G
GUINEA PIG    I  G  V  V  G
RAT           R  G  L  V  G
Conserved        G     V  G
```

The accessible surface areas of the two glycine residues in the Horse acylphosphatase structure (Saudek, Williams & Ramponi, 1989) are for Gly38, 41.4, and for Gly41, 0.4.

Gly38 is on the surface and Gly41 is buried. However, the NMR measurements (and hence the program) can not distinguish which is on the surface and which is buried because the amino acid composition of the surface is insensitive to this ambiguity. Different permutations of the assignments to surface and interior, within the set of positions containing absolutely conserved amino acids of the same type, are equally valid solutions of the equations.

It should be emphasized that not all absolutely conserved residues necessarily create such problems. For example, in the acylphosphatase sequences there are two conserved tryptophans, and two conserved prolines. But all of them are on the surface, and the method can detect this correctly.

3.1   An example: Acylphosphatase

As a test case we used acylphosphatase, for which the structure of the molecule from the Horse has been determined by NMR (Saudek, *et al.*, 1989), and for which 12 homologous sequences from mammals and birds are known (Table 1). The sequences were taken from the Protein Identification Resource data base (PIR) (George, Hunt & Barker, 1989). Table 1 shows their alignment. The longest sequences contain 102 residues. The percentage of identical residues between pairs of sequences ranges from 54 to 97%.

Our test was carried out using simulated experimental data. We calculated the amino acid composition of the surface of the known structure of Horse acylphosphatase using the method of Shrake & Rupley (1973). Usually, the distinction between buried and surface residues is drawn by setting a threshold of 15-20 A^2; however, because the probe in this case is larger than a water molecule, we have chosen a threshold of 30 A^2. We assumed that the positions of surface residues are identical in all of the twelve molecules, and asked whether we could recover these positions from the computed surface amino acid compositions.

The results using all 12 sequences are shown in Table 2. Here b represents a buried residue, S represents a surface residue, and * a position that cannot be specified. The results show that there are no errors, and no ambiguities other than those arising from combinations of conserved residues. The 30 ambiguities are all "essential" ones; that is, they all correspond to combinations of conserved residues.

We ran several combinations of sequences, to try to determine the minimum number of proteins required to determine the surface positions, in order to minimize the amount of experimental work required. Taking only three distantly related sequences in Table 1 — Chicken2, Human and Horse — there are no additional absolutely conserved positions, beyond those already appearing in Table 1. However these sequences all lack the N-terminal extension observed in Duck, Chicken and Turkey sequences.

In this case there were two solutions. One was the correct one, and the other corresponded to an interchange among four positions

	37	39	91	101
CHICKEN2	L	L	I	I
HUMAN	L	L	I	I
HORSE	I	V	I	V
Solution 1:	b	S	S	b
Solution 2:	S	b	b	S

Table 2: Results of calculation of surface/buried positions of residues in 12 acylphosphatase sequences, with threshold 30 A^2 (b = buried, S = surface, * = essential ambiguity in experiment).

```
                                  10         20         30
                                   |          |          |
Correct            bbbbSSSSSSSbbbbSbSbSSSSbSbSSSbSSSbS
Experiment         *b*bSSSSSSS**bbSbS*S****S*SSSbSSSbS
Conserved          S L         SVDYE  G VQGV FR       E


                                  45         55         65
                                   |          |          |
Correct            SSSbbbSbSSSSSSbbSbSbbbSSSSbSSbSSSbS
Experiment         SS*b**S*SS*SS***S**b**SSSS*SSbSSS*S
Conserved          G VGWV NT   GTV GQ QGP    V      WL


                                  80         90         100
                                   |          |          |
Correct            SbSSSSbSbSSbSSSSSSSbSSSSSSSbSSSS
Experiment         Sb**SS*SbSSbS*SSSSbSS*SSSS*SSSS
Conserved          GSP S I    F NEK I  L      F
```

Both solutions give the same surface amino acid compositions. If necessary, data from a single additional sequence would resolve these ambiguity; alternatively, in hindsight the choice of rat, horse and chicken2 produce the single solution of Table 2 from only three sequences.

These calculations were carried out assuming that there are no ambiguities in the identification of different types of residues. In practice, structurally-related residues may not be easily distinguishable. In this case, instead of determining the exact amino acid composition of the surface, it may be possible to state only the total number of surface residues within certain groups of amino acids. However, provided that among the sequences there are mutations into and out of these groups, the method described here may be robust enough to tolerate the loss of information.

As a test, we edited the twelve sequences of Table 1 to replace every C with an S (this corresponds to being unable to distinguish C from S in analysing the spectral data) and reran the calculation on the simulated data from all 12 sequences. This produced the same solution. Note that the acylphosphatase sequences contain conserved serines but no conserved cysteines. Running the calculation on the human, horse and chicken2 sequences, also edited to change every S to a C, again produced the same two solutions as the original sequences.

Without the assignment of the full spectrum or even of only the main chain, many residues, or groups — typically the aromatic side chains, the β-methyls of alanines, threonines — can be recognised easily. In some cases, although of course not in all, depending on the degree of overlap, it is also possible to trace each spin system by means of experiments.

It would be then possible to obtain a table that for each of the amino acid types lists the number of buried and exposed residues. Some ambiguities in the assignment may be tolerable. It is usually possible to recognize the location in the spectrum of at least one or two of these proton types. A coarse classification of the information obtained from different regions of the spectrum would provide a list of accessible areas such as the data assumed for acylphosphatase in the calculations reported here.

Table 3: Surface residues in Horse acylphosphatase: experiment and theory (b = buried, S = surface, * = essential ambiguity in experiment).

```
Amino acid sequence     STARPLKSVDYEVFGRVQGVCFRMYAEDEAR
Prediction              SSSSSbSSbSbSbbbSbbbbbbSbbbSSSbS
Errors in prediction        X X X   X XXX X  XX
Correct                 SSSSSSSbbbbSbSbSSSSbSbSSSbSSSbS
Experiment              SSSSSSS**bbSbS*S****S*SSSbSSSbS


Amino acid sequence     KIGVVGWVKNTSKGTVTGQVQGPEEKVNSMKSWLS
Prediction              SbbbbbbbSSbSSbSbSbSbSbSSSSbSSbSSSbbb
Errors in prediction    XX  X  X XX    X           X X
Correct                 SSSbbbSbSSSSSSbbSbSbbbSSSSbSSbSSSbS
Experiment              SS*b**S*SS*SS***S**b**SSSS*SSbSSS*S


Amino acid sequence     KVGSPSSRIDRTNFSNEKTISKLEYSNFSVRY
Prediction              SSSSSSSSbSSSSbSSSSSbbSbSSSSbbbSS
Errors in prediction     X    X    X X      X X     XX
Correct                 SbSSSSbSbSSbSSSSSSSbSSSSSSSbSSSS
Experiment              Sb**SS*SbSSbS*SSSSbSS*SSSS*SSSS
```

After obtaining similar results for several proteins of a family, the method outlined here may be applied. The number of proteins it is necessary to analyse may be as few as three, minimizing the experimental effort. In addition, because the whole procedure is based primarily on comparisons between spectra to detect disappearance of some protons, the analysis is straightforward and automation should be possible.

4　Combination of experimental results with theoretical predictions of surface residues

Holbrook, Muskal & Kim (1990) trained a neural network to predict the residues on the surface of a protein from the amino acid sequence. They claimed an average of 72% accuracy for a test set, using a two-state buried/exposed model. We were kindly given the predictions of this network for the three acylphosphatase sequences, and have compared the quantity and distribution of successes with those of the experimental procedure.

Table 3 shows the horse acylphosphatase sequence, together with the surface/buried predictions from simulated experiments with TEMPOL (see preceding section) and the neural network.

Of the 98 residues, 55 are determined correctly and unambiguously both from the simulated experiment and the neural network prediction. Of the others, 30 are essential ambiguities in the experimental determination, and 27 residues are incorrectly predicted by the neural network. However, only 12 residues are incorrectly determined by both methods.

If we apply the following rule: take the experimental determination except, when it is ambiguous take the neural network prediction, then the result will contain 12 errors: 18 out of the 30 ambiguities have been correctly resolved.

Table 4: Number of absolutely conserved residues buried and superficial determined from experiment and predicted.

Residue	Experiment		Prediction	
	buried	surface	buried	surface
G	4	4	7	1
S	3	1	4	0
T	1	1	1	1
V	6	1	7	0
L	2	1	3	0
Q	1	2	1	2
F	2	1	3	0

In principle, we might be able to do even better than this, because we know from the experiment *how many* residues of each type are on the surface. Table 4 shows, for each set of absolutely conserved residues of each type of amino acid that are ambiguously determined by the experiment, how many are buried and how many on the surface, and how many of them the neural network predicts to be buried and how many on the surface. For example, of the 8 absolutely conserved glycines, the simulated experimental results are that 4 are buried and 4 on the surface. The prediction is that 7 are buried and 1 on the surface. In fact the one predicted to be on the surface is predicted correctly. But of the 7 remaining, we know that only three of them are buried. If the surface-prediction method could reliably rank the predictions in order of confidence, it would be possible to use the predictions of the four that are most confidently predicted to be buried, and to assume that the other three are on the surface.

However, in fact the surface predictions are not that precise. And for this distribution, it is not possible to use *only* the knowledge that four of the eight are buried to make a better prediction, any more than, in roulette, it helps to know that red will come up approximately half the time: One will do as well betting red all the time as betting on red half the time and black half the time. A careful analysis of the relations between the number of residues indicated as buried by experiment and theory (Table 4) has not permitted an improvement in the quality of the final result.

5   Conclusions

We have described an experimental technique for determining the residues on the surface of a protein of unknown structure. Using acylphosphatase as a test case we have shown that by combining the experimental results with theoretical predictions we can achieve a rate of correct prediction of 86 residues out of 98, or 88%, better than is possible with either experimental results or the Holbrook, Muskal & Kim neural network alone.

The ability to determine the residues on the surface of a protein is of interest in its own right. It can be applied to studies of protein folding, and has been shown to be useful in energy refinement of proteins during structure determination (von Freyberg, Richmond, & Braun, 1993).

But our ultimate hope is to use this information to determine the folding pattern of the protein. We note that the method of Benner & Gerloff (1989, 1991) make use of a determination of the residues on the surface of the protein, and therefore our approach could supplement their methods, which have achieved a correct prediction, in a blind test, of the structure of the catalytic domain of protein kinase C. Other possibilities include the type of "threading" calculations that are currently being explored, although these will be limited to proteins with conformations similar to those of structures already known; or directed energy minimization with "push-pull" forces applied from the center of mass. We recognize that it has yet to be demonstrated that these methods will work.

Acknowledgements

We thank S. R. Holbrook, S.M. Muskal, S.-H. Kim and I. Dubchak for the surface predictions on acylphosphatase sequences. A.M.L. thanks The Kay Kendall Foundation for generous support.

References

Anglister, J., Frey, T. & McConnell, H. M. (1984a). Magnetic resonance of a monoclonal anti-spin label antibody. Biochemistry 23, 1138–1142..

Anglister, J., Frey, T. & McConnell, H. M. (1984b). Distance of tyrosine from a spin-label hapten in the combining site of a specific monoclonal antibody. Biochemistry 23, 5372–5375.

Arean, C. O., Moore, G. R., Williams, G. & Williams, R. J. P. (1988). Ion binding to cytochrome. Eur. J. Biochem. 173, 607–615.

Artymiuk, P. J. & Blake, C. C. F. (1981). Refinement of human lysozyme at 1.5 A resolution. An analysis of non-bonded and hydrogen bond interactions. J. Mol. Biol. 152, 737–762.

Benner, S. A. & Gerloff, D. (1989). Adv. Enzyme Regulation 28, 219-236.

Benner, S. A. & Gerloff, D. (1991). Adv. Enzyme Regulation 31, 121-181.

Berliner, L. J. (1976). Spin labelling: Theory and Applications. London: Academic Press.

Campbell, I. D., Dobson, C. M. & Williams, R. J. P. (1975). Assignment of the ^{1}H NMR spectra of proteins. Proc. Roy. Soc. Ser. A 345, 23–40.

Cassels, R., Dobson, C. M., Poulsen, F. M. & Williams, R. J. P. (1978). Study of the tryptophan residues of lysozyme using ^{1}H NMR. Eur. J. Biochem. 92, 81–97.

De Jong, E. A. M., Claesen, C. A. A., Daemen, C. J. M., Harmsen, B. J. M., Konings, R. N. H., Tesser, G. I. & Hilbers, C. W. (1988). Mapping of ligand binding sites on macromolecules by means of spin-labeled ligand and 2D difference spectroscopy. J. Magn. Res. 80, 197–213.

Draney, D. & Kingsbury, C. A. (1981). Free radical induced nuclear magnetic resonance shift: comments on contact shift mechanisms. J. Am. Chem. Soc. 103, 1041–1047.

Dwek, R. A., Knott, J. C. A., Marsh, D., McLaughlin, A. C., Press, E. M., Price, N. C. & White, A. I. (1975). Structural studies on combining site of myeloma protein MOPC315. Eur. J. Biochem. 53, 25–39.

Englander, S. W. & Kallenbach, N. R. (1984). Hydrogen exchange and structural dynamics of proteins and nucleic acids. Quart. Revs. Biophys. 16, 521–655.

Englander, S. W. & Wand, A. J. (1987). Biochemistry 26, 5953–5958.

Esposito, G., Molinari, H, Motta, A., & Niccolai, N. (1989). Proc. XXIII A 2D NMR delineation of the solvent exposure of protein nuclei from paramagnetic relaxation filtering. Convegno Nazionale GDRM, p. 11, Cagliari (Italy).

Esposito, G., Lesk, A. M., Molinari, H., Motta, A., Niccolai, N. & Pastore, A. (1992). J. Mol. Biol. 224, 659–670.

Esposito, G., Lesk, A. M., Molinari, H., Motta, A., Niccolai, N. & Pastore, A. (1993). Biopolymers, 33, 839-846 (1993).

George, D. G., Hunt, L. T. & Barker, W. C. (1989). In: Computational Molecular Biology: Sources and Methods for Sequence Analysis. A.M. Lesk, ed., Oxford, Oxford University Press, p. 17-26.

Hammer, P. & Rudeanu, S. (1968). Boolean methods in operations research and related areas. New York:Springer-Verlag.

Holbrook, S. R., Muskal, M. & Kim, S. H. (1990). Prot. Eng. 3, 659-665.

Imoto, T., Johnson, L. N., North, A. C. T., Phillips, D. C. & Rupley, J. A. (1972). The Enzymes, 3rd ed., vol. 7, p. 665.

Kaptein, R., Dijkstra, K. & Nicolay, K. (1978). Nature 273, 293-294.

Kosen, P. A., Scheek, R. M., Naderi, H., Basus, V. J., Manogaran, S., Schmidt, P. G., Oppenheimer, N. J. & Kuntz, I. D. (1986). Two dimensional ^{1}H NMR of three spin-labeled derivatives of BPTI. Biochemistry 25, 2356–2364.

Lesk, A. M. (1973). J. Comp. Phys., 12, 150-152.

Moonen, C. T. W., Scheek, R. M., Boelens, R. & Müller, F. (1984). The use of 2D NMR spectroscopy and 2D difference spectra in the elucidation of the active center of Megasphoena elsdenii flavodoxin. Eur. J. Biochem. 141, 323–330.

Morishima, I., Inubushi, T., Yonezawa, T. & Kyogoku, Y. (1977). Proton magnetic resonance studies of specific association of DTBN radical to probe affinity of hydrogen-bonding involved in complementary base-pairs. J. Am. Chem. Soc. 99, 4299–4305.

Morris, G. A. & Freeman, R. (1978). Selective excitation in Fourier transform nuclear magnetic resonance. J. Magn. Res. 29, 433–462.

Niccolai, N., Valensin, G., Rossi, C. & Gibbons, W.A. (1982). The stereochemistry and dynamics of natural products and biopolymers from proton relaxation spectroscopy: spin label delineation of inner and outer protons of gramicidin S including hydrogen bonds. J. Am. Chem. Soc. 104, 1534–1537.

Niccolai, N., Rossi, C., Valensin, G., Mascagni ,P. & Gibbons, W.A. (1984). An investigation of the mechanisms of notroxide. Induced proton relaxation enhancements in biopolymers. J. Phys. Chem. 88, 5689–5692.

Nishikawa, K. & Ooi, T. (1986). J. Biochem. 100, 1043-1047.

Petros, A. M., Mueller, L. & Kopple, K. D. (1990). NMR identification of protein surface using paramagnetic probes. Biochemistry 29, 10041–10048.

Redfield, C. & Dobson, C. M. (1988). Sequential ^{1}H NMR assignments and secondary structure of hen egg white lysozyme in solution. Biochemistry 27, 122–136.

Saudek, V., Williams, R. J. P. & Ramponi, G. (1989). FEBS Letters, 242, 225-232.

Schmidt, P. G. & Kuntz, I. D. (1984). Distance measurements in spin labeled lysozyme. Biochemistry 23, 4261–4266.

Shrake, A. & Rupley, J. A. (1973) Environment and exposure to the solvent of protein atoms: lysozyme and insulin. J. Mol. Biol. 79, 351–371.

von Freyberg, B., Richmond, T.J. & Braun, W. (1993). J. Mol. Biol. 233, 275-292.

Wien, R. W., Morrisett, J. D. & McConnell, H. M. (1972). Spin label induced nuclear relaxation. Distances between bond saccharides, histidine-15 and tryptophane-123 on lysozyme in solution. Biochemistry 11, 3707–3716.

Woodward, C. K., Simon, I., and Tüchsen, E. (1982). Hydrogen exchange and dynamics structure of proteins. Mol. Cell. Biochem. 48, 135–160.

Wüthrich, K. (1986). NMR of Proteins and Nucleic Acids. New York: John Wiley & Sons.

Zuiderweg, E. R. P., Hallenga, K. & Olejniczak, E. T. (1986). Improvement of 2D NOE spectra of biomacromolecules in H_2O solution by coherent suppression of the solvent resonance. J. Magn. Res. 70, 336–343.

Comparative Protein Modeling by Satisfaction of Spatial Restraints

Andrej Šali and Tom Blundell[†]

Department of Chemistry, Harvard University, Cambridge, MA 02138, USA
[†] ICRF Unit of Structural Molecular Biology, Department of Crystallography, Birkbeck College, London WC1E 7HX, England

Abstract

We describe a comparative protein modeling method designed to find the most probable structure for a sequence given its alignment with related structures. The three-dimensional (3D) model is obtained by optimally satisfying spatial restraints derived from the alignment and expressed as probability density functions (pdf's) for the features restrained. For example, the probabilities for main chain conformations of a modeled residue may be restrained by its residue type, main chain conformation of an equivalent residue in a related protein, and the local similarity between the two sequences. Several such pdf's are obtained from the correlations between structural features in 98 families of homologous proteins which have been aligned on the basis of their 3D structures. The pdf's restrain C^α–C^α distances, main chain N–O distances, main chain and side chain dihedral angles. A smoothing procedure is used in the derivation of these relationships to minimize the problem of a sparse database. The 3D model of a protein is obtained by optimization of the molecular pdf such that the model violates the input restraints as little as possible. The molecular pdf is derived as a combination of pdf's restraining individual spatial features of the whole molecule. The optimization procedure is a variable target function method that applies the conjugate gradients algorithm to positions of all non-hydrogen atoms. The method is automated and is illustrated by the modeling of trypsin from two other serine proteases.

1   Introduction

Comparative protein modeling uses experimentally determined protein structures to predict conformation of other proteins with similar amino acid sequences [for reviews see refs. (1) and (2)]. This is possible because a small change in the sequence usually results in a small change in the 3D structure (3,4). The accuracy of protein models obtained by comparative modeling compares favorably with that of models calculated by other theoretical methods. The comparative method produces models with an RMS error as low as 1 Å for sequences that have sufficiently similar homologues with known 3D structures (5). However, comparative modeling is restricted to sequences with closely related proteins with known 3D structures. Nevertheless, since 28% of the known sequences have at least a 25% residue identity with one of the known structures (6), we can estimate that an order of magnitude more sequences

	1	2	3	4	5	6	7
structure A	A	f	s̃	t	l	Ñ	t
structure B	A	f	s̃	s	i	**Ñ**	**t̃**
structure C	A	Ŷ	p	s	i	S̃	a
sequence X	G	F	D	T	I	T	T
extrapolation				↓			
structure X	G	f	d̃	t	i	T	t̃

Figure 1: Comparative protein modeling by satisfaction of spatial restraints. A 3D model of sequence X has to be calculated from the known homologous structures A, B and C. First, the known 3D structures are compared. In order to indicate spatial features of the known structures, residue codes in the resulting alignment are formatted using the convention of the JOY program (11): UPPER CASE, solvent inaccessible amino acid residues; lower case, solvent accessible amino acid residues; underline, hydrogen bond to main chain carbonyl; **bold type**, hydrogen bond to main chain nitrogen; tilde (˜), side chain – side chain H-bond; italic, positive main chain dihedral angle Φ. The sequence of the unknown is then aligned with the related structures. Next, the spatial features of the known structures are transferred to the sequence of the unknown; thus, a number of spatial restraints on its structure are obtained. For example, since there is a conserved hydrogen bond to the main chain carbonyl at position 6 in all three known structures, we assume that the equivalent hydrogen bond also occurs in the sequence of the unknown. Finally, these restraints are satisfied as well as possible to obtain the model for the 3D structure of the unknown.

can be modeled by comparative modeling than compared to the protein structures determined by experiment. This ratio is likely to increase as the fraction of the known structural motifs increases and the gap between numbers of the known sequences and 3D structures widens.

Future improvements of comparative modeling should aim to model proteins with lower homology to known structures, to increase the accuracy of the models, to make modeling fully automated, and to allow inclusion of many different types of information. In this paper, we attempt to achieve these goals by pursuing the following fundamental question: *What is the most probable structure for a certain sequence given its alignment with related structures?* Our approach, outlined in Figure 1, follows from the method for comparison of protein structures implemented in the program COMPARER (7,8,9). The modeling method was developed to use as many different types of data about the unknown as possible (2). The method consists of three stages: alignment of the sequence to be modeled with related protein structures and segments, extraction of spatial restraints on the sequence using the alignment, and satisfaction of the restraints to obtain a 3D model. This paper describes the procedures involved in the last two stages and illustrates the approach by application to modeling trypsin from two other serine proteases.

Table 1: Size of the alignments database.

Alignments	98
Protein structures	379
Homologous protein pairs	2,188
Residues	70,996
Equivalent residue pairs	412,910
Intra-molecular residue pairs	21,453,983
Equivalent intra-molecular residue pairs	117,959,350

2   Derivation of spatial Restraints

In this section, relatively simple restraints on the protein conformation are defined from the information about a related protein structure. A restraint is most precisely defined in terms of a pdf, $p(x/a, b, \ldots, c)$, for the feature x that is restrained. This is a conditional pdf and gives a probability density for x when $a, b, \ldots, c$ are specified. For example, $p(\chi_1/\text{residue type}, \Phi, \Psi)$ could be used to predict the side chain dihedral angle χ_1 from the type of a residue and its main chain dihedral angles Φ and Ψ.

In reality, it is not possible to obtain the true function p, but only its approximations:

$$p(x/a, b, \ldots, c) \approx W_{x,a,b,\ldots,c} \approx f(x, a, b, \ldots, c, \mathbf{q}) \tag{1}$$

where $W_{x,a,b,\ldots,c}$ is a table spanned by $x, a, b, \ldots, c$ that contains as its elements the observed relative frequencies for the occurrence of x given $a, b, \ldots, c$, and f is an analytic function whose parameters $\mathbf{q}$ are fitted to the observed $\mathbf{W}$. The multidimensional table of relative frequencies $\mathbf{W}$ is calculated from the absolute frequencies $\mathbf{W}'$ using

$$W_{x,a,b,\ldots,c} = \frac{W'_{x,a,b,\ldots,c}}{\sum_x W'_{x,a,b,\ldots,c}}. \tag{2}$$

The absolute frequencies, $\mathbf{W}'$, are obtained directly by counting the number of occurrences of each combination of $(x, a, b, \ldots, c)$ values in the sample. In this study, the sample is derived from a database of known protein structures and their alignments. Thus, before the restraints can be derived, a database of known protein structures, their features, and alignments must be constructed.

Initially, a small database of 17 family alignments was built (2,10) by the protein structure comparison program COMPARER (7,8). This small database was then gradually extended and used to obtain environment specific residue substitution tables (11,12), to improve homology modeling of loops (13), and to increase the sensitivity and accuracy of aligning sequences with structures (14). Recently, a large number of alignments was collected and presented (15). Currently, 378 members of 98 families of related proteins were extracted from the Brookhaven Protein Databank (16) and aligned both by COMPARER and MNYFIT (17) to obtain multiple alignments for each of the families (A. Šali & J. Overington, in preparation).[1] The size of the database is illustrated further in Table 2. A number of features of protein structures were also calculated and stored in the database (Table 2).

The program MDT was written to explore the alignments database and to derive the best pdf's for comparative modeling. The inputs to the program are names of features selected

---

[1]Several results in this paper were derived from the first smaller database of only 17 alignments.

Table 2: Features used in this paper that may be selected in MDT to span multi-dimensional frequency tables $\mathbf{W}'$. The first column lists the variable names that are used for these features. It also indicates whether an intra-molecular average or inter-molecular difference can be calculated. The $^-$ symbol indicates an average of the feature at two residue positions in the same protein, such as an average accessibility of a certain residue pair. Features that are not associated with two proteins can be used independently for two related proteins in a pairwise alignment or for three related proteins in a triple alignment. For example, a 2D table can be constructed that is spanned by a residue type r in one protein and a residue type r' at the equivalent position in a related protein; the prime is generally used to designate that the feature is from the second protein and two primes that it is from the third protein. The Δ symbol refers to the difference between features f and f': $\Delta f = f - f'$.

Variable	Feature
r	amino acid residue type
$\Phi, \Delta\Phi$	main chain dihedral angle Φ
Φ_c	main chain dihedral angle Φ class
$\Psi, \Delta\Psi$	main chain dihedral angle Ψ
Ψ_c	main chain dihedral angle Ψ class
$\omega, \Delta\omega$	main chain dihedral angle ω
ω_c	main chain dihedral angle ω class
χ_i	side chain dihedral angle χ_i, $i = 1, 2, 3, 4, 5$
c_i	side chain dihedral angle χ_i class, $i = 1, 2, 3, 4, 5$
t	secondary structure class of a residue (positive Φ, α, β, other)
M	main chain conformation class of a residue (24)
α	fractional content of residues in the main chain conformation class A
S	side chain conformation class (χ_1, χ_2)
$a, \bar{a}$	(fractional) contact solvent area of a residue
$s, \bar{s}$	residue neighborhood difference between two proteins
i	fractional sequence identity between two proteins
$d, \Delta d$	distance between two specified atom types
b	average residue isotropic temperature factor
R	resolution of X-ray analysis
n	number of atomic contacts with non-protein non-water atoms per residue
$g, \bar{g}$	distance of a residue from a gap in the alignment
l	number of residues in the protein
G	several residue type groups (*e.g.* hydrophobic/hydrophilic)

from Table 2, a list of discrete values for tabulating these features (numerical or symbolic), and the list of alignments. These are then used to calculate various multi-dimensional frequency tables $W'_{x,a,b,\dots,c}$ by counting the occurrences of all the required combinations of features $x, a, b, \dots, c$ in the alignments database. The tables $\mathbf{W}'$ are subsequently used as outlined above to calculate the relative frequency tables $\mathbf{W}$ and sometimes the corresponding pdf's f. For fitting pdf f to the observed relative frequencies $\mathbf{W}$, the Levenberg-Marquardt algorithm for non-constrained least-squares fitting of a non-linear multidimensional model (18) was implemented in the program LSQ. Since the database is sparse for some pdf's, smoothing of the pdf's was also performed as described in (2); the smoothing method is an extension of the procedure proposed in ref. (19).

2.1 Stereochemical Restraints

All stereochemical restraints are easily obtained from the amino acid sequence of a protein. Stereochemical restraints include bond distances, bond angles, torsional angles defined by three consecutive bonds, and improper dihedral angles as used in CHARMM (20). The corresponding pdf's are obtained from the CHARMM 22 parameter set (21), based on classical statistical mechanics (22).[2] For example, the pdf for the bond length is a Gaussian probability density function

$$p^b(b) = \frac{1}{\sigma_b\sqrt{2\pi}} \exp[-\frac{1}{2}(\frac{b-\bar{b}}{\sigma_b})^2] = N(\bar{b}, \sigma_b), \tag{3}$$

where $\sigma_b = \sqrt{kT/c}$; c and $\bar{b}$ are the CHARMM force constant and mean length, respectively. The following pdf is used for two atoms restrained by the van der Waals repulsion

$$p^v(d) = c \cdot \left\{ \begin{array}{ll} N(d_o, \sigma_w) \,; & d \leq d_o \\ \frac{1}{\sigma_w\sqrt{2\pi}} \,; & d_o < d < d_{max} \end{array} \right. \tag{4}$$

where d is the distance between the two atoms, d_o is the sum of their van der Waals radii and σ_w is the standard deviation of the Gaussian part of the whole pdf (usually 0.05Å). d_{max} is the maximal possible linear dimension of a protein and constant c is chosen so that $p^v(d)$ integrates to 1. This pdf does not differentiate between contact distances larger than d_o, but it does select against distances smaller than d_o. This is achieved by imposing a repulsive harmonic potential on atoms that are less than d_o apart.

2.2 Restraining a Distance between two C^α Atoms

The unknown feature is defined as the difference between two equivalent C^α–C^α distances, $d - d'$; d' is from the 'known' or template structure and d from the 'unknown' or target structure. Using the database of alignments described above, the distribution of $d - d'$ was found as a function of four independent variables: the corresponding C^α–C^α distance in the 'known' structure (d'), the fractional sequence identity of the two aligned sequences (i), the average of the fractional solvent accessibilities of the two residues spanning the distance in the 'known' structure ($\bar{a}'$), and the average number of positions that separate the two residues spanning the distance from the closest gap in the alignment ($\bar{g}$).

Examples of the histograms of probability distributions obtained by the MDT program are shown in Figures 2a–b. These histograms demonstrate that the conditional distribution of the distance differences may be approximated by a Gaussian function with a mean of zero and a standard deviation dependent on the values of the independent variables. Therefore, the pdf restraining a $C^\alpha - C^\alpha$ distance in the sequence of an unknown, given an alignment

---

[2]Several results in this paper were derived with an earlier version of our program that relied on the GROMOS86 IFP37C4 parameter set (23).

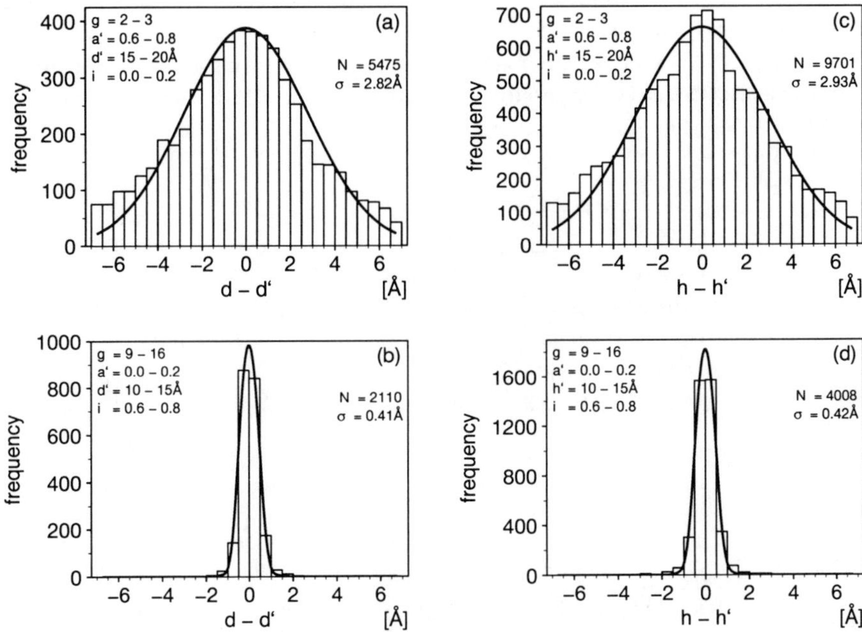

Figure 2: Distribution of the differences between two equivalent distances. The histograms show the frequency of the differences between two equivalent distances as observed by MDT in the alignments database. (a) and (b), C^α–C^α distances; (c) and (d), main chain N–O distances. The curves are fitted Gaussian models [Eqs. (5) and (8)]. The values of the dependent variables, the number of C^α–C^α distances in the database (N), and the standard deviation of the Gaussian model (σ) are shown for each histogram.

with a single related known structure, can be modeled as

$$p^d(d/\bar{g}, i, \bar{a}', d') = \frac{1}{\sigma(\bar{g},i,\bar{a}',d')\sqrt{2\pi}} \exp[-\tfrac{1}{2}(\tfrac{d-d'}{\sigma(\bar{g},i,\bar{a}',d')})^2]$$

$$\begin{aligned}
\sigma(\bar{g}, i, \bar{a}', d') = {}& \alpha_1 + \alpha_2\,\bar{g} + \alpha_3\,i + \alpha_4\,\bar{a}' + \alpha_5\,d' + \\
& \alpha_6\,\bar{g}^2 + \alpha_7\,\bar{g}i + \alpha_8\,\bar{g}\bar{a}' + \alpha_9\,\bar{g}d' + \alpha_{10}\,i^2 + \alpha_{11}\,i\bar{a}' + \\
& \alpha_{12}\,id' + \alpha_{13}\,\bar{a}'^2 + \alpha_{14}\,\bar{a}'d' + \alpha_{15}\,d'^2 + \\
& \alpha_{16}\,\bar{g}^3 + \alpha_{17}\,\bar{g}^2i + \alpha_{18}\,\bar{g}^2\bar{a}' + \alpha_{19}\,\bar{g}^2d' + \alpha_{20}\,\bar{g}i^2 + \alpha_{21}\,\bar{g}i\bar{a}' + \\
& \alpha_{22}\,\bar{g}id' + \alpha_{23}\,\bar{g}\bar{a}'^2 + \alpha_{24}\,\bar{g}\bar{a}'d' + \alpha_{25}\,\bar{g}d'^2 + \alpha_{26}\,i^3 + \\
& \alpha_{27}\,i^2\bar{a}' + \alpha_{28}\,i^2d' + \alpha_{29}i\bar{a}'^2 + \alpha_{30}\,i\bar{a}'d' + \alpha_{31}\,id'^2 + \\
& \alpha_{32}\,\bar{a}'^3 + \alpha_{33}\,\bar{a}'^2d' + \alpha_{34}\,\bar{a}'d'^2 + \alpha_{35}\,d'^3
\end{aligned}$$

$$(5)$$

In relation to Eq. (5), the four features can be seen as the measure for the degree of transferability of the distance from the 'known' to the 'unknown' structure; the distance in the 'unknown' is more likely to be closer to the equivalent distance in the 'known' when the distance is short, the two residues spanning the distance are buried, the two structures are similar overall, and the residues are distant from the gaps in the alignment. The remaining problem is to determine the best values of parameters α_i. This is

Table 3: The best parameters for restraining $C^\alpha - C^\alpha$ distances (d) and main chain N – O distances (h). Full expressions for the standard deviations of the Gaussian p models for **W** [Eqs. (5) and (8)] are given. If $\bar{g} > 20$, $\bar{g}$ is reset to 20. Before using $\bar{g}$, $\bar{a}'$, and i with the parameters shown, they have to be scaled by 0.1, 0.01, and 0.1, respectively. The RMS deviations between the p models and **W'** are 0.0524 and 0.0441 for d and h, respectively.

$$
\begin{aligned}
\sigma(\bar{g}, i, \bar{a}', d') \;=\; & 0.849 - 2.033\,\bar{g} - 1.227\,i + 0.971\,\bar{a}' + 1.467\,d' + \\
& 1.382\,\bar{g}^2 + 1.539\,\bar{g}i - 0.504\,\bar{g}\bar{a}' - 0.259\,\bar{g}d' + 2.412\,i^2 - 1.496\,i\bar{a}' - \\
& 3.094\,id' - 0.425\,\bar{a}'^2 + 0.670\,\bar{a}'d' - 0.159\,d'^2 - \\
& 0.307\,\bar{g}^3 - 0.213\,\bar{g}^2 i + 0.088\,\bar{g}^2\bar{a}' + 0.020\,\bar{g}^2 d' - 0.969\,\bar{g}i^2 + 0.453\,\bar{g}i\bar{a}' + \\
& 0.177\,\bar{g}id' - 0.058\,\bar{g}\bar{a}^2 - 0.042\,\bar{g}\bar{a}'d' + 0.020\,\bar{g}d'^2 - 0.847\,i^3 + \\
& 0.055\,i^2\bar{a} + 1.546\,i^2 d' + 0.527 i\bar{a}'^2 - 0.220\,i\bar{a}'d' + 0.254\,id'^2 + \\
& 0.066\,\bar{a}'^3 + 0.153\,\bar{a}^2 d' - 0.153\,\bar{a}'d'^2 - 0.0019\,d'^3
\end{aligned}
\tag{6}
$$

$$
\begin{aligned}
\sigma(\bar{g}, i, \bar{a}', h') \;=\; & 0.957 - 2.044\,\bar{g} - 1.078\,i + 0.995\,\bar{a}' + 1.477\,h' + \\
& 1.572\,\bar{g}^2 + 1.148\,\bar{g}i - 0.525\,\bar{g}\bar{a}' - 0.483\,\bar{g}h' + 1.505\,i^2 - 0.655\,i\bar{a}' - \\
& 2.849\,ih' - 0.625\,\bar{a}'^2 + 0.499\,\bar{a}'h' - 0.126\,h'^2 - \\
& 0.369\,\bar{g}^3 - 0.243\,\bar{g}^2 i + 0.121\,\bar{g}^2\bar{a}' + 0.067\,\bar{g}^2 h' - 0.592\,\bar{g}i^2 + 0.346\,\bar{g}i\bar{a}' + \\
& 0.276\,\bar{g}ih' - 0.032\,\bar{g}\bar{a}^2 - 0.061\,\bar{g}\bar{a}'h' + 0.036\,\bar{g}h'^2 - 0.329\,i^3 - \\
& 0.318\,i^2\bar{a} + 1.472\,i^2 h' + 0.284 i\bar{a}'^2 - 0.293\,i\bar{a}'h' + 0.198\,ih'^2 + \\
& 0.382\,\bar{a}'^3 + 0.110\,\bar{a}^2 h' - 0.079\,\bar{a}'h'^2 - 0.0095\,h'^3
\end{aligned}
\tag{7}
$$

achieved by least-squares fitting the model p^d in Eq. (5) to the histograms **W** obtained from the database scan (Table 3). The Gaussian conditional pdf's $p^d(d/\bar{g}, i, a, d')$, calculated from the least-squares parameters, are superposed on the experimental histograms in Figures 2a–b. These plots provide additional graphical evidence that the Gaussian model can describe the association between the unknown $C^\alpha - C^\alpha$ distance and the four independent variables included in this analysis.

2.3   Restraining a distance between main chain N and O atoms

The N – O distance in the target protein was modeled in the same way as the $C^\alpha - C^\alpha$ distance above (Figures 2c-d):

$$
p^h(h/\bar{g}, i, \bar{a}', h') = \frac{1}{\sigma(\bar{g}, i, \bar{a}', h')\sqrt{2\pi}} \exp\left[-\frac{1}{2}\left(\frac{h - h'}{\sigma(\bar{g}, i, \bar{a}', h')}\right)^2\right].
\tag{8}
$$

2.4   Restraining residue main chain conformation

The residue main chain conformation is defined by dividing the Ramachandran plot spanned by the Φ and Ψ main chain dihedral angles into six areas (Figure 3): A, B, P, L, G, and E

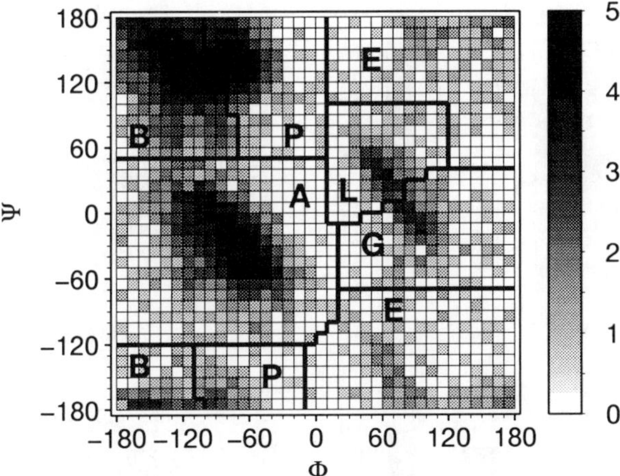

Figure 3: Definition of the main chain conformation classes. The plot shows $W'(\Phi, \Psi)$ determined from the proteins in the alignments database. It is divided into $10° \times 10°$ squares. The areas corresponding to the six characteristic peaks are delimited by thick lines. These areas define the residue main chain conformation classes A, B, P, L, G, and E (24). The scale on the right corresponds to $\ln[W'(\Phi, \Psi) + 1]$.

(24). Within each area, the distribution of the two dihedral angles is approximately Gaussian (2). Suppose we can predict the probability ω_i that the restrained residue is in the main chain conformation class i. Then the two pdf's restraining Φ and Ψ dihedral angles can be modeled as a weighted sum of six Gaussian functions, each function corresponding to one of the main chain conformation classes A–E and weighted by a probability that a residue is in the corresponding class:

$$p^m(\Phi) = \sum_{i=A,...,E} \omega_i \, N[\bar{\bar{\Phi}}_i, \sigma_i(\Phi)]$$

$$p^m(\Psi) = \sum_{i=A,...,E} \omega_i \, N[\bar{\bar{\Psi}}_i, \sigma_i(\Psi)]$$

$$(9)$$

where $N(\alpha, \sigma)$ stands for a Gaussian pdf with mean α and standard deviation σ (2). The remaining problem is to determine the probabilities ω_i of all six main chain conformation classes for each restrained residue. The database of alignments and the program MDT were used to obtain these weights.

The protein features that could correlate with the main chain conformation class of a restrained residue were selected from the list in Table 2. These are the types of the restrained (r) and equivalent (r') residues and the features that can be classified into the following three groups: main chain conformation of an equivalent residue (M', t', Φ', Ψ', α'), side chain conformation of an equivalent residue (c'_1, c'_2, c'_3), and variability measures (s, i, b', a', g, R'). Smoothed and non-smoothed pdf's of the form $p(M/a, b, \ldots, c)$ were derived from the alignments database for the 7249 possible combinations of up to five selected features $(a, b, \ldots, c)$ listed above. Each of the resulting pdf's was evaluated by predicting the most likely main chain conformation class for each residue in all the 5586 equivalent residue pairs in the test set of seven serine proteases and by comparing these predictions with the

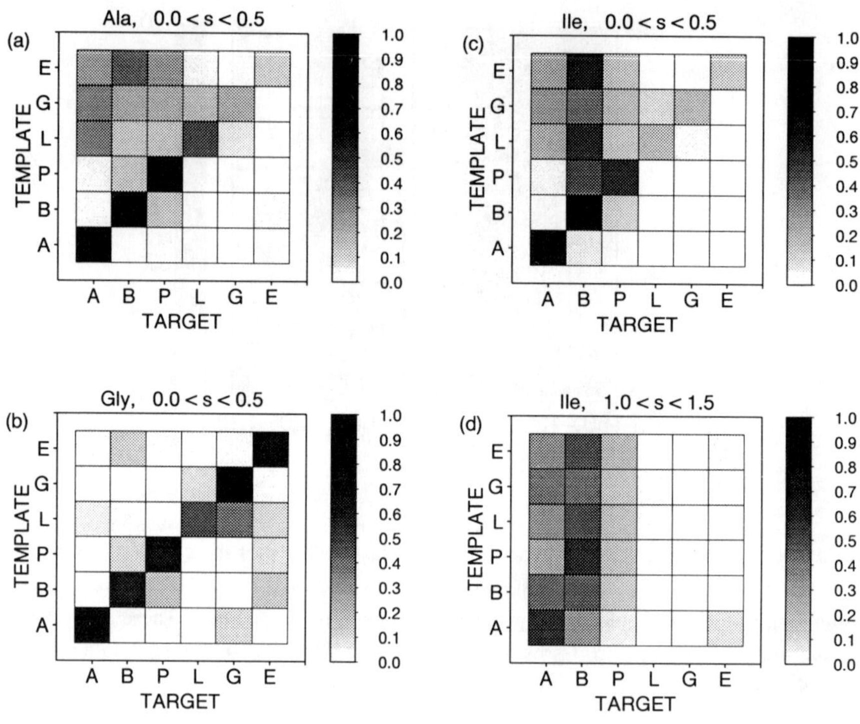

Figure 4: Sample cross-sections through the pdf for prediction of main chain conformation. The probabilities $W(M/M', r, s)$ for conformation classes A–E of a given residue type (horizontal row, M) are shown for each conformation class of an equivalent residue (vertical row, M'). The type of a restrained residue (r), and the residue neighborhood difference (s) are shown above each plot.

actual conformations found in the crystallographic structures.

A cross-section through the best pdf $p(M/r, M', s)$ is shown in Figure 4. Here, s is residue neighborhood difference that measures how different are the types of residues in the two spheres centered on two equivalent central residues; s depends only on the amino acid sequences and can be obtained from the alignment between the target and the template (2). The prediction success of this pdf on the test set of seven serine proteases is listed for the individual residue types in Table 4. The residues that are predicted most accurately (approximately 85%) are Trp, Gln, Pro, Phe, and Cys, whereas the least accurately predicted residues include Gly, Asn, Glu, and Leu (approximately 63%). This trend probably reflects the distribution of the various residue types in the core and on the surface of the molecule as well as the degree of restraint on the main chain provided by its side chain. The conformation of the core residues is expected to be more conserved, and therefore better predicted, than the conformation of the exposed residues. Likewise, the conformationally restrained residues, such as Pro, are predicted better than those that are more flexible, such as Gly. Leu is not predicted reliably because its intrinsic preferences for the A, B, and P classes are very similar. While the 73% prediction rate may seem low, many errors occur because of the swapping between the structurally similar (B, P) classes as well as between

Table 4: Success for the prediction of the main chain conformation class. Total number is the number of residue pairs in the test set that contain the residue being predicted; the numbers of the predicted residues are shown in parentheses. The smoothed pdf $p(M/M', r, s)$ was used for the prediction of the main chain conformation class M. The residue types are listed in descending order with respect to the success of the prediction. The bottom line gives the total number of equivalent residue pairs, the total number of residues with a defined main chain conformation state and the prediction success averaged over all residue types.

Residue type	Total number	% correctly predicted
W	219 (37)	90
Q	391 (67)	89
P	430 (79)	86
F	224 (38)	83
C	360 (62)	83
A	683 (123)	82
V	791 (136)	79
S	751 (139)	78
I	518 (88)	76
H	209 (36)	73
R	262 (46)	73
K	446 (77)	72
T	614 (108)	71
D	382 (69)	70
Y	330 (57)	69
M	137 (23)	68
G	918 (163)	66
N	481 (85)	62
E	319 (57)	62
L	685 (118)	57
	9150 (1608)	73

the (L, G) classes. When these two pairs are treated only as two classes, the prediction success increases to 87.4%.

2.5 Restraining residue side chain conformation

Side chain restraints are formulated in a similar way to the main chain conformation restraints. Most of the side chain dihedral angles are clustered in up to 3 characteristic intervals that span the range from $-180°$ to $180°$; this results in a small number of side chain rotamers (25,26). Thus, each dihedral angle can be described by a corresponding dihedral angle class within which the distribution of the dihedral angle is Gaussian (2). Similarly to the prediction of the main chain conformation class, the side chain dihedral angles χ_i are modeled as a weighted sum of Gaussians

$$p^s(\chi_i) = \sum_j \omega_{ij}\, N[\bar{\chi}_{ij}, \sigma_j(\chi_i)] \tag{10}$$

where ω_{ij} are the probabilities that the restrained side chain dihedral angle i is in class j, and $N(\alpha, \sigma)$ is a Gaussian pdf with mean α and standard deviation σ (2). The remaining problem is to determine the probabilities ω_{ij} of all side chain conformation classes for each restrained residue; the same approach is followed as for the derivation of the weights for the main chain conformation classes.

When $p(c_1/r)$ is used in the prediction of the χ_1 class, the prediction success is 57.4% because that many residues are in their most likely classes. When information about the type and χ_1 dihedral angle of an equivalent residue is added to obtain pdf $p(c_1/r, r', c_1')$, the prediction success increases for 6.4% to 63.8%. None of the remaining independent variables improves the prediction success of $p(c_1/r)$ or $p(c_1/r, r', c_1')$, irrespective of whether the variables are used on their own, in pairs, or in threes. The prediction successes of $p(c_1/r)$ and $p(c_1/r, c_1', r', s)$ are listed for the individual residue types in Table 5. The residues that are predicted most reliably (80%) by $p(c_1/r, c_1', r', s)$ tend to be large and buried (Trp, Cys, Leu, Val, and Tyr). The residues that are predicted least reliably (50%) tend to be small and exposed (Asn, Met, Arg, Glu, and Ser). The largest improvement as a result of using information about the equivalent side chain occurs for Trp (30%), His (23%), Asp (17%), Thr (12%), Tyr (10%), and Leu (10%). The amount of information provided by the type and χ_1 of an equivalent residue tends to be large for large or buried residues and small or non-existent for exposed residues. This improvement reflects the degree to which the side chain conformation of a residue is restrained by its environment. The restraints for χ_2, χ_3, and χ_4 dihedral angles were derived in a similar way. The prediction successes are summarized in Table 5.

3 Satisfaction of spatial restraints

It was shown in the previous Section how spatial restraints on the sequence to be modeled can be expressed as pdf's. These pdf's were obtained from stereochemical considerations and from a single homologous structure. In this Section, we describe how to combine the restraints from several homologous structures and how to use these restraints to derive a 3D model. The 3D model is obtained by an optimization of the molecular pdf which depends on the model and on the restraints.

3.1 The molecular probability density function

The molecular pdf is assembled from feature pdf's which in turn are obtained from basis pdf's.

3.1.1 Derivation of a feature pdf from basis pdf's

In general, every structural feature f can be restrained by several basis pdf's $p_k^f(f)$ for $k = 1, 2, \ldots$, such as those described in the preceding Section. A feature pdf, $p^F(f)$, is a pdf that combines all basis pdf's to use all the information about the possible values that the feature f can assume. The lowercase and uppercase superscripts are used for the basis and feature pdf's, respectively. The following example clarifies these definitions. The aim

Table 5: Success for the prediction of the side chain χ_i classes, c_i. Total number is the number of residue pairs in the test set that contain the residue being predicted; the numbers of predicted residues are listed in parentheses. The smoothed pdf's $p(c_1/r, c_1', r', s)$, $p(c_2/r, r', c_1', c_2')$, $p(c_3/r, c_3', r', t')$, and $p(c_4/r)$ were used for the prediction of χ_1, χ_2, χ_3, and χ_4 dihedral angle classes, respectively. The prediction successes of the smoothed pdf's $p(c_i/r)$ are shown in parentheses for $i = 1, 2, 3$. The residue types are listed in descending order with respect to the success of the c_1 prediction. The bottom line gives the total number of equivalent residue pairs tested by the pdf, the total number of residues with a defined χ_1 dihedral angle, and the c_i prediction successes averaged over all residue types that have defined χ_i dihedral angles.

Residue type	Total number	% correctly predicted			
		χ_1 class	χ_2 class	χ_3 class	χ_4 class
W	219 (37)	86.8 (56.8)	81.3 (62.2)	–	–
C	360 (62)	81.4 (77.4)	–	–	–
L	685 (118)	74.0 (64.4)	59.3 (55.9)	–	–
V	791 (136)	72.3 (72.1)	–	–	–
Y	330 (57)	69.4 (59.6)	100.0 (100.0)	–	–
I	518 (88)	68.9 (65.9)	73.9 (73.9)	–	–
K	446 (77)	65.2 (66.2)	63.5 (64.9)	76.0 (75.3)	71.4
F	224 (38)	64.7 (57.9)	100.0 (100.0)	–	–
D	382 (69)	64.7 (47.8)	100.0 (100.0)	–	–
H	209 (36)	61.7 (38.9)	62.2 (55.6)	–	–
Q	391 (85)	61.4 (64.2)	66.5 (62.7)	37.3 (35.8)	–
T	614 (108)	58.5 (46.3)	–	–	–
N	481 (85)	55.1 (52.9)	55.1 (56.5)	–	–
M	137 (23)	54.7 (52.2)	64.2 (69.6)	54.7 (21.7)	–
R	262 (46)	53.4 (52.2)	72.9 (73.9)	49.2 (54.3)	80.4
E	319 (57)	51.7 (49.1)	64.9 (63.2)	79.6 (80.7)	–
S	751 (139)	45.7 (40.3)	–	–	–
	7119 (1243)	64.4 (57.4)	72.3 (70.7)	60.6 (58.5)	74.8

is to construct a feature pdf for a particular C^α–C^α distance in a given sequence. Suppose two known related structures with equivalent distances are available; therefore, we have two corresponding basis pdf's for the C^α-C^α distance in the target sequence [Eq. (5)]. In addition, we also know that each of the two restraints has to comply with the van der Waals criterion, *i.e.* the distance has to be larger than the sum of the two van der Waals radii [Eq. (4)]. In order to combine all this information we have to combine the three basis pdf's into a single feature pdf. To find how to do that, we can use all possible alignments of three proteins in the alignments database. An example of the dependence of a C^α–C^α distance on the two equivalent distances from two related structures, $p(d/d', d'')$, is shown in Figure 5a. The histogram suggests that $p(d/d', d'')$ can be modeled as a weighted sum of the individual pdf's $p(d/d')$ and $p(d/d'')$:

$$p(d/d', d'', \bar{s}', \bar{s}'') = \omega(\bar{s}') \cdot p(d/d') + \omega(\bar{s}'') \cdot p(d/d''). \tag{11}$$

The weight ω of each term in this sum is proportional to the average residue neighborhood difference s between the corresponding structure and the sequence of the unknown. The

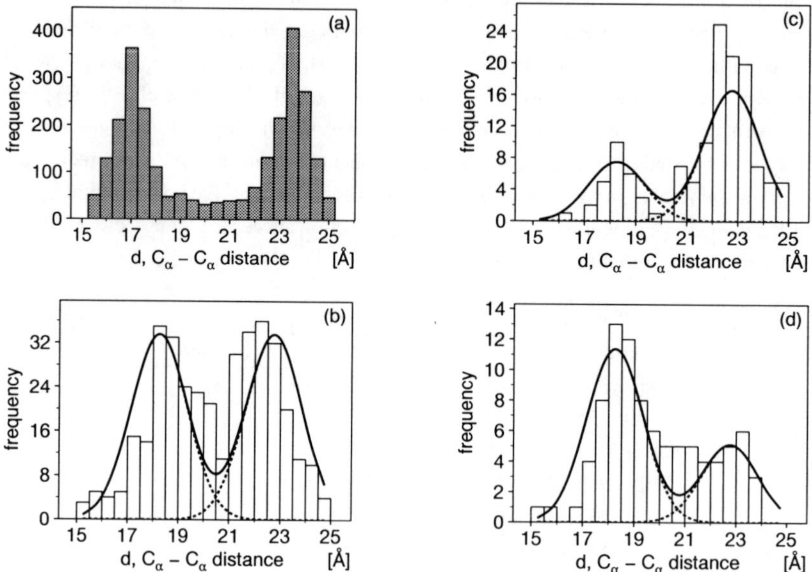

Figure 5: Derivation of a feature pdf from basis pdf's. In all plots, $18.0 < d < 18.5$ and $22.5 < d'' < 23.0$. (a) $W'(d/d', d'')$. (b) $W'(d/d', d'', \bar{s}', \bar{s}'')$, where $0.2 < \bar{s}' < 0.4$ and $0.2 < \bar{s}'' < 0.4$. (c) $W'(d/d', d'', \bar{s}', \bar{s}'')$, where $0.2 < \bar{s}' < 0.4$ and $0.4 < \bar{s}'' < 0.6$. (d) $W'(d/d', d'', \bar{s}', \bar{s}'')$, where $0.4 < \bar{s}' < 0.6$ and $0.2 < \bar{s}'' < 0.4$. The histograms are obtained by scanning the alignments database. The dashed lines are the basis pdf's $p^d(d/d', d'', \bar{s}', \bar{s}'')$ calculated from Eq. (5). The continuous lines are the feature pdf's $p^D(d/d', d'', \bar{s}', \bar{s}')$ calculated with Eqs. (11)–(12).

data can be fitted by the following model for $w(s)$:

$$\omega(s) = \frac{w(s)}{\sum_j w(s_j)} \quad \text{where} \quad w(s) = a + \exp(bs^c) \; ; \qquad \sum_j \omega(s_j) = 1 \, , \qquad (12)$$

where the best values for the parameters, as obtained by the LSQ program, are: $a = 0.0331 \pm 0.0025$, $b = -4.98 \pm 0.11$, and $s = 1.800 \pm 0.079$. The result is that the contribution of a structure to the 3D model of a related structure falls faster than linearly with the average residue neighborhood difference between the two sequences. Examples of histograms and analytical curves for the feature pdf corresponding to different weights are shown in Figures 5b–d.

The last step in the derivation of the feature pdf is to include the van der Waals restraint. Since all stereochemical restraints have to be satisfied in all structures, these restraints are multiplied into the feature pdf and we obtain the final feature pdf $p^D(d) = [\omega_1 p_1^d(d) + \omega_2 p_2^d(d)]p^v(d)$.

This simple approach to combining of two basis pdf's was used for any number of basis pdf's of the same type that were derived from related structures. When properties such as main chain and side chain conformation are predicted, average residue neighborhood difference is replaced by the residue neighborhood difference.

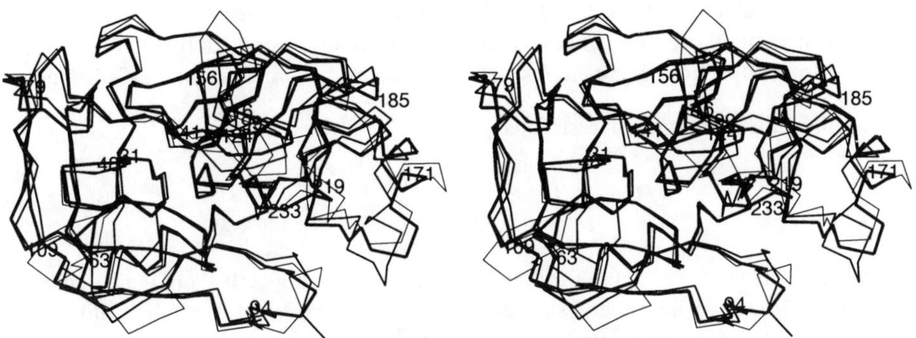

Figure 6: Comparison of trypsin, elastase, and tonin. The stereo plot shows the superposition of the C^α backbones of elastase (medium line) and tonin (thin line) on that of trypsin (thick line). The pairs of the C^α atoms that are aligned in the COMPARER alignment were used for the superpositions. Chymotrypsinogen numbering is used.

Definitions of all types of feature pdf's follow, with the basis pdf's on the right side of the equations as defined in Sections 2.1 – 2.5. The subscript i in the sum refers to the sequences with known structure that are aligned with the sequence of the unknown. The independent variables $a, b, \ldots$ refer to the features correlated with the restrained feature as described in Sections 2.2 – 2.5. The weights ω_i are determined from Eq. (12).

1. $C^\alpha - C^\alpha$ distance restraints:

$$p^D(d) = p^v(d) \sum_i \omega_i \, p_i^d(d/a, b, \ldots) \tag{13}$$

for all pairs of C^α atoms in the sequence of the unknown that satisfy the following three criteria: (1) there is at least one equivalent C^α-atom pair in the known structures, (2) there are at least N_α (usually 1) residues between the two residues spanning the distance in the sequence of the unknown and (3) at least one equivalent distance in the known structures is less than d_d (usually 20 Å). The sum runs over all known structures with an equivalent C^α pair present.

2. Main chain $N - O$ distance restraints:

$$p^H(h) = p^v(h) \sum_i \omega_i \, p_i^h(h/a, b, \ldots) \tag{14}$$

for all pairs of main chain N and O atoms in the sequence of the unknown that satisfy the following criteria: (1) there is at least one equivalent (N, O) pair in the known structures, (2) there are at least N_h (usually 2) residues between the two residues spanning the distance in the sequence of the unknown and (3) at least one equivalent distance in the known structures is less than d_h (usually 10 Å). The sum runs over all known structures with an equivalent $N - O$ pair present.

3. Stereochemical restraints:

$$p^E(e) = p^e(e) \ . \tag{15}$$

Feature e can be bond length, bond angle, torsion angle, improper dihedral angle, or van der Waals contact (Section 2.1). The feature pdf for van der Waals contacts, $p^V(v)$, restrains only those pairs of atoms that are not already restrained by the feature pdf's for the bond lengths, bond angles, C^α–C^α distances and main chain N–O distances.

4. Main chain conformation restraints:

$$p^M(\theta) = \begin{cases} \sum_{i=1}^n \omega_i \, p_i^m(\theta/a, b, \ldots) & n > 0 \\ p^m(\theta/R) & n = 0 \end{cases} \tag{16}$$

where θ stands for either Φ or Ψ main chain dihedral angle. If there is no equivalent residue in any of the related structures ($n = 0$), the restraint depending only on the residue type in the sequence of the unknown is applied.

5. χ_1, χ_2, χ_3, and χ_4 side chain dihedral angle restraints:

$$p^S(c) = \begin{cases} \sum_{i=1}^n \omega_i \, p_i^s(c/a, b, \ldots) & n > 0 \\ p^s(c/R) & n = 0 \end{cases} \tag{17}$$

where c stands for either χ_1, χ_2, χ_3, or χ_4 side chain dihedral angle. A rotamer library based only on the residue type is used when there is no equivalent residue in any of the available structures ($n = 0$).

3.1.2   Derivation of a molecular pdf from feature pdf's

The last stage in the derivation of a molecular pdf is to combine all feature pdf's into a molecular pdf. The 3D-structure of a protein is uniquely determined if a sufficiently large number of its spatial features, f_i, are specified. The goal is to find the 3D structure that is consistent with the most probable values of individual features f_i. The molecular pdf should give a probability for occurrence of any combination of these features simultaneously. Then, the model for the 3D structure of the unknown would correspond to the maximum of the molecular pdf. Assuming that feature pdf's are independent of each other, the molecular pdf is simply a product of feature pdf's defined in Eqs. (13)–(17):

$$P = \prod_i p^F(f_i). \tag{18}$$

Thus, by maximizing function P we find the most probable model for the 3D structure of the unknown given its alignment with the known structures.

3.2   Optimization of the molecular pdf

Derivation of restraints from an alignment and satisfaction of those restraints are implemented in the computer program MODELLER. The protein model may consist of all atoms or any subset such as all heavy atoms, mainchain atoms, or only C^α atoms. The function that is actually optimized is a transformation of the molecular pdf P:

$$F = -\ln(P) \tag{19}$$

where all the features are expressed in terms of atomic Cartesian coordinates. Function F is referred to as the objective function. The same Cartesian coordinates that maximize P also minimize F. To increase the radius of convergence, the variable target function approach is implemented in MODELLER. This method has been introduced by Braun and Gō in the DISMAN program for calculating protein 3D structures consistent with 2D-NMR constraints (27). The main difference between the original method and the present implementation is that the current optimization proceeds in the Cartesian space whereas the original procedure optimized the dihedral angles. Following the variable target function method, the optimum of the molecular pdf is found by successive optimizations of increasingly more complex 'target' functions, culminating in the true molecular pdf at the end. This series is obtained by starting with sequentially local restraints and then introducing more and more long range restraints, finally arriving at the true molecular pdf incorporating all the restraints. More precisely, the target function $P(\Delta r)$ is defined as a function of an integer variable $\Delta r = 1, \ldots, N$ where N is the number of residues in the sequence being modeled. The target function $P(\Delta r)$ is obtained in the same way as the molecular pdf, except that only those restraints whose atoms originate from residues not more than Δr residues apart in the sequence are included. The whole calculation consists of a number of conjugate gradient optimizations (18) of target functions $P(\Delta r)$ with increasing Δr values. The starting conformation for $P(1)$ optimization is either an extended structure or a conformation derived from an extended chain by rotation around the main chain and side chain dihedral angles. In the subsequent steps of the variable target function method, the starting conformation is the final model from the previous step. An ensemble of different final models is obtained by using different initial conformations.

4   Modeling of trypsin

To illustrate the method of comparative modeling by satisfaction of spatial restraints, this section describes the modeling of trypsin from two other serine proteases, elastase and tonin. The availability of the crystallographic 3D structure of trypsin allowed an evaluation of the model. Two other examples of application of MODELLER include modeling of ferredoxin (10) and of mouse mast cell chymases (28).

The 3D structures of trypsin [223 residues; (29)], elastase [240 residues; (30)], and tonin [227 residues; (31)] were compared using the program COMPARER (7) (Fig. 6). This program relies on many structural properties and relationships, such as positions of C^α atoms, local main chain conformation, solvent accessibility, and main chain hydrogen bonding patterns. When only those aligned C^α atoms that are less than 3.5Å apart from each other are considered, 217 pairs superpose with the RMS of 1.07Å in the more similar pair of trypsin and elastase, whereas only 209 pairs superpose with the higher RMS of 1.18Å in the superposition of trypsin and tonin. This trend is reversed for the sequence comparisons, where the sequence identity between elastase and trypsin is only 38%, and that between tonin and trypsin is 42%. There are only a few short gaps of up to 6 residues in the alignment. The structural alignment was used for extraction of spatial restraints on the sequence of trypsin as described in Section 2. The types of restraints and their numbers are listed in Table 6.

39 models of trypsin were calculated by optimizing the molecular pdf from 39 different initial conformations. These conformations were obtained by setting the main chain and side chain dihedral angles Φ, Ψ, and χ_i to random values between $-180°$ and $180°$. The progress of modeling was followed by monitoring the average atomic shifts and the value

Table 6: Spatial restraints used to model trypsin. [a]Lists a number of basis restraints of a given type that were used to model trypsin. [b]Lists a number of feature restraints of a given type that were assembled from the basis restraints. [c]For the best model, a number of the features that differ from the closest optimum in the feature pdf's by more than the cutoff in the parentheses is given. These cutoffs generally lie between one and two standard deviations of the corresponding basis pdf's. The best model is defined as the one with the lowest value of the molecular pdf. [d]RMS deviation between the actual values in the best model and the closest optimum. [e]RMS deviation between the actual values in the best model and the most likely optimum. [f]These dihedral angles restrain the planarity of peptide bonds and rings as well as chirality of the chiral carbon atoms. [g]All pairs of atoms that are not restrained by any of the bond or bond angle terms are restrained by the minimal contact distance. Only the number of pairs that violate this restraint in the final model is listed. [h]There are no *cis*-peptide bonds in trypsin. The only *cis*-peptide bond in tonin is at Pro 198 which is aligned with Gly in trypsin. Therefore, no *cis*-peptide bonds were imposed on trypsin.

Type	Basis pdf's[a]	Feature pdf's[b]	Violations[c]	RMS[d]	RMS[e]
bond lengths	1659	1659	0 (0.1Å)	0.005Å	0.005Å
bond angles	2250	2250	5 (10°)	2.00°	2.00°
dihedral angles[f]	919	919	1 (20°)	3.40°	3.40°
van der Waals contacts[g]	531	531	0 (0.2Å)	0.02Å	0.02Å
$C^\alpha - C^\alpha$ distances	23538	11914	26 (1.5Å)	0.22Å	0.47Å
main chain N – O distances	7480	3832	19 (1.5Å)	0.31Å	0.51Å
main chain Φ dihedral angles	1110	222	2 (20°)	10.8°	21.2°
main chain Ψ dihedral angles	1332	222	9 (20°)	10.6°	20.3°
side chain χ_1 dihedral angles	528	176	5 (25°)	8.4°	16.8°
side chain χ_2 dihedral angles	264	103	3 (25°)	10.2°	13.0°
side chain χ_3 dihedral angles	92	32	2 (25°)	11.9°	48.1°
side chain χ_4 dihedral angles	48	16	0 (25°)	4.5°	21.9°
disulfide bridge bonds	6	6	0 (0.1°)	0.007Å	0.007Å
disulfide bridge angles	12	12	0 (10°)	3.7°	3.7°
disulfide bridge dihedral angles	6	12	0 (20°)	10.0°	12.9°
cis-peptides[h]	0	0	—	—	—

of the objective function. The optimization schedule and a typical progress of optimization are shown in Figure 7. A total of 11 models with low values of the objective function were obtained (10293 ± 655). These models were close to the correct trypsin structure. The remaining 28 models were the mirror images of either the whole molecule or of a part of it. They all had a significantly higher value of the objective function (> 15000) and were thus easily identified as misfolded models. The model with the lowest value of the objective function (9388) among the 11 successful trials was taken to be the representative trypsin model (the best model). The violations of the restraints by this and other 10 models are small (Table 6). The stereochemistry of the models is comparable or better than that of the crystallographic trypsin structure refined at a high resolution.

The accuracy of the model is different for buried and exposed parts; thus, we will evaluate the model separately for the residues that have fractional side chain solvent accessibility

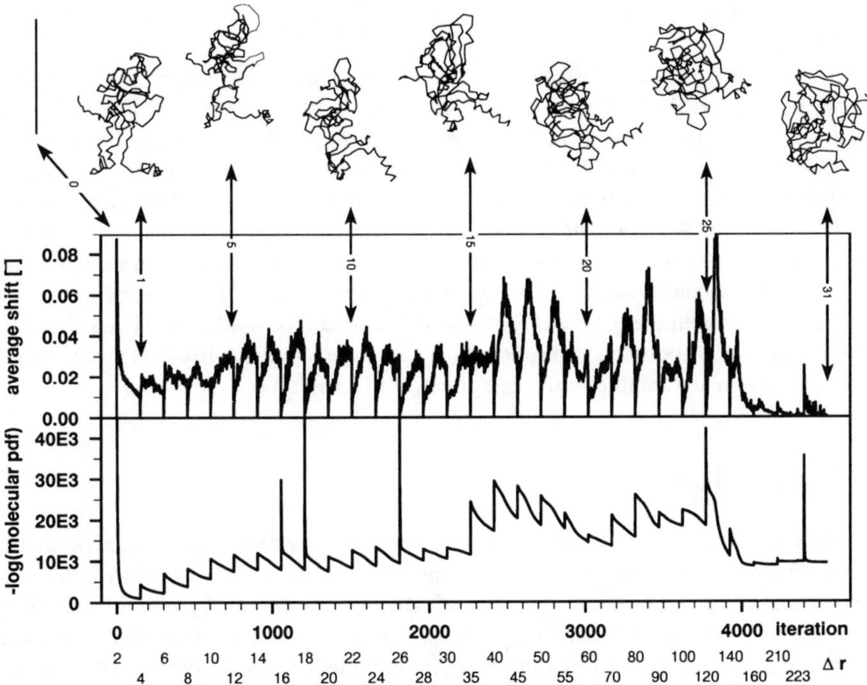

Figure 7: Schedule and progress of optimization. The optimization schedule is specified in the bottom three lines. The 'iteration' line counts the conjugate gradient steps. The bottom two lines show the changes in Δr: Δr is increased every 150 conjugate gradient steps or when the largest atomic shift is smaller than 0.005 Å. Each change in Δr corresponds to a step in a variable target function method. There are 31 such steps to get one model. The method starts with a few restraints that involve only the atoms from residues at most Δr residues apart and gradually incorporates all restraints (the final Δr equals the length of a sequence). The C^α traces of the evolving model at several stages during the refinement are shown on the top of the figure. The starting conformation in this case is an extended chain; generally, it is a chain with random Φ, Ψ, and χ_i dihedral angles. The van der Waals criterion was gradually introduced in the last five steps of the variable target function method by scaling the corresponding standard deviations by 8, 4, 2, 1, and 1. The data for the trial resulting in the model with the lowest value of the molecular pdf are shown. The CPU time needed to calculate one model is 30 minutes on a DEC Alphastation workstation.

less than 20% (buried residues) and for the remaining residues (exposed residues). Only 4 of the 107 buried C^α atoms are more than 3.5Å away from their correct positions whereas 6 out of 116 exposed C^α atoms are further than 3.5Å from their positions in the actual trypsin structure. There is no significant difference between the accuracies of the C^α atoms and all main chain atoms; the RMS error for buried main chain atoms is approximately 0.75Å, and for exposed main chain atoms, approximately 1.3Å.

Similarly to the main chain, buried side chains were modeled more accurately than exposed side chains. 82% of the buried χ_1 classes and 69% of the exposed classes were predicted correctly. For the χ_2 class, 79% of the buried residues and 80% of the exposed residues were modeled successfully. The average χ_3 prediction score for all χ_3 classes is

68%. There are no buried Arg and Lys residues; they are all exposed and predicted with 75% accuracy.

The modeling example described in this section is not a particularly difficult problem because of a relatively high similarity between the target sequence and the two template structures. There is no region in the target sequence that does not have aligned residues in at least one of the templates. If no equivalent residues in the template structures were available, MODELLER would use only the main chain dihedral angle restraints based on the residue type alone. We would not expect such weak restraints to result in an accurate model. Thus, structurally similar segments from the database of all known protein structures would have to be found and added to the alignment. In principle, filtering methods based on the distances between the gap flanking regions (32) could be used for this task, but general applicability of this approach is questionable (33). Another possibility may be an exhaustive conformational search employing energy criteria (34,35).

5   Discussion

The challenge for the future is to unify all the techniques for determination and prediction of protein structure into a single protocol, making the best use of all available information about the structure of a given protein, regardless of whether it is directly based on experiment, on the broader knowledge base, on empirical force potentials, or intuition (9). The methods that combine molecular dynamics and energy potentials with NMR derived constraints (36, 37) and X-ray data (37,38) to refine the initial models can be seen as the first step in this direction. Recently, the advantages of a joint crystallographic and NMR refinement were demonstrated (39).

Before we start prediction of the 3D structure of a protein, we know nothing about positions of the atoms. In the terminology of classical mechanics, the actual structure could be a point anywhere in the phase space spanned by the axes for the positions of all atoms. We can then imagine modeling as a process of reducing the volume of the phase space in which we know the actual structure is located. This is achieved by using various kinds of information. First, stereochemical restraints derived from the chemical connectivities can be used to remove some of the *a priori* available phase space. This can be pursued further by inclusion of experimental data, such as that from X-ray crystallography and NMR techniques. We can also add additional theoretical restraints originating from empirical energy potentials and known protein structures. Each of these kinds of information allows the model to be in a different area of the phase space with a different probability. The goal is to find the most probable conformation or a set of most probable conformations according to all types of information. All the information pooled together results in a smaller allowed volume of phase space than any of the methods can locate on their own.

The most useful representation of information is a pdf for the feature that is restrained. The present modeling method uses pdf's in a relatively general way. Thus, the method, even though it has so far been applied only to comparative modeling, could possibly be extended to include other types of information, such as NMR-derived constraints and coarse-grained potentials of mean-force describing residue–residue interactions.

6   Conclusions

1. A database of family alignments for proteins with known structures was constructed.

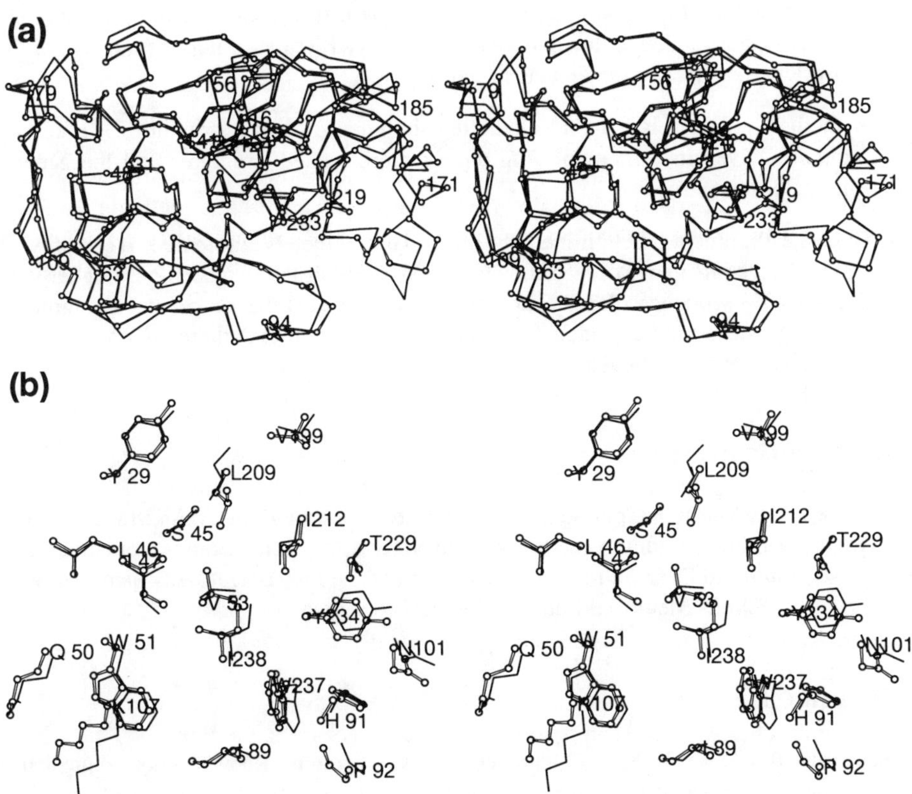

Figure 8: Comparison of the best trypsin model with trypsin. Comparison is obtained by superposing all C^α atoms. Chymotrypsinogen numbering is used. Trypsin (open bonds with circles), trypsin model (line). (a) Comparison of the C^α traces. (b) Comparison of side chains in a mostly buried region.

2. It was shown how pdf's and other tools can be used to explore quantitatively various relationships between features in individual proteins and in families of proteins.

3. The usefulness of the pdf's was improved by a new smoothing procedure that minimized the problems of a sparse data set.

4. Using these tools and the alignments database, the best pdf's for comparative modeling of a side chain conformation of a given residue were constructed. They relied mainly on its type, on the side chain conformation of the equivalent residue and on the similarity between the two local environments.

5. The best possible pdf for modeling the main chain conformation from the main chain of a homologue was found. It was based on the main chain conformation of the

equivalent residue and on the similarity between the two local environments.

6. The pdf's for restraining the C^α–C^α distances and the main chain $N - O$ distances on the basis of homologous structures were calculated. It was shown that the most likely distance corresponded to that in one of the related structures, not to the average of the equivalent distances in the related structures.

7. A method was developed for calculating the most probable structure for a certain sequence, given its alignment with one or more related structures and the general rules of protein structure.

8. Once the alignment is determined, the method is completely automated. It can provide a 3D model equivalent to a medium resolution X-ray structure when homologues with at least 40% sequence identity are known. This means that an order of magnitude more sequences can be modeled at a medium resolution than there are entries in the Brookhaven Protein Databank.

Acknowledgments

We are indebted to Martin Karplus and Alex MacKerell for providing CHARMM 22 parameters. We thank John Overington and Mark Johnson for discussions about protein modeling. We are also grateful to Daša Šali for critical reading of the manuscript. A.Š. is a Fellow of The Jane Coffin Childs Memorial Fund for Medical Research.

References

1. Blundell, T. L., B. L. Sibanda, M. J. E. Sternberg & J. M. Thornton, "Knowledge-based prediction of protein structures and the design of novel molecules," *Nature* **326**, 347–352 (1987).

2. Šali, A. & T. L. Blundell, "Comparative protein modelling by satisfaction of spatial restraints," *J. Mol. Biol.* **234**, 779–815 (1993).

3. Lesk, A. M. & C. H. Chothia, "The response of protein structures to amino-acid sequence changes," *Phil. Trans. Roy. Soc.* **317**, 345–356 (1986).

4. Hubbard, T. J. P. & T. L. Blundell, "Comparison of solvent inaccessible cores of homologous proteins: Definitions useful for protein modelling," *Prot. Eng.* **1**, 159–171 (1987).

5. Srinivasan, N. & T. L. Blundell, "An evaluation of the performance of an automated procedure for comparative modelling of protein tertiary structure," *Prot. Eng.* **6**, 501–512 (1993).

6. Chothia, C., "One thousand families for the molecular biologist," *Nature* **360**, 543–544 (1992).

7. Šali, A. & T. L. Blundell, "Definition of general topological equivalence in protein structures: A procedure involving comparison of properties and relationships through simulated annealing and dynamic programming," *J. Mol. Biol.* **212**, 403–428 (1990).

8. Zhu, Z. -Y., A. Šali & T. L. Blundell, "A variable gap penalty function and feature weights for protein 3-D structure comparisons," *Prot. Eng.* **5**, 43–51 (1992).

9. Šali, A., J. P. Overington, M. S. Johnson & T. L. Blundell, "From comparisons of protein sequences and structures to protein modelling and design," *TIBS* **15**, 235–240 (1990).

10. Šali, A., "Modelling three-dimensional structure of proteins from their sequence of amino acid residues," in *PhD Thesis*, University of London, London, 1991.

11. Overington, J., M. S. Johnson, A. Šali & T. L. Blundell, "Tertiary structural constraints on protein evolutionary diversity; Templates, key residues and structure prediction," *Proc. Roy. Soc. Lond.* **B 241**, 132–145 (1990).

12. Overington, J., D. Donnelly, M. S. Johnson, A. Šali & T. L. Blundell, "Environment-specific amino acid substitution tables: Tertiary templates and prediction of protein folds," *Protein Sci.* **1**, 216–226 (1992).

13. Topham, C. M., A. McLeod, F. Eisenmenger, J. P. Overington, M. S. Johnson & T. L. Blundell, "Fragment ranking in modelling of protein structure. Conformationally constrained environmental amino acid substitution tables," *J. Mol. Biol.* **229**, 194–220 (1993).

14. Johnson, M. S., J. P. Overington & T. L. Blundell, "Alignment and searching for common protein folds using a data bank of structural templates," *J. Mol. Biol.* **231**, 735–752 (1993).

15. Overington, J. P., Z. -Y. Zhu, A. Šali, M. S. Johnson, R. Sowdhamini, G. V. Louie & T. L. Blundell, "Molecular recognition in protein families: A database of aligned three-dimensional structures of related proteins," *Biochem. Soc. Trans.* **21**, 597–604 (1993).

16. Abola, E. E., F. C. Bernstein, S. H. Bryant, T. F. Koetzle & J. Weng, "Protein Data Bank," in *Crystallographic databases — Information, content, software systems, scientific applications*, F.H. Allen, G. Bergerhoff & R. Sievers, eds., Data Commission of the International Union of Crystallography, Bonn/Cambridge/Chester, 1987, 107–132.

17. Sutcliffe, M. J., I. Haneef, D. Carney & T. L. Blundell, "Knowledge based modelling of homologous proteins, Part I: Three dimensional frameworks derived from the simultaneous superposition of multiple structures," *Prot. Eng.* **1**, 377–384 (1987).

18. Press, W. H., B. P. Flannery, S. A. Teukolsky & W. T. Vetterling, *Numerical Recipes*, Cambridge University Press, Cambridge, 1986.

19. Sippl, M. J., "Calculation of conformational ensembles from potentials of mean force. An approach to the knowledge-based prediction of local structures in globular proteins.," *J. Mol. Biol.* **213**, 859–883 (1990).

20. Brooks, B. R., R. E. Bruccoleri, B. D. Olafson, D. J. States, S. Swaminathan & M. Karplus, "CHARMM: A program for macromolecular energy minimization and dynamics calculations," *J. Comp. Chem.* **4**, 187–217 (1983).

21. MacKerell Jr., A. D., D. Bashford, M. Bellott, R. L. Dunbrack Jr., M. J. Field, S. Fischer, J. Gao, H. Guo, S. Ha, D. Joseph, L. Kuchnir, K. Kuczera, F. T. K. Lau, C. Mattos, S. Michnick, T. Ngo, D. T. Nguyen, B. Prodhom, B. Roux, M. Schlenkrich, J. Smith, R. Stote, J. Straub, M. Watanabe, J. Wiorkiewicz-Kuczera & M. Karplus, *in preparation*.

22. Hill, T. L., *An introduction to statistical thermodynamics*, Addison-Wesley Publishing Company, Reading, Massachusetts, 1960.

23. Berendsen, H. J. C., J. P. M. Postma, W. F. van Gunsteren, A. DiNola & J. R. Haak, "Molecular dynamics with coupling to an external bath," *J. Chem. Phys.* **81**, 3684–3690 (1984).

24. Wilmot, C. M. & J. M. Thornton, "β-turns and their distortions: a proposed new nomenclature," *Prot. Eng.* **3**, 479–493 (1990).

25. Janin, J., S. Wodak, M. Levitt & B. Maigret, "Conformation of amino acid side-chains in proteins," *J. Mol. Biol.* **125**, 357–386 (1978).

26. Ponder, J. W. & F. M. Richards, "Tertiary templates for proteins: Use of packing criteria in the enumeration of allowed sequences for different structural classes," *J. Mol. Biol.* **193**, 775–791 (1987).

27. Braun, W. & N. Gō, "Calculation of protein conformations by proton-proton distance constraints: A new efficient algorithm," *J. Mol. Biol.* **186**, 611–626 (1985).

28. Šali, A., R. Matsumoto, H. P. McNeil, M. Karplus & R. L. Stevens, "Three-dimensional models of four mouse mast cell chymases. Identification of proteoglycan-binding regions and protease-specific antigenic epitopes," *J. Biol. Chem.* **268**, 9023–9034 (1993).

29. Walter, J., W. Steigemann, T. P. Singh, H. Bartunik, W. Bode & R. Huber, "On the disordered activation domain in trypsinogen. Chemical labelling and low-temperature crystallography," *Acta Crystallogr.* **B38**, 1462–1472 (1982).

30. Meyer, E., G. Cole, R. Radakrishnan & O. Epp, "Structure of native porcine pancreatic elastase at 1.65 Å resolution," *Acta Crystallogr.* **B44**, 26–38 (1988).

31. Fujinaga, M. & M. N. G. James, "Rat submaxillary gland serine protease, tonin. Structure solution and refinement at 1.8 Å resolution," *J. Mol. Biol.* **195**, 373–396 (1987).

32. Jones, T. H. & S. Thirup, "Using known substructures in protein model building and crystallography," *EMBO J.* **5**, 819–822 (1986).

33. Tramontano, A. & A. M. Lesk, "Common features of the conformations of antigen-binding loops in immunoglobulins and application to modeling loop conformations," *Proteins* **13**, 231–245 (1992).

34. Moult, J. & M. N. G. James, "An algorithm for determining the conformation of polypeptide segments in proteins by systematic search," *Proteins* **1**, 146–163 (1986).

35. Bruccoleri, R. E. & M. Karplus, "Prediction of the folding of short polypeptide segments by uniform conformational sampling," *Biopolymers* **26**, 137–168 (1987).

36. Clore, G. M., A. T. Brünger, M. Karplus & A. M. Gronenborn, "Application of molecular dynamics with interproton distance restraints to 3D protein structure determination," *J. Mol. Biol.* **191**, 523–551 (1986).

37. Brünger, A. T., R. L. Campbell, G. M. Clore, A. M. Gronenborn, M. Karplus, G. A. Petsko & M. M. Teeter, "Solution of a protein crystal structure with a model obtained from NMR interproton distance restraints," *Science* **235**, 1049–1053 (1987).

38. Brünger, A. T., J. Kuriyan & M. Karplus, "Crystallographic R-factor refinement by molecular dynamics," *Science* **235**, 458–460 (1987).

39. Shaanan, B., A. M. Gronenborn, G. H. Cohen, G. L. Gilliland, B. Veerapandian, D. R. Davies & G. M. Clore, "Combining experimental information from crystal and solution studies: Joint X-ray and NMR refinement," *Science* **257**, 961–964 (1992).

Recurrent Neural Networks for Protein Distance Matrix Prediction

Martin Reczko and Henrik Bohr

DKFZ, German Cancer Research Center, Department of Molecular Biophysics, Im Neuenheimer Feld 280, 69120 Heidelberg, Germany

Abstract

The distances between C_α atoms of globular proteins are predicted from the amino acid sequence using recurrent neural networks. Due to the ability of these networks to store information about different patterns using not only synaptic weights but also internal activation values, details of the sequence and the structure can be accumulated while the network reads in the primary sequence. The three dimensional backbone structure for the protein is then constructed through an abstract energy minimization procedure using the distance and hydrophobic constraints to achieve globular structures. As an example the structure of rubredoxin (1RDG) being 64% homologous to a rubredoxin (6RXN) in the training set was predicted with an accuracy of 3.9 Ångstrom.

1 Introduction

The determination of the amino acid sequence of proteins is still at least one order of magnitude faster than the determination of the three-dimensional structure of a protein using crystallographic or NMR techniques. The structure depends only on the sequence of amino acids of the protein and the surrounding conditions. Therefore a fundamental task in biosciences is to develop methods for predicting the unknown tertiary structure of a protein with known sequence. This task is only solved partly in the form of homology modelling [1, 2] approaches. In this modelling a unknown protein structure is assembled from parts of other known protein structures whose sequences are similar to the sequence parts of the protein in question. It is obvious that the accuracy of these methods will deteriorate as the smallest sequence identity between the sequence of the unknown protein and the sequence of the parts of the proteins of known structure drops.

Artificial neural networks have been used to associate the sequence of of homologous proteins with their structure [3]. The structure to be associated was represented by binary band distance matrices for distance inequalities between the C_α atom positions on the protein backbone. A distance inequality for two atomic positions would tell whether the distance between them was greater or smaller than a certain threshold. An ordinary feedforward neural network could, when trained on known protein structures within a given homology class, fairly well predict new protein structures in the form of these binary distance matrices on the basis of the sequence of the novel protein provided that the sequence had large sequence similarity to the proteins in the homologous training set. Once a binary distance matrix was predicted (usually within a narrow band along the diagonal of the distance matrix) a minimization algorithm could reproduce the real 3-dimensional protein backbone structure through the distance constraints from the binary distance matrix. Apart from the strong requirements of sequence similarity between the given test protein and the proteins in the trainingset the band width of the band along the diagonal (the correlation length in the distance matrix prediction) was also a limiting factor.

We shall here try to employ roughly the same methodology and see if we can overcome the mentioned limitations of sequence similarity and yet obtain somewhat higher accuracy in predicting correct structures of protein backbones. The most obvious limitation given by the size of the input window, and resulting in a limited band of output values, can be overcome when employing recurrent neural network with the abillity of a time regression through feedback loops.

Concerning the other limitation, for proteins with very little homology to other proteins there exists no method that can predict those protein's 3-dimensional structure to high accuracy just from there sequence data. One can, however, in the case of low homology, divide the task up in several parts, one being first to determine what homology class (fold-class) a protein belongs to, and then next predict the distance matrix from the training on that particular homology class. We shall in the following report on experiments with the distance matrix prediction part of such a procedure while the fold class determination part is described in a previous chapter.

A distance matrix prediction will, due to the structurally homologous training set, be accurately enough for constructing the 3-dimensional backbone structure for the protein through an abstract energy minimization procedure [4] and subsequently position the side-chains in order to obtain a full protein structure to high accuracy.

In the following we shall first introduce our methodology and then report on some results on predicting distance matrices with subsequent minimization and generation of 3-dimensional protein structures.

2   Methodology

Distance matrix prediction may be optimized either for accuracy or generality. The most servere limitation in the previous application of a feedforward neural network for distance matrix prediction [3] is the requirement of more than 70% sequence identity between a test protein and the most homologous training protein for accurate predictions. The test protein has to have this level of homology since for a large input window there are not enough input sequences to sample the space of all possible input patterns to provide good statistics. The input window to the network was required to be at least twice as large as the output window holding information since the network needs the information about the residue type at each position in the output window in order to enable prediction of the distance to this residue. The output window however has to be as large as possible to provide enough information for the structure optimization step.

This approach tries to weaken the homology limitation in two ways. The first is to predict information about interresidue distances not as binary constraints but as a real valued transformed distance. It was shown [4] that the optimization of the backbone geometry can be performed with substantially less number of real valued distances than binary constraints. Therefore a smaller output window containing real valued information may be used to achieve the same accuracy as with a large binary output window. A smaller output window faciliates the use of a smaller input window which weakens the possibility for the network to overgeneralize non-general features in a large input window during prediction which is the

Table 1: Proteins used for training.

PDB-code	Name
1ake	ADENYLATE KINASE
1ald	LYASE (ALDEHYDE)
1bbp	BILIN BINDING
1cob	OXIDOREDUCTASE
1cox	OXIDOREDUCTASE(WITH OXYGEN AS RECEPTOR)
1crn	PLANT SEED PROTEIN
1cse	COMPLEX(SERINE PROTEINASE-INHIBITOR)
1dfn	DEFENSIN
1ecd	OXYGEN TRANSPORT
1f3g	PHOSPHOCARRIER
1fkf	/FK506$ BINDING PROTEIN
1gky	GUANYLATE KINASE
1gp1	GLUTATHIONE PEROXIDASE
1hip	HIGH POTENTIAL IRON PROTEIN
1ifb	FATTY ACID-BINDING PROTEIN
1l26	LYSOZYME
1lh3	LEGHEMOGLOBIN
1nxb	NEUROTOXIN (POST-SYNAPTIC)
1ova	OVALBUMIN
1pal	PARVALBUMIN
1ppd	2 - HYDROXYETHYLTHIOPAPAIN
1ppt	PANCREATIC POLYPEPTIDE
1rbp	RETINOL BINDING PROTEIN
1rns	HYDROLASE (PHOSPHORIC DI-ESTER, RNA)
1rnt	HYDROLASE (ENDORIBONUCLEASE)
1rop	ROP: COLE1 REPRESSOR OF PRIMER
1sar	HYDROLASE (STREPTOMYCES $AUREOFACIENS)
1tgs	TRYPSINOGEN COMPLEX, CHAIN Z
1tgs	TRYPSINOGEN COMPLEX, CHAIN I

PDB-code	Name
1yeb	B-2036 COMPOSITE CYTOCHROME
2blm	BETA-LACTAMASE (PENICILLINASE)
2cab	CARBONIC ANHYDRASE FORM B
2ccy	CYTOCHROME (HEME PROTEIN)
2fcr	FLAVODOXIN
2gbp	D-GALACTOSE/D-GLUCOSE BINDING PROTEIN
2hmz	HEMERYTHRIN (ADIYZOMET)
2lhb	HEMOGLOBIN V (SEA LAMPREY)
2ltn	LECTIN
2pab	PREALBUMIN
2scp	SARCOPLASMIC CALCIUM BINDING PROTEIN
2snm	STAPHYLOCOCCAL NUCLEASE
2utg	UTEROGLOBIN
2zta	LEUCINE ZIPPER
3cbh	CELLOBIOHYDROLASE
3cpa	CARBOXYPEPTIDASE
3csc	CITRATE SYNTHASE COENZYME
3dfr	DIHYDROFOLATE REDUCTASE
4fgf	BASIC FIBROBLAST GROWTH FACTOR
3grs	GLUTATHIONE REDUCTASE
4fxn	FLAVODOXIN (SEMIQUINONE FORM)
4il b	INTERLEUKIN-1BETA
5acn	ACONITASE
5tnc	TROPONIN-C
6pcy	PLASTOCYANIN
6rxn	RUBREDOXIN
9ins	INSULIN
9wga	WHEAT GERM AGGLUTININ

main reason for the requirement of highly homologous test sequences for accurate distance matrices.

Secondly the input window can be further reduced if the neural network has a possiblilty to store relevant information about the input sequence also during prediction time. In a feedforward network information about patterns can only be stored during the training phase by adapting the weights. If a network is processing a sequence of patterns and can use the activations of some of its neurons from the processing of the previous pattern these neurons may hold information about the sequence that can be used in the processing of patterns appearing much later in the sequence. A network with cycles in the connectivity graph is called recurrent and has this possibility of developing a short term memory.

A learning algorithm for non-fixpoint recurrent neural networks 'Truncated Backpropagation Through Time' (TBPTT) [5] is extended by using the 'Quickprop'[6] rule for weight update. In benchmarks the algorithm is shown to use at least one order of magnitude less epochs as an exact algorithm as 'Realtime Recurrent Learning' (RTRL)[12] to reach the same error level. The time complexity for each epoch scales as $O(N^2L)$, where N is the number of units in the net and L is the length of an activity queue stored for each unit. This compares favourably with the $O(N^4)$ epoch complexity of RTRL and $O(N^3)$ of improvements thereof. An important feature of TBPTT is the possibility to use the activity queue

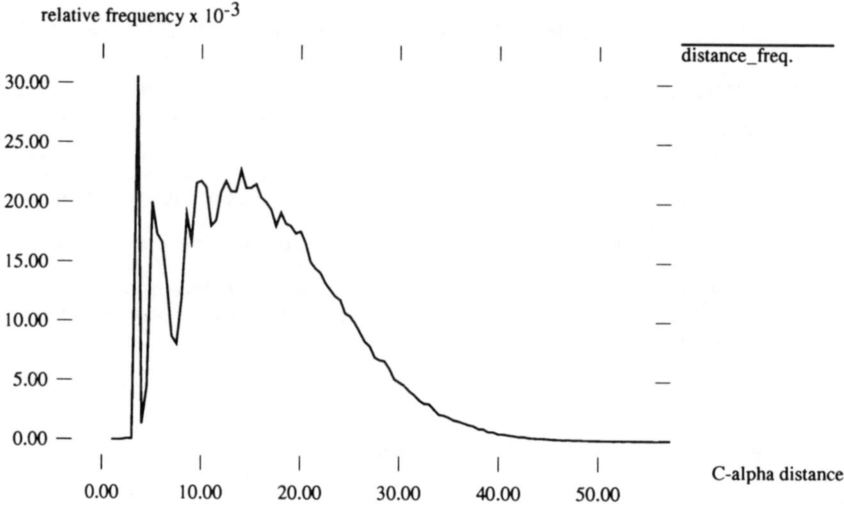

Figure 1: Relative frequency of the occurrence of a C_α distance in all proteins of the training set.

length L to control the transition of one network from being feedforward ($L = 1$) to Elman partially recurrent ($L = 2$) to fully recurrent (L equal to the length of the longest training sequence). Learning complex dynamic systems is improved by performing these transitions during the training process.

3 Optimizing backbone structure from distance constraints

The task of generating a three dimensional backbone structure compatible with the predicted distances between the C_α atoms is closely related to the problem of generating the spatial structure of a molecule from informations about contacts between hydrogen atoms obtained in NMR experiments. The sophisticated methods of distance geometry[8] should therefore be applied whenever possible. An important difference in the applicability of distance geometry in distance matrix predictions and NMR is the number and quality of the distance information used as input. The prediction of the band of distances along the diagonal can not identify long range contacts seperated by more than *outband* residues in the protein sequence. Due to the limited number of weights in the network the distances can only be predicted with a finite accuracy. The sigmoidal transformation of the distances prevents most distances above 20 from being predicted accurate which would otherwise waste most of the memory capacity of the neral network. The different quality of the distances being predicted should be taken into account in their subsequent use. This is achieved by a 'triangular smoothing' of the predicted distances prior to the optimization step. The triangular inequality is used to reduce distances that are predicted as being too large. Since the input to the network is a window that is local in sequence it is assumed that distances between residues that are close in sequence are predicted more accurately than distances with larger seperations in sequence. So for each possible triple (a, b, c) ordered linearly on the seqence with a positioned before b and b positioned before c, the triangle inequality

$$d(a, b) + d(b, c) \geq d(a, c) \tag{1}$$

is verified. The total number of residues between a and c will be called the triangle sequence length $trilen$. If 1 is violated, it is assumed that the accuracies of the predictions for $d(a, b)$ and $d(b, c)$ are both better than the accuracy of the prediction for $d(a, c)$. Thus the value of the larger distance is set to

$$d(a, c) = d(a, b) + d(b, c) \tag{2}$$

The new value is a prediction for the upper bound of the real distance. In order to propagate the constraint 1 consistently through the distance matrix the correction step 2 is first applied for all triples (a, b, c) violating 1 with a triangle sequence length $trilen = 1$. Using the verified distances that were seperated in sequence by one residue the process is repeated for all triples with $trilen = 2$. The triangle sequence length is increased this way by one residue until it reaches $nres - 2$ residues. This procedure is called triangle smoothing and is applied to the distance matrix after all unknown distances are set to infinity. In this way upper bounds are obtained for all unknown distances.

The smoothed distances and upper bounds can now be converted to binary distance constraints and used for the gradient descent optimization method for distance constraints described in [4]. The valuable information about the real valued distance is discarded this way and new quantization errors for the distances close to the threshold θ_d are introduced. Hence the use of a pseudo potential using predictions of real valued distances for structure optimization is more appropriate.

The distance matrix of a structure and its mirrored counterpart are identical. Therefore the optimisation of distance constraints often gets stuck in a local minimum of the pseudo energy function if one part of the structure is found in a locally optimal right handed solution while simultaneaously another part develops into a locally optimal left handed structure assuming that the globally optimal solution is left handed. A local optimization method like gradient descent has to be restarted with a new random conformation in these situations. Global optimization method like genetic algorithms[9] or simulated annealing seem more appropriate for this application. An annealing method for continous parameters [10] is modified to be efficient for large optimisation problems as the $3nres$-dimensional space of C_α coordinates.

Simulated annealing [11] was originally developed for combinatorical optimization problems. For a spin system $\mathbf{x} = \{x_i\}$ of dimension N with $x_i \in (-1, +1)$, an energy $E(\mathbf{x})$ is defined and has to be minimized. The method for generating a new solution $\mathbf{x}'$ for a given solution $\mathbf{x}$ is simply to choose a random position i and flip the spin x_i. The new energy $E(\mathbf{x}')$ is evaluated and accepted as a better solution if $E(\mathbf{x}') \leq E(\mathbf{x})$. If $E(\mathbf{x}') > E(\mathbf{x})$ the solution $\mathbf{x}'$ is accepted with a probability

$$p = \exp((E(\mathbf{x}) - E(\mathbf{x}'))/T), \tag{3}$$

where T is an acceptance temperature. If steps to higher energy are accepted the path in the space of possible solutions may escape from local minima. The acceptance temperature $T(n)$ changes with the number of generated states and is set to a high value at the start of the annealing process so that almost any solution is accepted. It is gradually lowered to $T(\infty) = 0$ in which case only steps to better energy are accenpted.

In the case of an energy defined for a vector $\mathbf{x}$ of continous valued parameters an appropriate state generation method has to be defined. The state generation should also be temperature dependent with large steps in the beginning of the annealing process in order to explore the parameter space with large energy differences and smaller steps at lower temperature to prevent all new solutions from being rejected. The state generation starts with a vector $\mathbf{u} = \{u_i\}$ of random numbers distributed uniformly in the interval

$u_i \in [-\sqrt{3}, \sqrt{3}]$. Each u_i has mean 0 and variance 1. Each component of this vector is multiplied with a parameter temperature matrix $\mathbf{P} = \{p_{ij}\}$ to give the state offset $\Delta\mathbf{x} = \mathbf{P} \cdot \mathbf{u}$ used to generate a new state $\mathbf{x}' = \mathbf{x} + \Delta\mathbf{x}$. The diagonal elements p_{ii} of the parameter temperature matrix determine the variance of each parameter offset Δx_i if all off diagonal elements $p_{ij}, i \neq j$ are 0. This variance acts like a independent temperature for each parameter. In order to adapt each of the parameter temperatures p_{ii} independently N_{test} new solution vectors are generated and evaluated at a fixed acceptance temperature $T(n)$. This is called an equilibrium phase. For each parameter the variance v_i of all parameter offsets Δx leading to a new state accepted with $T(n)$ is recorded. The extent of fluctuations leading to lower energies in all parameter dimensions is measured this way. For the next equilibrium phase at acceptence temperature $T(n + 1)$ the parameter temperatures are set to $t_{ii} = 1.5 \cdot v_i$. Since the variances for generating parameter offsets in the next eqilibrium phase will be larger than the variance of accepted parameter offsets, the process is able to explore the parameter space but uses information about the topology of the energy function to ensure efficient stepwidths for the generation of new solutions. An extended version of this algorithm using also the covariances of accepted parameter offset pairs for generating off-diagonal elements in the parameter temperature martix is described in [10]. For high dimensional optimization problems the calculation of these covariances between all parameters is very time consuming and in this application the approximation with using independent parameter temperatures was found to be efficient.

The energy function used for optimizing the backbone geometry is essentially the same that has been used for using steepest descent minimization with binary distance constraints [4].

4   Data selection and representation

In a first experiment a recurrent network is trained on a set of 54 proteins with less than 30% pairwise sequence identity and a crystallographic resolution better than 2.0. These are selected using the PDB-Select [12] database and are listed in table 1.

For each protein the matrix of pairwise C_α atom distances is calculated. The distances have to be transformed into activation values that can be generated as an output of the neural network. This transformation should map distances that occur with a higher probability on activation values that can be reproduced more accurately. In order to determine such a transformation the relative frequency of the occurrence of a C_α distance in all proteins of the training set is shown in figure 1. If these distances have to be mapped into activities bound to an interval is is clearly advantageaous if this mapping had the largest slope for distances close to 16 Åsince most distances appear in this range. The integral from 0 to d of the frequency of occurrence gives the probability for a distance being smaller than d. This probability distribution is shown in figure 2 and has the desired property of mapping a distance ranging from 0 to ∞ to the interval $[0, 1]$ with a slope of the mapping proportional to the frequency of occurrence of that distance. The probability distribution is approximated by a sigmoidal function

$$a(d) = \frac{0.99}{1 + \exp(0.31(16 - d))} + 0.05. \qquad (4)$$

This mapping encoporates the additional constraint of having an activitiy of 0.05 for the first neighbour distance of 3.8 Åand an activitiy of 0.95 for distances going to infinity. Since the normal sigmoidal function having values in $(0, 1)$ is used as an activation function for

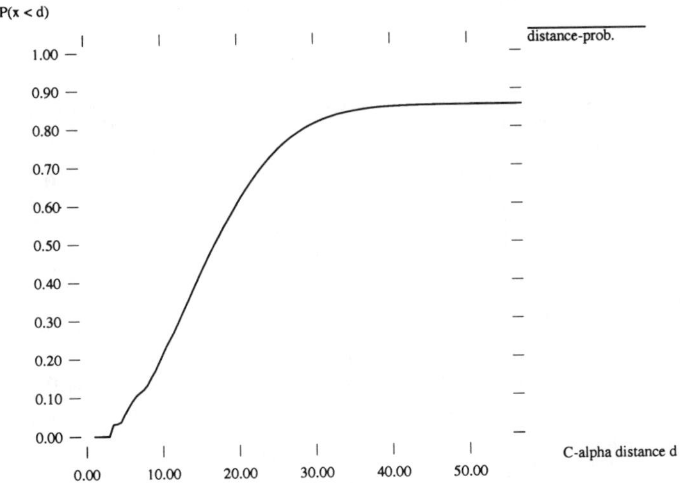

Figure 2: Probability of C_α distances being smaller than d.

the output neurons this constraint prevents the weights leading into an output neuron from growing to infinity if the smallest occuring distance has to be reproduced by that neuron.

The input to the network represents a window of 31 residues shifted over the primary sequence of a protein with a translation of one residue for subsequent patterns. Each residue is represented by an activity vector containing 7 physical properties listed in table 2. For each input window the distances between the C_α atom of the residue in the middle of the input window and 30 residues directly preceeding this residue has to be predicted. The beginning and end of each protein sequence is padded so that missing parts in the input window are filled with activity vectors containing only zeros.

5   Results

As a test protein the sequence of rubredoxin (1RDG) being 64% homologous to a rubredoxin (6RXN) in the training set was converted into a sequence of input windows. The resulting band of the distance matrix may be compared with the correct distance matrix of 1RDG if it is converted to binary distance information being 1 if a given distance is below 16 Åand 0 else. The bands along the diagonal of the native and predicted binary distrance matrix is shown in figure 3.

Although the binary distances are only 72.3% correct the average of the squares of all differences between the transformed distances and the predicted transformed distances is only 0.0236. If the predicted transformed distances are mapped back to distances that are then used in the optimization of the backbone geometry the RMS difference between the resulting structure and the native strucutre is 3.9 Ångstrom. The predicted and native structure are superimposed and shown in figure 4.

Table 2: Physical properties used for representing amino acids.

residue	hydro-phobicity	size	polarity	charged	amphipatic	aromatic	proline
A	0.25	1.69	0	0	0	0	0
R	-1.8	1.9	1	1	0	0	0
N	-0.64	1.82	1	0	0	0	0
D	-0.72	1.79	1	-1	0	0	0
C	0.04	1.69	0	0	0	0	0
Q	-0.69	1.88	1	0	0	0	0
E	-0.62	1.91	1	-1	0	0	0
G	0.16	1.52	0	0	0	0	0
H	-0.4	1.76	1	1	0	1	0
I	0.73	1.9	0	0	1	0	0
L	0.53	1.91	0	1	1	0	0
K	-1.1	1.99	1	0	0	0	0
M	0.26	1.84	0	0	0	0	0
F	0.61	1.7	0	0	0	1	0
P	-0.07	1.71	0	0	0	0	1
S	-0.26	1.72	1	0	0	0	0
T	-0.18	1.8	1	0	0	0	0
W	0.37	1.7	1	0	0	1	0
Y	0.02	1.72	1	0	0	1	0
V	0.54	1.88	0	0	1	0	0

6 Discussion

A preliminary result with the application of a recurrent neural network for protein distance matrix prediction shows that it is possible to predict protein backbone structures with a lower sequence homology to the training set proteins than usually required for feedforward networks. The structures still have the comparable threedimensional accuracy. If the recurrent network will be trained only for members within one fold class it is expected that more accurate results can be obtained even at lower levels of sequence identity since the large space of possible distance matrix configurations is much more confined for one fold class. Complemented with the prediction of the fold class from a protein sequence described in a previous chapter a general system for improved protein structure predictions will be studied further.

7 Implementation

The recurrent neural network for predicting distance matrices is implemented in the SNNS (Stuttgart Neural Network Simulator) environment[13] and is freely avaliable for non-commercial purposes.

Figure 3: Binary bands of 30 residues along the diagonal of the native (left) and predicted (right) distance matrices of Rubredoxin (1RDG). A '1' indicates a C_α distance smaller than 16.

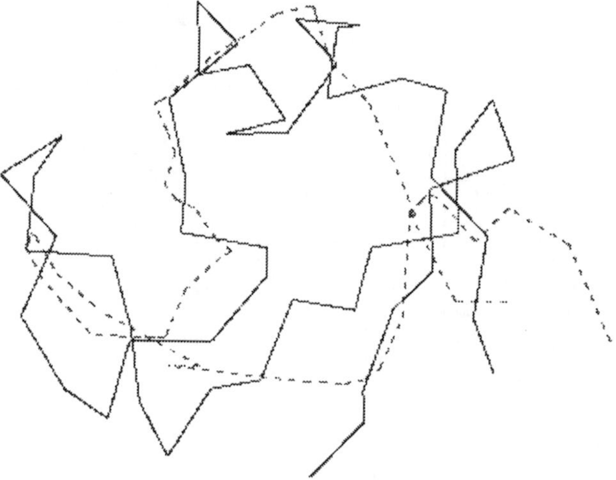

Figure 4: Picture of the minimized structure of $1RDG$ on the basis of a 30 residue long band matrix predicted by the neural network. The native backbone structure is superimposed with the solid line.

Acknowledgements

We should like to thank the staff and students of the physical chemistry department at the University of Illinois as well as the molecular bio-physics group at the DKFZ institute at Heidelberg for technical assistance and stimulating discussions.

References

[1] T. L. Blundell, B. L. Sibanda, M. J. Sternberg, and J. M. Thornton. *Nature*, **326**, 343–352 (1987).

[2] J. Greer. *Methods Enzymol.*, **202**, 239–252 (1991).

[3] H. Bohr, J. Bohr, S. Brunak, R.M.J. Cotterill, H. Fredholm, B. Lautrupt, and S.B. Petersen. A novel approach to prediction of the 3-dimensional structures of protein backbones by neural networks. *FEBS-Letters*, **261**(1), 43–46 (1990).

[4] J. Bohr, H. Bohr, S. Brunak, R. M. J. Cotterill, B. Lautrup, and S. B. Petersen. *J. Mol. Biol.*, **231**, 861–871 (1993).

[5] D. Zipser. Subgrouping reduces compexity and speeds up learning in recurrent networks. In D.S. Touretzky, editor, *Advances in Neural Information Processing systems II*, pages 638–641, San Mateo, California, 1990. Morgan Kaufmann.

[6] Scott E. Fahlman. Faster-learning variations on back-propagation: An empirical study. In T. J. Sejnowski G. E. Hinton and D. S. Touretzky, editors, *1988 Connectionist Models Summer School*, San Mateo, CA, 1988. Morgan Kaufmann.

[7] R. J. Williams and D. Zipser. A learning algorithm for continually running fully recurrent neural networks. *Neural Computation*, **1**, 270–280 (1989).

[8] T. Havel, I. Kuntz, and G. Crippen. *Bulletin of Mathematical Biology*, **45**(5), 665–720 (1983).

[9] J. H. Holland. *Adaptation in Natural and Artificial Systems*. Univ. of Michigan Press, Ann Arbor, Mich., 1975.

[10] D. Vanderbilt and S.C. Louie. A monte carlo simulated annealing approach to optimization over continuous variables. *Journal of Computational Physics*, **36**, 259–271 (1984).

[11] Scott Kirkpatrick. Optimization by simulated annealing: Quantitative studies. *Journal of Statistical Physics*, **34**, 975–986 (1984).

[12] U. Hobohm, M. Scharf, R. Schneider, and C.Sander. *Protein Science*, **3** (1992).

[13] A. Zell, N. Mache, T. Sommer, and T. Korb. In *Proc. Applications of Neural Networks Conf., SPIE, Aerospace Sensing Intl. Symposium, Orlando Florida*, 708–719, 1991.

Growth of Domains in Distance Geometry through Protein Folding

Henrik Bohr, Jin Wang[†] and Peter Wolynes[†]

Center for Biological Sequences Analysis, The Technical University of Denmark,
DK-2800 Lyngby, Denmark
[†] University of Illinois at Urbana-Champaign, Urbana IL 61801, USA

Abstract

A model study is presented of how domains of contacts in proteins in the distance geometry picture could evolve predominantly in the early stages of protein folding. The model is making use of spin dynamics and in such statistical mechanical view it is analogous to an anti-ferromagnetic model in an external field and with a random term. Basically we distinguish between domains of close contact or no contacts. By omitting the random term we obtain power law growth and with the random term logarithmic growth. We can reasonably well describe formation of domains of different secondary structure patterns.

1   Introduction

The main aim of the present work is to study the formation of contacts and structures in protein folding in the language of statistical mechanics of domain growth. Important for such a description is the apparent analogy between protein folding and Ising spin systems with randomness.

We shall first construct a general basis for studying domain growth by chosing an appropriate representation for the patterns of 3-dimensional protein structure. Such representation is given by the distance matrix geometry in which contacts between any pair of residues are the fundamental variables. Earlier we were able[1] to describe development of contacts between residues in biopolymers during the early stages of folding. Here we shall extend that study to domain growth and somewhat beyond the early stages.

In a distance matrix geometry one can study growth of domains of contacts which in turn stands for specific 3-dimensional configurations of protein structures. The energy function used for describing contact formation in protein molecules is constructed from polymer dynamics containing entropy and interactions among smooth strings[2, 3, 4] and with side-chain interactions contained in memory terms of known protein structures.

We can describe early stages of protein folding by a set of first order differential equations standing for down-hill energy processes. These equations can be solved analytically if we assume that they are only weakly coupled but else, in the general case of a fully coupled system of equations, we will have to resort to numerical computer simulations.

The pattern of these contacts can easily be translated into the corresponding 3-dimensional structure of the protein. The contacts are usually represented by a 2-dimensional distance matrix with the columns and rows designating the sequence of residues, as shown in figure

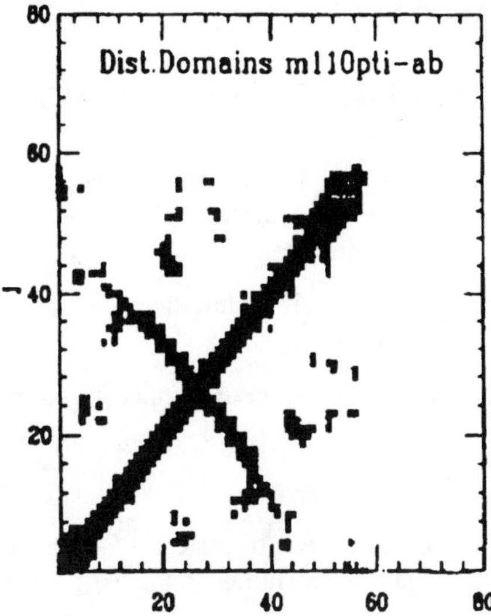

Figure 1: The contact map of the native folded protein 6PTI obtained from the Crystallographic data. The picture is clearly dominated by the apparent secondary structures, an anti-parallel β-turn crossing the diagonal in the middle and two smaller α-helices along the diagonal in the beginning and the end.

1 in the case of the native structure of the protein 6PTI. This 2-dimensional distance matrix representation of residue contacts is an ideal framework for describing growth of domains of contacts. As in a lattice gas a "positive" specific contact corresponds to a lattice site (i.e. distance matrix element) being occupied and no contact to an empty site. In turn the growth of domains of positive contacts signifies the formation of a given substructure in the protein, which is relatively easy to study in terms of domain growth laws and scaling. Thus by applying such domain growth techniques we can analyze how substructures in the protein get formed and thereby how the protein folds. Furthermore the analogy to neural networks and chemical activation processes can be used to steer the domain growth since the neural network methodology gives prescriptions for how to include and use protein "memories" as a scaffolding and nucleation of protein structure formation.

The main results coming out of the domain growth analysis and through a numerical study is that the domain growth of protein substructures in the early stages of the folding is governed by power laws of rapid growth while later stages exhibit a slower logarithmic growth. Such a scenario makes good sense compared to results of protein folding experiments[5] where secondary structures are rapidly formed in the early stages of folding followed by slower processes of docking and rearrangements of secondary structures. Furthermore the nucleation in the start is predominantly of local nature which means that the earliest contacts are formed between neighbouring residues. These facts resemble the dynamics of first order phase transitions in disordered media where bubles (here contact domains) grow slowly in time. What concerns the formation of secondary structures the helical patterns are starting to form earlier but once the sheet patterns are nucleated they grow faster.

We shall first introduce the basic framework for studying protein contacts and construct the theory for evolution of these contacts.

2   Dynamics of contacts

Let us first introduce the basic framework of contacts in biopolymers. We consider a string of monomers representing residues on a protein back-bone and where the variables σ_{ij} describe contacts between residue i and j as

$$\sigma_{ij}^{\rho} = \begin{cases} 1 & \text{if j is within distance } \rho \text{ of i} \\ 0 & \text{if j is not} \end{cases}$$

In ref. 1 the free energy for the formation of contact structures was written as

$$\mathcal{F} = k_B T [\sum_{ij} \sigma_{ij} \log \sigma_{ij} + (1-\sigma_{ij}) \log(1-\sigma_{ij})] + \sum_{ij} W_{ij}^o \sigma_{ij} + \sum_{ij} \sum_{kl} W_{ijkl}^o \sigma_{ij} \sigma_{kl} + \sum_{ij} W_{ij}^I \sigma_{ij} + \dots$$

$$(1)$$

where the first logarithmic terms are pure entropy terms for a Boltzmann gas and the factors W_{ij}^o, W_{ijkl}^o measure the energy gain for forming contact (i, j) alone and contact (i, j) in the presence of contact (k, l), respectively. The term W_{ij}^I is an interaction term describing the interaction between the residues. In references[1, 2] it was used as an "instructed" term which could be written as

$$W_{ij}^I \sigma_{ij} = \sum_{\alpha} \gamma^{\alpha} (q_i^{\alpha} q_j^{\alpha} q_i q_j) \sigma_{ij}^{\alpha} \sigma_{ij}, \tag{2}$$

where α denotes particular stored protein memories and γ^{α} is a generalized residue "charge" being a function of the residue properties q_i.

Hence in such a picture we instruct the interactions among protein residues from known interaction patterns stored as protein memories.

It is possible to derive the expression above for the free energy by means of functional methods where the entropy term $\sigma \log \sigma$ will appear as a term proportional to a free propagator. However, here we shall just go on with the application of the free energy functional.

2.1   Determination of contact factors from polymer kinetics

It is important to bear in mind that the factors W^o in principle can be derived from polymer kinetics. As explained in ref. 3 the effect of a contact on a polymer chain can be measured by a reduction R in the conformational freedom of the chains. If $\Omega(N, i, j)$ stands for the number of conformations of a chain with N units and an (i, j) contact and $\Omega_o(N)$ is the total number of accessible conformations, the reduction factor $R(N, i, j)$ is given by

$$R(N, i, j) = \frac{\Omega(N, i, j)}{\Omega_o(N)} \tag{3}$$

and in terms of our energy contact factor

$$W_{ij}^o = \log(R(N, i, j)). \tag{4}$$

In the Jacobson and Stockmayer limit [3] of "random flight statistics" and without including excluded volume effects the reduction factor can be written as

$$R(N, i, j) = (\Delta v)[\frac{d}{2\pi \mid i - j \mid}]^{d/2}; (N \geq \mid i - j \mid), \tag{5}$$

where d is the dimensionality of the configurations and $\triangle v$ is a tolerance volume factor, actually the volume of a sphere of radius ρ in which the residues qualify as being in contact.

Similarly the effect of two contacts (i, j) and (k, l) can be measured by a reduction factor $R(N, i, j, k, l)$ [4] and one can define a topological correlation factor R_2 as

$$R_{2_{k_1,k_2}}(L) = \frac{R(N, i, j, k, l)}{R(N, i, j)R(N, k, l)},\tag{6}$$

where k_1 and k_2 are the two contact orders: $(k_1 = \mid i - j \mid)$ and $(k_2 = \mid k - l \mid)$, respectively. L measures the distance between the two contacts: $(L = \mid j - l \mid)$. Correspondingly our 2-contact factor W_{ijkl} is given by $W_{ijkl} = \log(R_{2_{k_1,k_2}}(L))$. In the limit of "random flight statistics" we can again obtain a fairly easy expression for the correlation factor R_2.

First we realize that there are only 3 types of possibilities for the index configurations: 1. case: $0 \leq L \leq k_2 - k_1$, 2. case: $k_2 - k_1 < L < k_2$, 3. case: $L \geq k_2$, (when we assume positive separation $L > 0$ and in these cases the correlation factor can be derived to be

$$R_2(L) = \begin{cases} [k_2/(k_2 - k_1)]^{d/2} & \text{if } 0 \leq L \leq k_2 - k_1 \\ [1 - (k_2 - L)^2/(k_2 - k_1)]^{-d/2} & \text{if } k_2 - k_1 < L < k_2 \\ 1 & \text{if } L \geq k_2 \end{cases}$$

If the contact-orders are equal, and hence $k_1 = k_2$, case 1 is absent. The importance of the above classification is that it is complete. In the case of periodic structures it is clear that antiparallel β-sheet types are of case 1, α-helices, π-helices and parallel β- sheets of case 2, while random coil (periodic or not) are of case 3. For negative L, the case can be transformed into the former case by the symmetry

$$R_2(L) = R_2(k_2 - k_1 + \mid L \mid)\tag{7}$$

Thus, there is a natural relation between these cases and the secondary structures. The periodicity in the helical structure means that the orders for the non-zero contacts are equal, i.e. $k_1 = k_2$, and that itself exclude case 1. For regular sheet structures the non-zero contact orders are regular series of numbers such that $k_1 = k_2 + k = k_3 + 2k$ etc. occurring in case 1. The question about parallel and anti-parallel structures are determined by a twist on the contact loops.

Including excluded volume effects basically mounts to multiplying R_2 with a factor D whose coefficient f can be determined from Feynman rules[4]: $D = (1 - \nu_o/2\pi f)$.

The equation of motion for fast down-hill energy processes is

$$\frac{d\sigma_{i,j}}{dt} = \frac{\delta F}{\delta \sigma_{i,j}} \cdot \Gamma,\tag{8}$$

which leads to the following evolution equation, quite similar to the equation for the evolution of a Hopfield neural network [6]

$$\sigma_{ij}(t + 1) = \tanh[(k_B T)^{-1}(\sum_{kl} W_{ijkl}\sigma_{kl}(t) + W_{ij}^o + W_{ij}^I)]\tag{9}$$

where the contacts σ correspond to the neuron states and W to the synaptic weights and thresholds in such a network.

3   Domain growth

In order to make a good semiquantitative illustration of domain growth we consider a continuous representation of the contact variable σ denoted as a field ϕ being a real valued function of the coordinates x, y that are the continuous realizations of the discrete indices i, j. This is a reasonable description for proteins when considering the backbone as a long smooth polymer chain consisting of just C_α-atoms representing each residue. Thus in our continuous picture the number of residues is large compared to the length of the chain. The continuous version serves here as a pedagogical guideline for the more realistic discrete case.

Basically we are interested in the dynamical equation for fast down-hill processes and we consider a kind of Langevin equation without the noise term. A similar equation arises when we are to study domain growth which is a non-equilibrium thermodynamical process. The corresponding non-equilibrium equation of motion is derived from phenomenological thermodynamics by using a suitable order parameter ϕ and equate its displacement with the present thermodynamical force. We therefore write the equation for the non-equilibrium thermodynamical domain growth as

$$\frac{d\phi}{dt} = -\Gamma \frac{\delta \mathcal{F}}{\delta \phi} \tag{10}$$

with $\frac{d\mathcal{F}}{dt} < 0$ and where Γ controls the time scale of the system.

The energy functional or free energy in the equation above is written in terms of ϕ as

$$\mathcal{F} = \int [\frac{1}{2}(\nabla \phi)^2 + \frac{1}{2}\mu\phi^2 + \frac{1}{4}b\phi^4 + H\phi + h\phi]dxdy, \tag{11}$$

where μ corresponds to W^o_{ijkl}. Actually it is here an average value over different structures and represents an approximation to the real quantity that is non-local. Hence the term $\frac{1}{2}\mu\phi^2$ should be understood as an approximation to $\int \phi(x')\phi(y')dx'dy'$. We shall for simplicity consider only the local case in the continuous version thus neglecting various non-local interaction terms. Part of the ϕ^2 and ϕ^4 terms are due to the entropy contribution and the other part of ϕ^4 and the gradient term are from the energy part of the free energy. The factor H corresponds to W^o_{ij}, and a small part of the instructed interaction term W^I_{ij} plus an extra constant and h corresponds to the rest of W^I_{ij} (the non-condensed part). The term with h is to be regarded as a random term, arising as the contribution from the non-condensed memories exactly as in a spin glass memory model. H is of course to be regarded as a constant external field. It is important to notice that the specific properties of protein chemistry is buried in the factors μ, H and predominantly h but not in b. By protein chemistry dependence is ment a dependence on the protein side-chains through the coordinates x and y that are taking the place of the side-chain (residue) indices. Thus the model above is manifestly a 2 dimensional theory. It is analogous to an anti-ferromagnetic, frustrated system with randomness.

Next we need a useful quantity for studying the growth of domains. We shall here use the domain size of contacts as a sort of radial parameter

$$R^2 = \int \phi(x, y)dxdy. \tag{12}$$

3.1　Fast growth

In this section we study the very first stage of the process where we have no random term h. We consider our energy functional which in the continuous case becomes

$$F = \frac{1}{2}\nabla\phi^2 + \frac{\mu}{2}\phi^2 + \frac{b}{4}\phi^4 + H\phi,\tag{13}$$

in the case of an Ising ferromagnetic system, and the kinetic equation in d dimensions is

$$\frac{d\phi}{dt} = \Gamma[K\frac{\partial^2\phi}{\partial r^2} + K\frac{(d-1)}{r}\frac{\partial\phi}{\partial r} - \frac{\partial\mathcal{F}}{\partial\phi} + H],\tag{14}$$

which is derived from the equation: $\frac{d\phi(r)}{dt} = -\Gamma\frac{\partial\mathcal{F}}{\partial\phi}$ and where ϕ again is the order parameter (the magnetization) and where $\phi \sim \phi(r - R(t))$, (we can put $\Gamma = 1$).

Our equation in two dimensions, with $\phi = \phi(r - vt)$, is (in the spherical symmetric case)

$$K\frac{d^2\phi}{dr^2} + (K\frac{(d-1)}{R} + \frac{v}{\Gamma})\frac{d\phi}{dr} - \mu\phi - b\phi^3 - H = 0\tag{15}$$

and by integration we get

$$\triangle(f - \phi H) = 2\phi_s H = \frac{v}{\Gamma}\frac{\sigma}{K}\frac{(d-1)}{R} + \frac{K(d-1)}{R^*}\tag{16}$$

In these equations $\triangle$ is the change in the argument when crossing the interface, σ is the equilibrium surface tension and ϕ_s the minimum of f, $\phi_s = \sqrt{-\frac{\mu}{b}}$. We have introduced an effective radius where the energy is maximized

$$R^* = \frac{1}{2}(d-1)\frac{\sigma}{\phi_s H} = \frac{(d-1)}{c}.\tag{17}$$

We can now follow the prescription from the standard theory of domain growth [7] and write the growth equation for the domain boundary as follows

$$v = \frac{dR}{dt} = \Gamma(\frac{1}{R^*} - \frac{1}{R}) \;\Rightarrow$$
$$\frac{R^* R dR}{R - R^*} = \Gamma dt \;\Rightarrow$$
$$R^*(R + R^* \log(R - R^*)) = \Gamma t + c_o\tag{18}$$

In the case of large R: $R > \log R$ we have: $R^* R = \Gamma t$, i.e. linear growth. In the case of small R: $R <| \log R |$ we have

$$(R^*)^2 \log(R - R^*) = \Gamma(t - t_o)\tag{19}$$

so in this limit

$$R - R^* = e^{\frac{\Gamma}{(R^*)^2}(t-t_o)}.\tag{20}$$

and the behaviour of ϕ will have power law growth and the field ϕ will spread out and grow like a spherical bubble.

So far we have been treating the cases of spherical symmetry, but if we also want to extend the study to aspherical cases we could replace the term $\frac{1}{R}$ in the equation above with the mean curvature K for the curve that is describing the profile of our domain of contacts.

3.2 Slow growth

Next we would like to examine the growth when we have included the random term $\eta(x, y, t)$. We shall also like to comment on the physical scenario of what happens when a random term is introduced, see also [9, 10].

Concerning barrier crossing, if we include the random term h in the energy function we can write the barrier height as a surface potential energy term and a volume term

$$\Delta = -JR^{d-2}\delta - h(\delta R^{d-1})^{\frac{1}{2}}, \tag{21}$$

and minimization with respect to δ (the amount the radius is allowed to shrink) gives

$$\Delta = \frac{Rh^2}{J}. \tag{22}$$

Thus, the time τ it takes to overcome the barrier is

$$\tau = \tau_o exp((\beta)[\frac{Rh^2}{J}]) \tag{23}$$

and

$$R \simeq \mu \frac{Jk_BT}{h^2} \log(\frac{t}{t_o}). \tag{24}$$

Hence we have $R \sim \log(\tau)$, i.e. logarithmic growth. The specific protein chemistry is contained in the constant J that basically is a function of μ and hence our contact factor W^o_{ijkl}. It is very important that the power law growth of contact domains in the first stages of the protein folding process, that we are to describe, in the later stages is slowed down and becomes a logarithmic growth that eventually will stop. This is of course due to the noise term from the spurious memory terms that becomes more important in the later stages of the process. This conclusion is true for all dimensions $d \geq 2$.

4 Domain wall dynamics and the jump condition in the discrete case

We start with the free functional

$$F = W_{ij}\sigma_{ij} + W_{ijkl}\sigma_{ij}\sigma_{kl} + k_BT\left[\sigma_{ij}\log\sigma_{ij} + (1 - \sigma_{ij})\log(1 - \sigma_{ij})\right] + W^I_{ij}\sigma_{ij} \tag{25}$$

We use the Einstein convention for a repeated index, meaning summation over the same index. The dynamical equation for the contacts is

$$\frac{d\sigma}{dt} = W_{ij} + W_{ijkl}\sigma_{kl} + k_BT\log\frac{\sigma_{ij}}{1 - \sigma_{ij}} + W^I_{ij} \tag{26}$$

We would like to impose the thin wall ansatz to the problem we study. Therefore in the proper coordinates our current two dimensional complex problem can be simplified to a one dimensional form

$$\sigma_{ij} = \sigma(i - V_i t). \tag{27}$$

We choose this particular form for the wall ansatz, because α helix and β sheets in our contact coordinates are approximately all one dimensional perpendicular to each other. The indices i and j are not independent quantities. They are linearly related to each other in the helix and sheets cases. After substituting this ansatz into the dynamical equation of contacts, we obtain

$$-V_i(\sigma_{i+1} - \sigma i) = W_i + W_{ikl}\sigma_{kl} + k_B T \log(\frac{\sigma_i}{1 - \sigma_i}) + W_{ij}^I. \tag{28}$$

We multiply each side of the equation and then sum over i, and obtain a "jump" condition for the velocity across the boundary of the contact domain

$$V_i\sigma_0 = \sigma_i^0(W_i + W_i^I). \tag{29}$$

Here σ_i^0 is the solution of the stationary equation $\frac{\delta F}{\delta \sigma_i} = 0$, while σ_0 is the surface energy of the wall

$$\sigma_0 = \sum_i (\sigma_{i+1} - \sigma)^2 + F(\sigma_{ij}) - F(\sigma_{ij}^0) \tag{30}$$

Since this is an approximate one dimensional problem, we obtain in the case of no random source term

$$X_i = A_i t + B_i. \tag{31}$$

A_i and B_i are constant terms depending on W_i. X_i is the position of the wall in sequence space. When we include the random term, we obtain

$$X_i = A_i(\ln t)^2 + B_i. \tag{32}$$

So we see that in contact space, the domain of contacts (helices and sheets) is growing linearly in time when there is no random source term, and it is growing logarithmically in time when including the random source term. This might be approximately the domain growth law of the early stages of folding. In this derivation we have not been using any spherical symmetry assumptions, so in this sense the discrete case is more general and more appropriate for protein applications.

5   Numerical studies

In this chapter we present a numerical study of domain growth of protein structures described by the evolution equations. These first order differential equations were derived in the previous sections and represented the time evolution of contacts between residues in protein during the earlier stages of protein folding. Thus it is in the 2-dimensional plane of contacts, where each point signify a contact σ_{ij} between residues i and j ordered along the axis of the plane, that the domain growth is studied. Thus each domain is consisting of points (i, j) that stand for a close contact between residue i and j, i.e. residues having a distance between them being less than a given threshold. In figure 1 is pictured a typical 2-dimensional distance matrix of contacts σ_{ij} for the folded structure of the test protein 6PTI. We are mainly concerned with contacts related to the most common secondary structures, the helical and the beta strand structures, but the analysis can easily be carried over to other structures. The reason for being primarily concerned with these two types of structures is that they are easy to recognize in the contact space where the helical structures are forming close to, and along, the diagonal ($i = j$), while the parallel beta-strands are far away and parallel to the diagonal and the anti-parallel beta-strands are structures orthogonal to the diagonal. These circumstances are easily verified in figure 1.

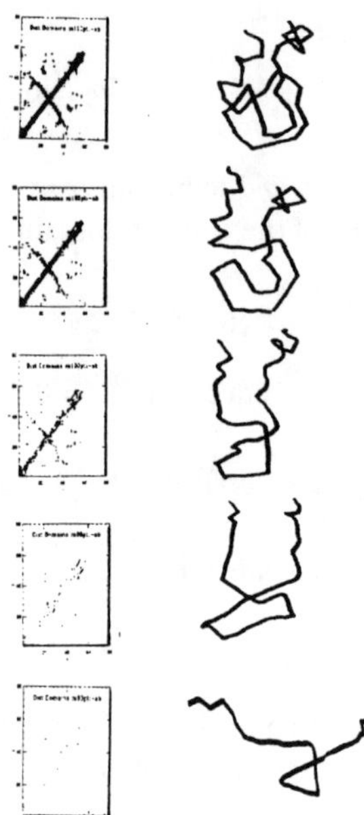

Figure 2: The figure shows to the left a set of pictures of domain growth patterns at different stages of the "folding" in the distance matrix representation, where the residues are numbered along each axis and where a cross at (i, j) in the plot stands for a contact between the corresponding residues i and j. To the right are snap-shots of the 3-dimensional structure of the protein backbone corresponding to the distance matrix plots.

In the evolution equation we have included the polymer factors corresponding to these types of structures. The evolution equations for the contact variables were in the earlier sections derived from the dynamical equation for energy down-hill processes

$$\frac{d\sigma_{ij}}{dt} = \frac{\partial F}{\partial \sigma_{ij}} \cdot \Gamma, \tag{33}$$

where F is the energy functional

$$F = \sum_{ij} W_{ij}\sigma_{ij} + \sum_{kl} W_{ijkl}\sigma_{ij}\sigma_{kl} + k_B T \left[\sigma_{ij} \log \sigma ij + (1 - \sigma_{ij}) \log(1 - \sigma_{ij})\right] + W_{ij}^I \sigma_{ij}. \tag{34}$$

The 2-loop factors W_{ijkl} can be derived[2, 4] for the specific secondary structures, $\alpha, \beta,$ under consideration

$$W_{ijkl}^\alpha = \log([1 - (\mid n - L \mid /n)^2](\frac{\triangle V)^2}{(n2\pi)^d}); d = 3, L = l - j, n = 3; \tag{35}$$

and

$$W_{ijkl}^\beta = \log[(\frac{(j - i)}{(k - l - j + i)})^{d/2}(\frac{\triangle V}{(j - i)2\pi})] \tag{36}$$

while the 1-loop factors basically behave like $W_{ij} \sim \log[(\frac{1}{2\pi|i-j|})^{d/2}]$, d being the dimension.

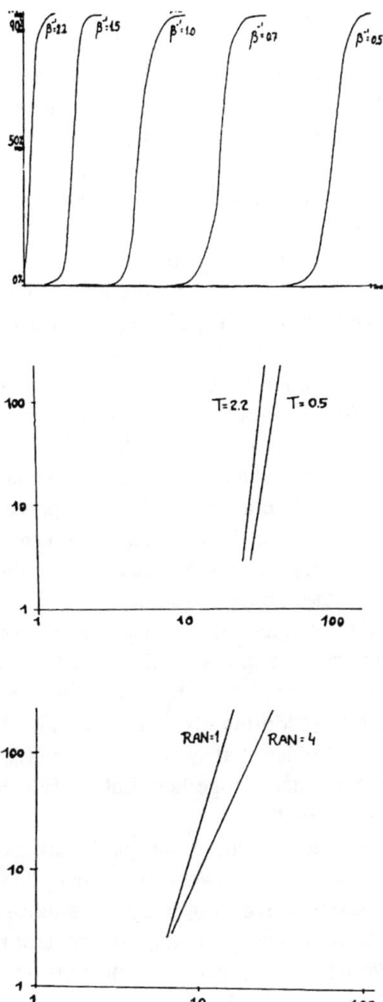

Figure 3: The domain growth as function of the varying temperature and random factor. First (top) the domain growth is monitored for different temperatures (β^{-1}) by measuring the number of contact points, where 0% is the start configuration with no contacts and 100% the full native structure of contact points. On logarithmic scale (middle and bottom) is pictured the growth laws where the sizes of the domains are monitored for progressing integration steps.

The evolution equation can then, with the stochiometric assumption that there are only α and β secondary structures (i.e. helices and sheets), be written as

$$\frac{d\sigma_{ij}}{dt} = \sum_{kl} W_{ijkl}^{\alpha}\sigma_{kl}^{\alpha} + W_{ijkl}^{\beta}\sigma_{kl}^{\beta} + W_{ij} + \log\left(\frac{\sigma_{ij}}{1 - \sigma_{ij}^{\alpha} - \sigma_{ij}^{\beta}}\right) + W_{ij}^{I} + .. \tag{37}$$

and then be approximated by a finite difference equation and integrated on both sides to become

$$\sigma_{ij}(t+1) = \sigma_{ij}(t) + \triangle(k_B T)[\log(\frac{\sigma_{ij}}{1 - \sigma^{\alpha}{}_{ij} - \sigma^{\beta}{}_{ij}}) + W_{ij} + W_{ij}^I + ..] \qquad (38)$$

These equations are solved numerically by calculating the contact variables at a finite time step and integrating. The integration is split up in m integration steps, typically of the order of $m = 100$ to $m = 500$ steps. The integration steps become the time parameter in the evolution of contacts.

We start out with zero contacts at time $t = 0$ and then construct the contacts for all i and j residues at the next time $t + 1$. Thus we are able to follow the evolution of contacts up to the time when the desired patterns have converged to a final stage determined by the contact patterns enforced in the equation through the interaction term W^I in the equation. These enforced patterns can be fully folded realistic protein structures or it can be patterns constructed from energy functions calculated in a more general framework of spin glass theories[11]. Basically enforcing patterns from such a framework give the same result with respect to domain growth as adding contacts from real native proteins.

Both a "desired" pattern of contacts we wish the trial structure to attain, as well as other structures functioning as noise and a random number generator term, are added to the evolution equation in order to simulate a realistic folding process. The desired pattern is emphasized with a slightly larger factor than the other patterns. We chose the Pancreatic Trypsin Inhibitor, 6PTI, as a good test protein for the desired pattern since it had reasonably well-defined and clearly confined secondary structures.

In figure 2 is shown a series of integration steps as evolutionary stages of the domain growth in the distance matrix representation together with snap-shots of the related 3-dimensional protein backbone structure. After a clear stage of nucleation happening along the diagonal of the contact plot predominantly in the middle of the protein stages follow of domain growth of elongated bubles that grow mostly along the diagonal towards the boundaries of each other and then merge together. Later, after the boundaries are marked, each domain is filled out or completed.

On figure 3 the growth laws are depictured both on linear and logarithmic scales of the full protein at different temperature and with varying intensity of the random generator. It is clear from figure 3, and also expected from our analytical studies in the proceeding section, that the growth laws of the middle stages of domain formation (i.e. not the nucleation or completion stages) is governed by a power law, decreasing in power with decreasing temperature and increasing random factor.

6   Acknowledgement

We would like to thank the members of the physical chemistry group at the Noyes Laboratory at Urbana for helpful discussions and especially Drs R. Goldstein, Z. Schulten and S. Scofield are acknowledged for help during the project. Also the use of computer resources at NCSA are acknowledged.

References

[1] H. Bohr and P. G. Wolynes: "Protein folding: A physical view of Neural Network Approaches", in Neural Networks: From Biology to High Energy Physics, 261-276, ETS Editrice, Pisa (1992).

[2] H. Bohr and P. G. Wolynes: The Initial Events of Protein Folding from an Informational Viewpoint", *Phys. Rev.*, **A46**, 5242 (1992).

[3] H. Jacobson and W. J. Stockmayer, *J. Chem. Phys.*, **18**, 1600 (1950).

[4] H. S. Chan and K. A. Dill, *J. Chem. Phys.*, **92**, 3118 (1990).

[5] O. B. Ptitsyn, R. H. Pain, G. V. Semisotnov, E. Zerovnik and O. I. Razgulgaev, *FEBS LETT.*, **262**, 20 (1990).

[6] J. J. Hopfield, *Proc. Natl. Acad. Sci. USA*, **79**, 2554 (1982).

[7] J. S. Langer, "An Introduction to the Kinetics of First Order Phase Transitions", in Solids far from Equilibrium, (Cambridge Uni. Press) (1991).

[8] R. Bruinsma, "Statics and Dynamics of the Random Field Ising Model (Theory)", in Condensed Matter (Springer Verlag) (1986).

[9] S. Coleman, *Phys. Rev. D*, **15**, 2929 (1977).

[10] T. Nattermann and J. Villain, "Random-Field Ising Systems: A survey of Current Theoretical Views" in Phase Transitions, Vol. 11, (Gordon and Breach Scientific Pub. (1988).

[11] R. Goldstein, Z. L. Schulten and P. G. Wolynes, *Proc. Natl. Acad. Sci. USA*, **89**, 9029 (1992).

Distance–based Energy Function Methods for Protein Structure

Predictive Power of Mean Force Pair Potentials

Manfred J. Sippl and Markus Jaritz

Center for Applied Molecular Engineering, Institute for Chemistry and Biochemistry,
University of Salzburg, Austria

Abstract

A long standing goal in protein structure theory is the development of force fields and energy functions for protein solvent systems, which can be used to calculate or predict native folds from amino acid sequences. The most essential property of reasonable models of protein solvent systems is that they must be able to recognize native folds of amino acid sequences among a set of alternative conformations. Otherwise any attempt to locate the native fold will necessarily fail. The goal of the present study is the development of computer experiments which can be used to investigate the predictive power of mean force potentials in terms of the ability to recognize the native folds of proteins among a large number of non-native decoys. The technique is based on the construction of a polyprotein from the known structures in the data base. The native fold of a given sequence is hidden somewhere in the polyprotein. When the sequence is combined with all possible fragments along the polyprotein the question is, whether or not the fragment corresponding to the native fold has the lowest energy. The version of the knowledge based mean field investigated in this study is based on C^{β}-C^{β} interactions only. This is the most simple model conceivable. In spite of its incompleteness the mean field is able to recognize most native folds in the current release of the Brookhaven data base, including virus coat proteins and membrane associated proteins. In most cases failures to recognize experimentally determined folds are confined to low resolution structures or folds which are known to be incorrect. The polyprotein technique is used to investigate the stability of the results as a function of global force field parameters, like cut off distances for energy calculations, the distance grid size used to sample mean force potentials, or the size of the data base used to compile the mean field. The predictive power of the mean field is at its optimum when the potentials are calculated to distances up to $20 - 30$.

1   Introduction

The protein folding problem belongs among the most fascinating and important problems in contemporary biology. Driven by forces originating from interactions of the polypeptide chain with itself and the surroundings, proteins fold into unique three-dimensional structures. In general the information required for folding is entirely specified by the amino acid sequence and the physico-chemical properties of its amino acids residues in physiological environments [1]. This is the basic tenet in protein folding and in spite of the discovery that chaperons are often required to accelerate folding and to prevent unfolded proteins from aggregation it is still the generally accepted view in protein structure theory.

If we accept this view physics immediately provides a powerful strategy for the solution of the folding puzzle which would enable the calculation of native folds from the information contained in amino acid sequences alone. In physiological conditions native states are stable states of the protein solvent system and according to thermodynamics this is only possible if they are in a state of equilibrium where the free energy is at its minimum. Hence the minimum energy principle is also applicable to protein folding and all that we need in order to calculate protein folds from sequences are two ingredients. First we have to construct a reasonable energy model of protein solvent systems. Reasonable means that the energy function must have the global minimum equals native fold property and that the energy surface is well behaved in the sense that the function is smooth as compared to a chaotic function which changes erratically in response to small changes in the variables. This requirement is vital for the second step. Once a reasonable energy model of the system is found we have to locate the global minimum thus identifying the native fold. The search must be completed in acceptable time using present day computer power. If the search takes too long, then we are still unable to calculate native folds from amino acid sequences and our solution, although possibly correct, is impracticable.

The recipe is simple and straightforward. Its use as a guiding principle has a long tradition in protein structure theory. Based on this prescription semi-empirical force fields and simplified energy functions have been developed over the last decades [2-7] but it is still impossible to calculate native folds from amino acid sequences. There are still severe theoretical problems as well as technical difficulties which have to be surmounted. The interactions and forces that guide the polypeptide to its unique three dimensional structure and stabilize the native fold are complex. Deduction of the functional form of these interactions from first principles is demanding if not impossible. And the minimization or optimization of functions of many variables is in itself a largely unsolved problem unless the function is very well behaved.

An alternative route to the construction of energetic models for protein solvent systems relies on the data base of known protein structures determined by X-ray analysis and nuclear magnetic resonance spectroscopy. The solved structures contain a tremendous amount of information on the forces and interactions which stabilize native folds in solution. Using Boltzmann's principle this information can be extracted from the data base in the form of potentials of mean force [8-11]. The result is a knowledge based mean field, a toolbox of potentials, describing the interactions of individual protein atoms relative to the average backgound of a dense medium comprised of protein and solvent molecules. The total energy function for an arbitrary protein of known or unknown structure is assembled from the toolbox of mean force potentials as a function of the amino acid sequence.

The design of knowledge based force fields and data base derived parameters has become an active research area in recent years (reviewed by Wodak and Rooman [12], Bowie and Eisenberg [13], and Sippl [9]). Following the demonstration of the feasibility and predictive value of the mean force approach one of our main goals has been the improvement, completion and refinement of the prototype implementation [8,9].

Force field construction involves phases of redesign, improvement and optimization since we cannot expect to derive a complete and optimized energy model in a single strike. A most important component in force field design therefore, concerns the critical evaluation of the quality and predictive power of the force field at each stage of development. It is clear that the prediction of three dimensional folds of proteins from amino acid sequences

Table 1:

Protein structure data sets[a]

Mean field data base									
1FC2-C	1RDG	1CHO-I	1FDX	HMBM-I	1AAP-A	1TGS-I	1HDD-C	1FXD	1CDT-A
2SNI-I	1SN3	2CRO	1CTF	3IL8	1BOV-C	1PGX	2CTX	2HIP-A	3ICB
1UBQ	1CCD	2KAI-A	1FXB	1GRD-B	351C	1CC5	1HIP	1ZAA-C	3B5C
1ABA	1TPK-A	1TEN	1LRD-4	1GMR-A	3FXC	HPPST-I	1PCY	3HLA-B	5HVP-B
1TLK	1CMB-A	1RNT	256B-A	4FD1	1ACX	1FKF	2CDV	2SSI	2TRX-A
5CPV	HCYS-I	1RNB	2MSB-A	2C2C	1CD8	2PAB-A	1CY3	2MHR	1GMF-B
1PAZ	1STP	1BP2	8RUB-S	1RSM	1BBK-D	2CCY-A	CHEY-1	1RCB	2AZA-A
1LZ1	2HMB	SBLP	1SNC	1ECA	1END	1BAR-B	4FXN	1ITH-A	1LE2
1SDH-A	2HHB-B	7AT1-D	2HBG	2LHB	1AAK	1SNV	2SOD-B	4I1B	1TNF-C
2KAI-B	1LH1	1MBD	2RN2	1MUP	1GPR	1OVB	4DFR-A	5TNC	1CPC-K
3DFR	1L01	RAS1	1OFV	1FHA	9WGA-B	1CPC-L	1BBP-D	1GCR	1RBP
2SCP-A	2CD4	1SAS	1GKY	8DFR	1COL-B	1ABM-A	1P07-A	1FC1-A	HGPF-A
3GAP-A	1CLA	2FBJ-L	1AKE-A	1GST-B	1PPG-E	2ACT	1SGT	1FB4-H	3PGM
2CNA	1RVE-A	1CGJ-E	1YPI-A	1CA2	3BLM	2TSC-A	4TGL-A	1APG	2HHM-B
3HLA-A	1DRI	2SNI-E	1PYP	1NIP-A	1RHD	2CYP	1EZM	1ABP	3CPA
2GBP	7AT1-C	2CMD	1TRB	1AVR	4TMN-E	4PFK	2APR	2HMG-A	1LDM
4MDH-A	1GD1-P	2CPK-E	9API-A	1IPD	2LBP	1BBK-C	1MPP	1VSG-A	1ALD
1OMP	ACT1-A	8ADH	1WSY-B	2BAT	2PHH	4XIA-A	2AAT	HCRTN-B	2CPP
4ICD	3PGK	4ENL	2NPX	1PII	9RUB-B	8RUB-L	1PGD	1GLY	3LAD-A
2TPR-B	8CAT-B	1COX	2PMG-A	1GAL	1LFG	1TMD	6ACN	3GPB	
Continued									

Miscelaneous proteins subjected to hide and seek									
4SGB-I	ROP2-A	1TRC-B	3FIS-A	2ABX-A	1CPB	1TRO-G	2WRP-R	1CY3	2HMG-B
1PTE	1MDA-C	1NCD-N	1NSB-B	RUBC-S	3FGF				

Viral proteins									
1MS2-B	2TMV-P	2BPA-1	2BPA-2	1BMV-A	1BMV-B	1BMV-C	2STV	1BBT-1	1BBT-2
1BBT-3	1TME-1	1TME-2	1TME-3	4SBV-B	2MEV-1	2MEV-2	2MEV-3	1TMF-1	1TMF-2
1TMF-3	2PLV-1	2PLV-2	2PLV-3	1R09-1	1R09-2	1R09-3	2R06-1	2R06-2	2R06-3
1R1A-1	1R1A-2	1R1A-3	1MEC-1	2DPV					

Membrane proteins									
4RCR-L	4RCR-H	4RCR-M	2RCR-L	2RCR-H	1PRC-L	1PRC-H	1PRC-M	1PRC-C	2POR

Folds known to be incorrect		
RUBW-S	2GN5	1THI

a, Brookhaven data bank codes. If available chain identifiers are appended. HMBM-I, HPPST-I, HCYS-I, HGPF-A, HCRTN-B were obtained from Robert Huber, SBLP from Paul Sigler and RUBC-S, RUBW-S are courtesy of David Eisenberg. ROP2-A, RAS1, ACT1-A were obtained from the EMBL file server.

will fail if the force field employed is unable to identify a native fold among a set of decoys. Hence, the ability to recognize native folds among a large number of alternative structures is the main theme we adopt in force field testing.

The emphasis here is on the development of techniques which support force field development and the assessment of their predictive power and on the reasoning behind the particular approach chosen. The technique is then applied to a most simple knowledge based mean field constructed solely from C^β-C^β interactions, the evaluation of the predictive power of this variant and the investigation of the stability of the results obtained in terms of several global parameters like cut off distances for the truncation of potentials, data base size, etc.

2   Strategy of Force Field Testing

A most important role of energy in protein structure prediction is that it can be used as a guiding principle. If the equivalence between global minima of energy functions and native structures holds, then, at least in theory, native folds of proteins can be identified and calculated by minimizing the energy as a function of the conformational variables. If the condition is violated the search for the native fold necessarily goes astray.

The 'global minimum equals native state' property is hard to verify for a given energy model. A rigorous proof requires the generation of all possible conformations of a single amino acid chain, the calculation of the associated conformational energies and the comparison of these energies to the energy of the native fold. The number of possible folds of small proteins is in the order of 10^{100}. Generation of these structures is far beyond the capabilities of present day computer technology.

On the other hand, if we find structures of lower energy as compared to the native state, the equivalence of global minimum and native fold is falsified. This is the principle we adopt. We try to disqualify the knowledge based mean field by pretending that we have a large number of possible candidates for the native fold, but we do not know the correct one. We then ask whether the force field will be able to identify the native fold. If it is confused by one or several folds, then the correspondence between global minimum and native state cannot hold. If the force field succeeds, we still have no rigorous proof of this correspondence, but at least we get some estimate of the predictive power of the force field.

Novotny et al. [14,15] have used this principle in their investigation of the predictive power of semi-empirical force fields in its most basic form. They used the native fold and one grossly misfolded structure and found that the force field had difficulties in distinguishing correct from incorrect folds.

In our earlier studies [11] a set of alternative conformations was prepared from the data base of known structures and the ability of the knowledge based mean field to recognize the native fold was investigated for a large fraction of proteins in the Brookhaven data base [16]. By shifting a sequence of length l through a given fold of length h, $h - l + 1$ possible conformations can be formed. It is clear that decoys can be obtained from a fold only if $h \geq l$. Hence, for short sequences l, a large set of decoys is obtained from the data base, but for longer sequences the yield is small. Consequently, the result that the force field is able to recognize the native fold of a large protein is insignificant and this prevented the investigation of the performance of the force field when applied to the larger proteins in the data base.

A quality check of force fields should cover all proteins whose structures are available. Besides the insignificance of the results obtained for large proteins an additional drawback is that results obtained for proteins of different sequence length are incomparable, since the number of decoys obtained from the data base depends on the sequence length of the protein under study. This prevents the definition of a meaningful performance measure over the whole test set.

When the structures in the current data base are combined to a single polyprotein these problems can be avoided. Using 200 proteins of known structure the total length L of the resulting polyprotein is in the order of $50,000$ residues and for any sequence l we have $l \ll L$ so that for practical purposes the number of decoys $L - l + 1$ is independent of the particular length of the test protein. For any protein we have approximately the same number of decoys and the test is comparably hard for all proteins in the data base. Individual results can be compared and the average predictive power of the force field over the whole data base can be estimated.

In the construction of the polyprotein it is important to model the joining regions between successive protein modules carefully. If linker regions have unusual properties polyprotein fragments containing such regions will be easily recognized as unfavourable folds. The basic requirements are that linker regions have good stereochemistry and that there are no steric overlaps between linker regions and protein modules or between adjacent protein modules along the polyprotein.

3 Polyprotein Construction

The basic operational task in the construction of a polyprotein is to connect two successive protein structures. It is usually impossible to join two structures directly, since generally the conformation of the C-terminus of one molecule will not be structurally compatible with the N-terminus of a second molecule. Protein modules therefore, have to be bridged by linker regions resembling connections as found in multidomain proteins.

Linker regions can be derived from the data base of known structures. This ensures that linkers obey general rules of protein folds. To establish an acceptable connection P_i-L_i-P_{i+1} between two successive proteins P_i and P_{i+1} the linker L_i has to meet several criteria. The passage from the C-terminus of protein P_i to the N-terminus of linker L_i has to be 'smooth', devoid of steric strain. In other words the C-terminal region of P_i and the N-terminal region of L_i have to be superimposable with a small error and there must not be any inacceptable distortion of bond lengths and valence angles. The same applies to the passage from the C-terminus of L_i to the N-terminus of P_{i+1}. In addition steric overlaps between any of the structures P_i, L_i, and P_{i+1} have to be avoided.

We construct a connection between protein P_i and linker L_i (or L_i and P_{i+1}) by superimposition of the respective two terminal residues using the technique reported by Sippl & Stegbuchner [17]. To have a general notation we label the two C-terminal residues $N - 1$ and N of P_i (where N is the sequence length of P_i) by 1 and 2 and the two N-terminal residues 1 and 2 of L_i by $1'$ and $2'$. The first step is to superimpose the nitrogen atoms N_2 and $N_{2'}$ by a translation. With the two nitrogens tied together an optimal superimposition of the remaining backbone atoms including C^β is performed where equivalent atoms are $N_1 \leftrightarrow N_{1'}$, $C_1^\alpha \leftrightarrow C_{1'}^\alpha$, and so on. Since the C-terminal oxygen atom O_2 is part of a carboxyl group but $O_{2'}$ is a member of the peptide bond superimposition is performed without the pair $O_2 \leftrightarrow O_{2'}$. Finally the surplus atoms $N_{1'}$, $C_{1'}^\alpha$, $C_{1'}^\beta$, $C'_{1'}$ and $O_{1'}$ of L_i and N_2, C_2^α, C_2^β, C'_2 and O_2 of P_i are removed. The individual steps in the construction of the passage are depicted graphically in Figure 1.

A connection is acceptable only if the root mean square error of superimposition is small. Otherwise, the connection may contain unacceptable distortions of bond lengths and bond angles. In addition a small error of superimposition guarantees that the ϕ-ψ angles around the point of connection are those observed experimentally.

If the connection is accepted the next step is to search for steric overlaps between P_i and L_i. If close contacts are found (which happens quite frequently) the linker L_i is rejected. If not, the linker is connected to protein P_{i+1} using the two C-terminal residues of L_i and the two N-terminal residues of P_{i+1} in the same manner as described above. Again the linker is rejected if close contacts are found between L_i and P_{i+1} after the connection is established. Now the mutual orientation of P_i and P_{i+1} is fixed by the linker and if steric overlaps are found between P_i and P_{i+1} the linker is again rejected. A linker is accepted only if it is able to pass all these tests successfully. Figure 2 depicts a successful link construction and the steps involved.

These are quite stringent requirements and it is clear that in general it will not be easy to find an acceptable linker in a data base of known structures. In addition the operations involved are computationally demanding. A workable solution of the polyprotein construction problem therefore requires an efficient search strategy and a careful implementation of the procedures used. Our strategy consists of the following steps.

1. The set of proteins used to build the polyprotein are sorted in arbitrary order.

2. A second set of proteins is prepared which is used as a source for linker peptides.

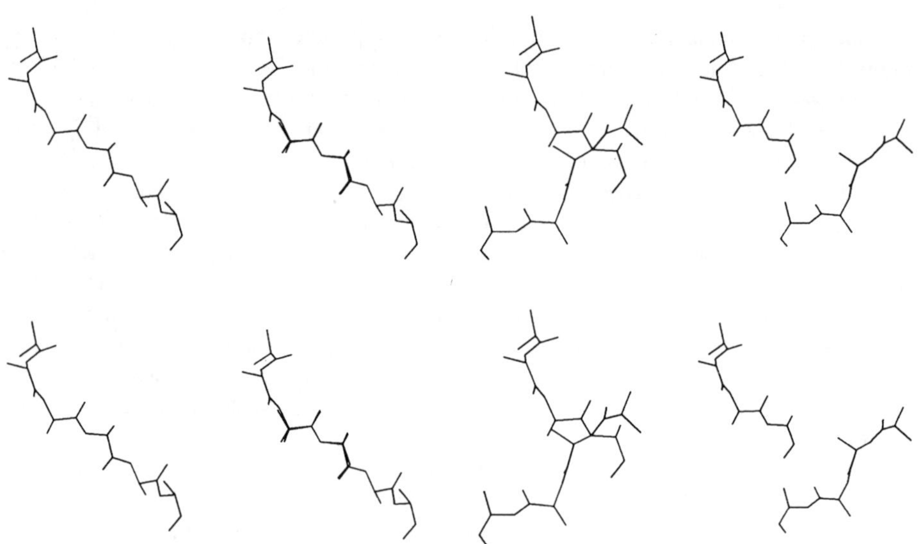

Figure 1: Stero pair drawings of the operations involved in the construction of connections between protein modules and linkers. Top: C-termin us of a protein module (right) and N-terminus of a linker peptide in arbitrary orientation. In the first step one molecule is translated, so that the N-atoms of the terminal residues coincide. Then the molecules are superimposed in the overlapping regions. Finally surplus atoms are removed.

For practical purposes this set may be identical to or it may be an enlarged set of data base 1.

3. Buildup starts with the first two proteins of set 1 and is repeated for any successive pairs P_i and P_{i+1} of set 1 by searching set 2 for acceptable linkers. This involves

    (a) Search for a dipeptide in set 2 which can be connected to the C-terminus of P_i with an rms error of superimposition of less than s The first superimposable dipeptide found constitutes the N-terminus of a possible linker. We denote the start of this linker by $Q_{j,k}$, where Q_j denotes protein j of set 2 and k corresponds to the start of a possible linker in Q_j.

    (b) The next step is the search for a possible C-terminus of the linker. Since joining of protein modules and linkers results in the removal of one residue from each chain a minimum linker size of four residues is appropriate. On the other hand to suppress insertion of entire protein modules as linkers we restrict the linker size to a length $\leq t$ residues. The shortest possible linker therefore, extends from $Q_{j,k}$ to $Q_{j,k+3}$ and the largest from $Q_{j,k}$ to $Q_{j,k+t-1}$.

    (c) Starting with the shortest candidate the C-terminus of the linker is superimposed on the N-terminus of P_{i+1} until an appropriate linker fragment is found which fits and at the same time is free of steric overlaps between protein modules and linker. At this point an acceptable linker

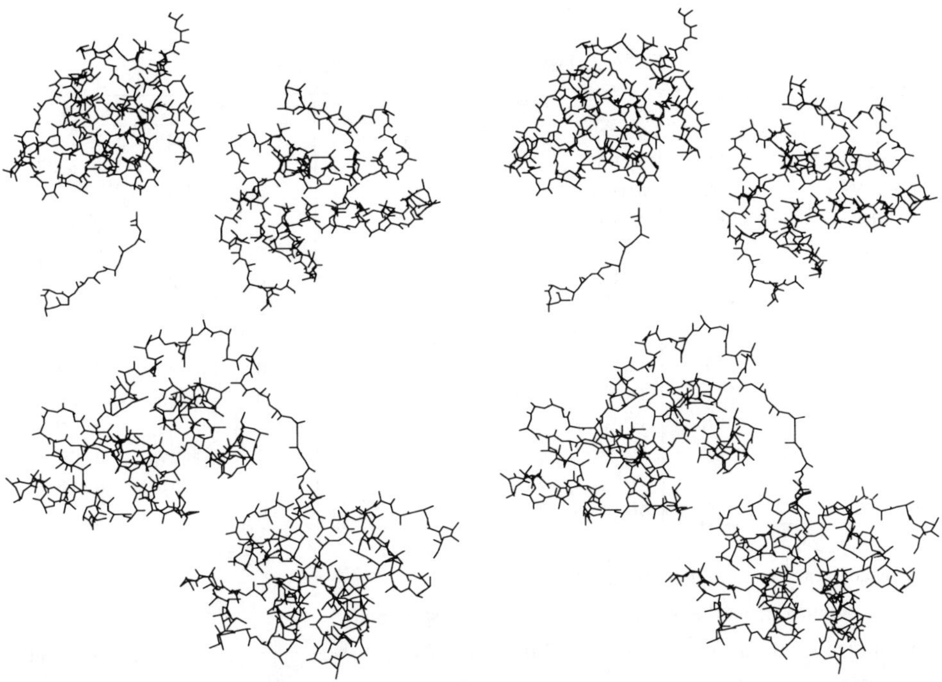

Figure 2: Example of a successful linker construction between 1ALC and 1BP2 (stereo pairs). The linker corresponds to fragment 149-160 of 1CHO-E.

has been found, the nascent polyprotein has grown by one additional linker and protein module, and the construction proceeds with the next protein in set 1 (step 3). If none of the fragments are acceptable the search for a new N-terminus is resumed (step 3.a).

There is however, no guarantee that the algorithm terminates successfully. Depending on the geometry at the N- and C-termini of protein folds it sometimes happens that continuation of the chain inevitably results in steric clashes. Hence, if the search for an acceptable linker fails one residue is removed from the N- or C-terminus of P_i or P_{i+1}, respectively and step 3 is repeated. If this fails again then more and more residues are removed from P_i and P_{i+1} one at a time until a solution is found. This is usually accomplished after removal of only a few residues. We found that this fully automatic procedure is generally successful.

The algorithm depends on two parameters: s, the cutoff value for sufficient similarity of linkers and protein modules in their joint regions and t the maximum linker length. It is clear that the efficiency of the procedure depends on s. If s is too small, then acceptable linkers will rarely be found. If it is too large then the stereochemistry at the connections may be bad. A value of $s = 0.3$ ensures efficiency of polyprotein construction and at the same time guarantees that the connections are of high stereochemical quality. The maximum linker length is less critical. Values of $t = 30$ to 40 perform satisfactorily.

An additional parameter concerns the length of polyprotein fragments which are free of overlaps. Overlap checking extends over a prescribed range of the polyprotein, instead of the length of P_i as implied above. A connection is accepted only if the linker and protein module P_{i+1} do not overlap with any of the m C-terminal residues of the nascent

Table 2: **Details of Polyprotein Buildup**

connection[a]			linker[b]				r-C[c]	r-N[d]	fit-C[e]	fit-N[f]
ACT1-A	↔	1NTP	1ECA	46	-	74	0	2	0.29	0.24
1NTP	↔	1BMV-A	3CPA	31	-	57	1	0	0.23	0.25
1BMV-A	↔	1BMV-C	ACT1-D	222	-	226	1	0	0.07	0.18
1BMV-C	↔	1BMV-B	ACT1-A	160	-	170	0	0	0.20	0.17
1BMV-B	↔	1CHO-I	2PFK-B	102	-	126	1	0	0.19	0.30
1CHO-I	↔	1P07-A	1NTP	93	-	103	0	0	0.29	0.23
1P07-A	↔	1ALC	1BMV-A	116	-	122	0	0	0.19	0.26
1ALC	↔	1BP2	1CHO-E	149	-	160	2	1	0.14	0.11
1BP2	↔	1CA2	2STV	126	-	141	0	0	0.29	0.23
1CA2	↔	1CCR	1PP2-R	19	-	37	0	0	0.28	0.26
1CCR	↔	1CRN	1CHO-E	23	-	27	0	0	0.18	0.17
1CRN	↔	5CPV	5CPV	74	-	84	4	2	0.27	0.20
5CPV	↔	5CTS	2PLV-2	218	-	231	0	4	0.30	0.18
5CTS	↔	2SEC-I	ACT1-A	116	-	137	0	0	0.20	0.23

a, two successive protein modules P_i-P_{i+1}.

b, linker obtained from the data base search (protein, start, end)

c, number of residues removed from the C-terminus of P_i before an appropriate linker was found.

d, number of residues removed from the N-terminus of P_{i+1} before an appropriate linker was found.

e, root mean square error of optimal superimposition of the two C-terminal residues of P_i and the two N-terminal residues of linker L_i.

f, root mean square error of optimal superimposition of the two N-terminal residues of P_{i+1} and the two C-terminal residues of linker L_i.

polyprotein. A search depth of $m = 2,000$ residues was used for the constructions reported in the present study. As an example Figure 3 shows a polyprotein generated from 18 protein modules with a total length of $2,803$ amino acid residues. The details of the construction of this polyprotein are found in Table 2.

There is still no guarantee that the algorithm will succeed, since this depends on the quality of the structures used to build the polyprotein. In low resolution structures with bad stereochemistry it may happen, that superimposition of linker and protein termini may fail in all cases. Acceptance can be increased by raising s so that the algorithm will work but at the price of loosing stereochemical control in the protein linker connections. However, using the parameter settings recommended above and a set of acceptable structures, the algorithm is able to generate polyproteins of arbitrary length quite efficiently. On a typical workstation construction of a polyprotein consisting in the order of 200 protein modules is accomplished within a few minutes of CPU time.

4   Hide and Seek on a polyprotein

Polyproteins contain the backbone atoms N, C^α, C' and O and the C^β atom of each residue, but no additional side chain atoms. If the original residue is a glycine, then a virtual C^β atom is constructed as reported previously [11]. The polyprotein therefore, is equivalent to a polyalanine chain of length L and the original sequence of the individual protein modules

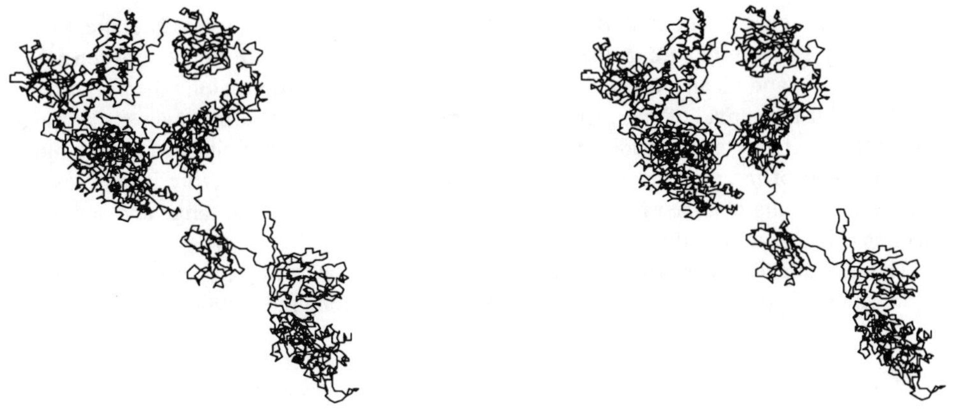

Figure 3: Stereo pair drawing of a polyprotein constructed from 18 protein modules (Table 2).

cannot be deduced from the polyprotein conformation.

The polyprotein can be employed to challenge the predictive power of a force field in the following way. The known native fold of a given amino sequence of length l is hidden somewhere along the polyprotein. The question is whether the knowledge based mean field is able to identify the native fold, which is achieved only if the respective polyprotein fragment has the lowest energy as compared to all other alternatives encountered along the polyprotein chain. Starting at the polyprotein's N-terminus the sequence is combined with the first possible fragment, corresponding to conformation C_1 (residues 1 to l of the polyprotein), and the associated conformational energy E_1 is evaluated and recorded. When finished the sequence is displaced by one residue towards the C-terminus and the associated energy is calculated. The process is repeated for all possible fragments $C_i, i = 1, \ldots, L - l + 1$. Finally the conformations are sorted by their respective energies E_i. The native fold C_n is successfully identified if its energy is lowest among all alternatives, i.e. $E_n < E_i$ for all $i \neq n$.

Given a particular amino acid sequence the fragments along the polyprotein represent the conformational space accessible in the hide and seek game. This is only a tiny fraction of the complete set of possible conformations, but the sample covers important regions of conformation space. Many of the conformations encountered along the polyprotein are compact and globular and have the characteristics of genuine native protein folds.

To avoid steric clashes in the linker regions it is sometimes necessary to remove one or more residues from the N or C-termini of individual protein modules. In most cases only a few residues are affected but the consequence is that modules in the polyprotein are no longer identical to the original native folds. But it does not really matter whether the native fold is hidden in the polyprotein or not. The role of the polyprotein is to provide a set of alternative conformations whose energies can be compared to the energy of the native fold. Consequently, for practical purposes, the energy of the original native fold is evaluated independently and compared to the energies obtained from the polyprotein. As a corollary it is therefore not neccessary to construct a new polyprotein in order to perform a hide and seek test on a protein whose fold is not contained in the polyprotein.

A thorough investigation of the predictive power of a force field involves the repeated application of the hide and seek test for all proteins of known structure and the calculation of a performance measure over the whole test set. The ability to recognize the native fold of a given sequence is one important feature. A second is the significance and strength of recognition. The significance or discriminating power is high if the difference between native fold energy E_n and mean energy $\bar{E}$, averaged over all conformations along the polyprotein is large and if the spread of energies σ around the mean value is small. However, the range of energies encountered depends on the size of a protein. Large proteins can form a much larger number of interactions as compared to small proteins and therefore it is necessary to scale the results obtained for different proteins with respect to the range and spread of energies calculated from the polyprotein.

This is achieved by transforming energies to z-scores. The average energy of a given sequence with respect to the conformations encountered along the polyprotein is

$$\bar{E} = \frac{1}{N} \sum_{i=1}^{N} E_i \tag{1}$$

and the associated standard deviation

$$\sigma = \sqrt{\frac{1}{N} \sum_{i=1}^{N} E_i^2 - \bar{E}}. \tag{2}$$

Given the energy E of some fold the associated z-score is

$$z = (E - \bar{E})/\sigma \tag{3}$$

which is a measure of the distance of energy E from the mean value in units of standard deviation. A large negative z-score calculated from the energy E_n, the energy of the native fold, indicates a high significance in the recognition of the native fold of the respective sequence.

5   Lysozyme — a case study

Before we apply the hide and seek technique we have to discuss two points of general importance in force field testing. These concern the introduction of gaps in sequence structure alignment and the effect of hidden correlation. Another question, specific for the polyprotein test strategy, concerns the quality of the linker regions.

We use the results obtained in a single hide and seek test applied to the turkey lysozyme sequence, Brookhaven code 1LZ3, as summarized in Table 3, to illustrate these points. The energies are calculated from C^β-C^β interactions only. The native fold has the lowest energy as compared to all other conformations encountered by the 1LZ3 sequence along the polyprotein, and the associated score indicates a significance of 8 units of standard deviation. The most favourable non native fold is 1GMF-B whose z-score of -4.41 is much lower.

The third position is occupied by 1LZ1, a human lysozyme, which is homologous to 1LZ3. The structures of 1LZ3 and 1LZ1 are very similar, but their optimal alignment and spatial superimposition requires a gap of one residue (Gly48 of 1LZ1 has to be deleted for optimal alignment with 1LZ3). In the hide and seek test gaps are not allowed and the combination of the 1LZ3 sequence with the 1LZ1 conformation is out of register for almost one half of the molecule, resulting in a rather high energy and corresponding score of low significance. However, if the gap is allowed in the 1LZ3 sequence then the z-score obtained from the combination with the 1LZ1 conformation has a significance of -6.88 indicating a native like fold.

Table 3: Hide and seek result for the sequence of lysozyme 1LZ3

Rank	fragment[a]	position[b]	z-score	energy[c]
1	1LZ3	0	-8.01	-81.2
2	1GMF-B	46	-4.41	-33.7
3	1LZ1	1	-4.36	-33.1
4	1XIM-C	299	-3.99	-28.2
5	8CAT-B	152	-3.86	-26.4

a, protein module in polyprotein.

b, relative position of first residue in sequence to the start of the module in the polyprotein (e.g.
1LZ1 1 means that the sequence of 1LZ3 starts one position downstream of the N-terminus of the
1LZ1 module in the polyprotein).

c, in units of E/kT

It is clear that optimal combinations of sequences and structures in general require the
introduction of gaps. In fact, this is a most important requirement in fold recognition. If
gaps are allowed, then the number of possible combinations of sequences and structures
is raised enormously. Every insertion of a gap or deletion of a residue generates a new
sequence structure pair and successful recognition of native folds is much more demanding.
In force field testing our goal must be to make identification of native folds as hard as
possible and a technique allowing gaps may seem to be preferable. However, the introduction of gaps changes the sequence. These changes have largely unpredictable effects
on the conformational preferences of the altered sequence. Deletion of key residues or
the introduction of gaps may have dramatic consequences on the native fold. A striking
experimental example is α_1-antitrypsin [18,19]. The molecule is deactivated by cleavage
of a single peptide bond (a process comparable to gap opening), resulting in large structural
rearrangements. After cleavage a formerly exposed loop inserts into a beta sheet and the
two residues at the cleavage site become separated by more than 70. In other words, if
a break (gap) in the backbone is allowed, the molecule relaxes to a more favourable state
which however, is inaccessible to the uncleaved molecule.

If we are interested in the predictive power of a force field it is necessary to compare
different conformational states of a unique amino acid sequence. If the sequence is changed
by substitutions, insertions and/or deletions, then the molecule may still fold to the same
native structure, but it also may adopt a completely different conformation. We simply
do not know. Keeping the sequence constant we know the native fold from experiment,
but if we allow changes in the sequence we lose control and the results obtained may be
misleading.

Our second point concerns the effect of hidden correlations in the potentials. The force
field employed to calculate the results of Table 3 did not contain information on lysozymes
or other related molecules (like α-lactalbumin, for example). If 1LZ3 is included in the data
base used to compile the mean force potentials the results change considerably (Table 4).
The significance of the 1LZ3 score rises by almost 3 units. Similarily, the score for 1LZ1
becomes more significant and the score for the more distantly related 1ALC (α-lactalbumin)
conformation improves considerably. It is noteworthy, that only the scores of 1LZ3-related
proteins are affected. The score of the unrelated 1GMF-B fragment is insensitive to the
presence or absence of 1LZ3 in the mean force data base.

Table 4: **Hide and seek result for the sequence of lysozyme 1LZ3 with 1LZ3 included in the mean field data base.**

Rank	fragment^a	position	z-score	energy
1	1LZ3	0	-10.79	-119.2
2	1LZ1	1	-5.45	-47.8
3	1GMF-B	46	-4.43	-34.2
4	1ALC	-3	-4.10	-29.9
5	1XIM-C	299	-3.97	-28.2

a, see legend to Table 3

In summary, a fair judgement of predictive power demands that the force field does not contain specific information on the molecule under study. With the exception of Table 4 all results reported in the present study were obtained from a jack knife test (e.g. Wodak and Rooman 1993). The molecule under investigation is removed from the data base and mean force potentials are recompiled before the hide and seek test is performed.

The result obtained for 1LZ3 shows that if information on the native fold is included in the mean field then the z-score of 1LZ3 is raised considerably, implying that the features of the mean field strongly depend on particular proteins in the data base. This however, is not the case. The result obtained for 1LZ3 does not change significantly if unrelated proteins are removed or added to the mean force data base (the result is similar to the score obtained for 1GMF-B, which is not affected by the addition of 1LZ3).

A major advantage of the hide and seek test presented here is the enlarged set of decoys obtained from the polyprotein. But the consequence is that a substantial fraction of fragments encountered by a sequence in this test contain linker regions. The question remains whether or not the linker-containing fragments are non-realistic as compared to fragments solely derived from protein modules (i.e. the set of decoys used by Hendlich et al. [11]). If a large fraction of linker containing fragments have high energies then the stringency of the test would be greatly reduced. The effect would be towfold. (1) The number of genuine decoys would be approximately the same as in the test strategy employed by Hendlich et al.[11]. (2) The scores obtained for the native folds would be artificially raised due to the large average energy and (possibly) small standard deviation calculated from the polyprotein.

To investigate this points we analyse the distribution of energies in the conformation space represented by the polyprotein. If linker-free and linker-containing fragments are distinct then the distribution of energies (i.e. the number of conformations as a function of energy) should have at least two peaks: one peak at low energies corresponding to linker-free fragments and a second peak representing linker-containing fragments at higher energies. Figure 4 shows the distribution obtained for lysozyme 1LZ3. The distribution of energies is gaussian. The energy distributions obtained for other protein sequences are similar to 1LZ3. In all cases investigated the distribution is gaussian. The result shows that the energies of linker-free and linker containing conformations are in the same range. Hence, the set of decoys derived from the polyprotein is homogeneous in the sense that linker-free and linker-containing conformations are, on average, energetically indistinguishable.

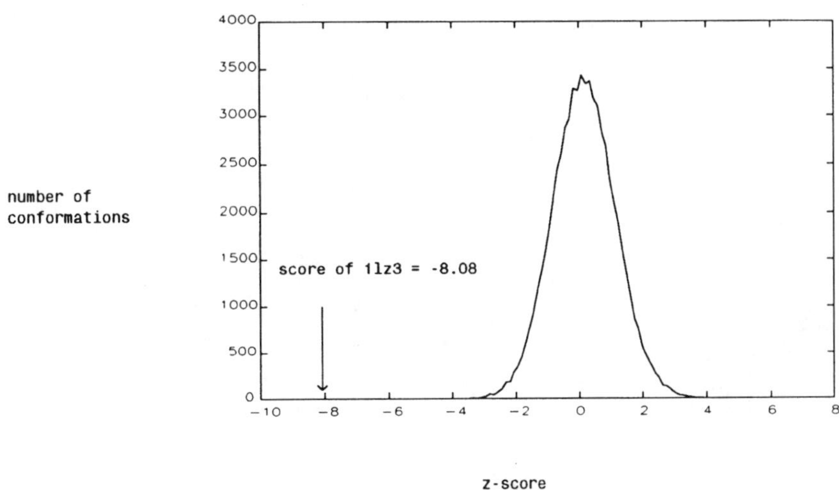

Figure 4: Energy distribution of all conformations encountered along the polyprotein with respect to the 1LZ3 sequence (energy interval versus number of conformations whose energy is in this interval). The energy scale is shown in units of standard deviation (z-score) ($z = (E - \bar{E})/\sigma$). Average energy and standard deviation of 1LZ3 are 22.4 and 13.2, respectively. The graph is constructed in the following way. The energy range (from -81.16 to 86.49) is devided in 100 intervals of equal size and the number of conformations in each interval is counted.

6 Predictive power of the knowledge based mean field

Our goal is to obtain an estimate of the predictive power of the C^β based mean field. It is clear that side chain-side chain interactions are very important for folding and stability of native states, as are interactions with prosthetic groups, co-factors and ions. It is however, technically cumbersome, if not impossible, to construct a complete force field containing all these interactions in a single effort. In the present study we therefore, concentrate on the performance and optimization of C^β-C^β interactions. It is clear that a model based on C^β-C^β interactions alone is the most reduced energy model of protein solvent systems conceivable. If the model has predictive power then the prospects for further improvement are encouraging and the results obtained will be relevant for further refinement and completion of the force field.

The data base used to compile the mean force potentials consists of soluble globular proteins (Table 1). The data base has been prepared so that the sequence homology between any two proteins in the data base is low (the three largest homolgies are: 1LFG:1OVB, sequence lengths 691 and 159, respectively, 56% identities; 4FD1:1FDX, (106, 54), 41%; 1PPG-E:2KAI-A, (218, 80), 34%; percentage of identities are calculated with respect to the shorter sequence).

Our intention is to evaluate the performance of the mean field for as many experimentally determined folds as possible. Therefore, the hide and seek test was performed on all proteins available from the current release and prereleases of the Brookhaven data base (avoiding strongly homologous cases), including low resolution structures.

Figure 5 reports the scores obtained for the proteins assembled in Table 1 and a few

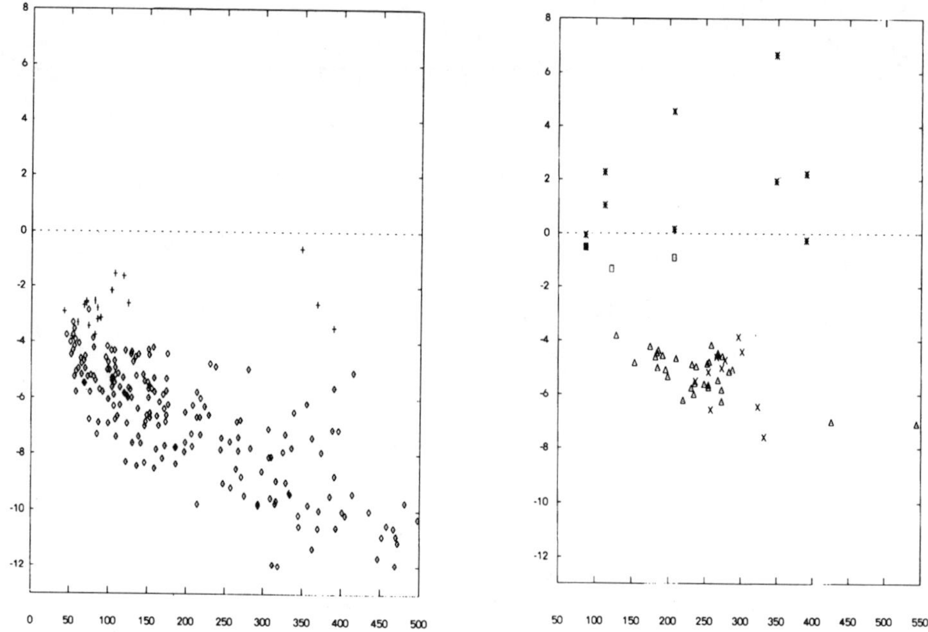

Figure 5: (a) and (b). Scores of experimentally determined protein folds obtained from hide and seek as a function of sequence length for several classes of proteins. $\diamond$ proteins recognized by the force field, $+$ proteins not recognized by the force field, $\triangle$ virus coat proteins, $\times$ membrane proteins, $\square$ experimentally determined folds, known to be incorrect and $*$ deliberately misfolded proteins.

deliberately misfolded proteins in terms of z_n, the z-scores of native folds versus sequence length. The polyprotein used for the calculation of z-scores was constructed from the 199 proteins of the mean force data base which resulted in a total length of $44,736$ residues and hence the number of non-native decoys for each protein is of the same order of magnitude.

Figure 5a shows that in most cases the native folds have lowest energy and are thus identified successfully as the most favourable folds for the respective sequences ($\diamond$). Proteins whose native fold is not identified as the most favourable conformation are marked by $+$ and the results on these proteins are summarized in Table 5. There are 18 cases of 260 (not counting deliberately misfolded and incorrect folds) where the native fold does not have lowest energy and in all these cases the corresponding scores z_n have a significance of less than four units of standard deviation (in absolute value).

Figure 5b collects the results obtained for virus coat proteins ($\triangle$), membrane associated proteins ($\times$) and, as a control, the results obtained for experimentally determined folds which are known to be incorrect ($\square$) and deliberately misfolded proteins ($*$). Deliberately misfolded proteins are obtained by arbitrarily combining sequences and structures. In particular the sequence structure combinations shown in Figure 5b are 1TPK-2GN5 (i.e sequence of 1TPK combined with structure of 2GN5), 2GN5-1TPK, 2MCP_VL-1HMQ, 1HMQ-2MCP_VL, 1GAP-1THI, 1THI-1GAP, 1MLE-1PTE, 1PTE-1MLE, 1NSB-1PHH and 1PHH-1NSB.

The monomers of virus coat proteins interact strongly to form a protective shell covering the genetic material. The shell is reminiscent of a membrane composed of strongly interacting protein units and hence, it is likely that the features of viral coat proteins differ

Table 5: **Proteins whose reported fold is not identified**

Protein	length	rank	z_n	z_{opt}
1PTE	348	12116	-0.62	-3.83
2CDV	107	2640	-1.56	-4.26
1CY3	118	2191	-1.63	-4.02
1MDA-F	103	578	-2.17	-3.92
1CPB	82	126	-2.55	-4.41
2HIP-A	71	172	-2.57	-3.72
2CTX	71	5	-2.61	-2.67
1MDA-C	368	204	-2.63	-4.05
1TRC-B	68	242	-2.69	-6.43
1HIP	85	152	-2.78	-4.36
3FGF	124	69	-2.90	-3.91
1FC2-C	43	78	-2.92	-4.24
1TEN	89	8	-3.16	-3.86
1ZAA-C	85	18	-3.21	-3.56
1CDT-A	60	7	-3.32	-3.93
1HOE	74	6	-3.45	-3.72
351C	82	6	-3.76	-4.16
1CHO-I	53	3	-3.82	-4.11

to some extent from those of soluble globular proteins. Nevertheless, the force field derived from a set of soluble globular proteins is able to recognize virus coat proteins as the most favourable folds in all cases (Figure 5b), but the significance of the scores z_n obtained for these proteins is generally lower compared to soluble proteins.

The native folds of membrane proteins are stable in hydrophobic environments. Therefore, it seems rather unlikely that a mean field derived from soluble globular proteins should be able to recognize the native folds of membrane proteins. However, in all cases investigated (Table 6) the native folds of membrane proteins are recognized in the hide and seek test, although the scores obtained for the native folds of these proteins are significantly lower as compared to soluble proteins. The result can be interpreted in the sense that the mean field based on intramolecular interactions reasonably models important aspects of the energetic architecture of membrane proteins but at the same time several components specific for this class of proteins are missing.

Figure 5b shows that the scores of folds which are known to be incorrect are insignificant. This is obvious for deliberately misfolded proteins, whose scores scatter around the zero-baseline or have strong positive scores like the 1PTE-sequence combined with the 1MLE-fold, producing a score of +6.66. Large positive scores indicate that the sequence is totally incompatible with the respective fold, as compared to random combinations of sequences and folds.

Similar results are obtained for the examples of incorrect structure determinations. The X-ray structure of gene 5 DNA binding protein (2GN5) is in conflict with recent data obtained from nuclear magnetic resonance studies [20]. The score z_n of -0.52 obtained for this structure indicates that the 2GN5 structure does not correspond to the native fold of the 2GN5 sequence. There are $13,692$ folds encountered along the polyprotein which have lower energy. It is noteworthy however, that the structure is still represented as a regular entry in the Brookhaven data base and that no corrected X-ray version has been published.

Table 6: **Pair potential z-scores of membrane proteins**

Protein	length	rank	z_n	z_{opt}
4RCR-M	296	1	-3.87	-3.67
2POR	301	1	-4.42	-4.07
4RCR-L	266	1	-4.61	-3.63
2RCR-L	278	1	-4.73	-3.71
1PRC-L	273	1	-5.03	-4.28
2RCR-H	255	1	-5.18	-4.13
4RCR-H	237	1	-5.51	-4.17
1PRC-H	258	1	-6.57	-4.66
1PRC-M	323	1	-6.46	-3.64
1PRC-C	332	1	-7.60	-3.85

A second example is the small subunit of Rubisco RUBW-S. The score z_n of the incorrect fold of -0.99 clearly points to a non native sequence structure combination and $7,086$ folds in the polyprotein are more favourable. On the other hand the correct structure RUBC-S is identified yielding $z_n = -7.00$, typical for a genuine native fold.

A third example of an experimentally determined fold which has subsequently been corrected is thaumatin 1THI. The current Brookhaven data bank entry (1THI) has been determined to 3.1 resolution. The polyprotein contains $8,320$ fragments having lower energy as 1THI and the z-score of -0.92 points to a problematic structure. Thaumatin has recently been refined to a resolution of 1.65 [21]. During refinement several frameshift errors have been detected in the low resolution model (1THI) and several parts of the structure have been remodeled. The refined structure is presently not available so that we are unable to report the respective scores for the high resolution structure.

The 1THI Brookhaven entry contains only the C^α coordinates. The results reported here were calculated from a set of C^β coordinates obtained from Chris Sander's group. The results obtained for the original 1THI C^α coordinates calculated from a force field based on C^α-C^α interactions are virtually identical to those for the C^β-C^β interactions discussed above, i.e. the z-score of 1THI reflects the frameshift errors in the low resolution structure determination and is not a consequence of the C^β buildup procedure.

The native folds of 18 proteins in Figure 5a and Table 5 are not the most favourable structures. Most of these folds correspond to small proteins of less than 100 residues. In addition several of these proteins have unique and unusual features, like 2CTX containing six disulfide bonds in its amino acid chain of 71 residues. The high potential iron proteins (2HIP-A, 1HIP) contain iron-sulfur clusters, which are important for the stability of these folds. Another example is 2CDV whose amino acid chain (107 residues) wraps around four heme groups. There are however, several examples of large proteins whose scores are striking. These are 1PTE ($z_n = -0.59$, rank $12,610$) and 1MDA-C ($z_n = -2.29$, rank 504). For the moment we note that the scores of these structures are not native-like but we postpone a discussion of the special featurs of these structures to a more detailed analysis.

7 Predictive power as function of force field parameters

The knowledge based mean field depends on several global parameters which affect the quality of the results. Therefore, an important question is whether the results obtained depend on a particular set of parameter values or if the results are stable over a broad range of values. And in applications it is, of course, important to use a set of parameter values which maximizes the predictive power of the force field.

One of these parameters, for example, is the cut off distance used to sample mean force potentials. Intra- and intermolecular interactions are usually thought to be rather short ranged. The functional forms used to model these interactions quickly approach zero with increasing distance. In current models of macromolecular force fields the most far reaching term is the electrostatic contribution which approaches zero as a function of r^{-1}. On the other hand mean force potentials can be calculated for large spatial separations and it is not clear in advance whether energies calculated for distances beyond, say, 10 increase or decrease its predictive power.

Our strategy to investigate this and related points is based on the predictive power of the knowledge base mean field expressed in terms of the average z-score $\bar{z}$ calculated for the known native folds in the data base

$$\bar{z} = \frac{1}{M} \sum_i^M z_i. \tag{4}$$

where, z_i is the z-score of the native fold of protein i obtained from the hide and seek test. The average z-score is a performance measure for the ability of the mean field to recognize native folds over the whole data base so that the predictive power is raised when the average significance of native fold recognition increases. The knowledge based mean field contains only a few global parameters which affect all potentials in the same manner. Such parameters are the distance cut off beyond which energy contributions are truncated, the distance interval size used to sample frequencies, and the size of the data base used to compile mean force potentials.

The systematic variation of several force field parameters is time consuming and requires considerable computing resorces, since for every combination of parameter values the hide and seek test has to be performed for all proteins in the data base. For the investigation of parameter sets we therefore used a subset of 68 proteins chosen randomly from the complete data base of 199 protein chains. In addition a shorter variant of the polyprotein (69 structures; 10,252 residues) was used to calculate z-scores. However, in all cases the mean field was derived from the complete set of 199 proteins.

The dependency of the average z-score $\bar{z}$ on distance truncation is shown in Table 7. The predictive power reaches a maximum at a distance cut off of 30. The predictive power is considerably diminished if the potentials are truncated at small distances (< 10). For cut off distances > 10 the average score increases slowly but steadily up to 30 where the predictive power reaches a maximum. In the range of $20 - 50$ the function is rather flat. In terms of successful identification of native folds truncation at 20 yields the best result. There is only one protein in 68 (cobratoxin 2CTX) whose native conformation is not identified as the most favourable fold for its amino acid sequence.

There is no distinct optimum value for the distance cut off but the range of mean force potentials extends to 20-30. In terms of mean force potentials this is the range where, on average, two amino acids interact and feel each other. In other words the information contained in mean force potentials is useful for predictive purposes up to this range. This is confirmed by calculations where the energies are defined in the range between 15 and

Table 7: **Dependency of predictive power on distance cutoff**

cutoff ()	$\bar{z}^a$	identified[b]	(%)
5.0	-2.18	10	14.7
7.5	-4.13	49	72.1
10.0	-5.05	59	86.8
15.0	-5.28	65	95.6
20.0	-5.46	67	98.5
30.0	-5.72	64	94.1
40.0	-5.69	61	89.7
50.0	-5.43	58	85.2

a, average score obtained for the native folds of a test set of 69 proteins.

b, number of proteins whose native fold is successfully identified (maximum is 69).

Table 8: **Dependency of predictive power on interval width**[a]

interval width ()	$\bar{z}$	identified	(%)
0.5	-5.48	65	95.6
1.0	-5.72	64	94.1
1.5	-5.70	64	94.1
2.0	-5.61	62	91.1
5.0	-5.30	60	88.2
10.0	-4.96	56	82.3
15.0	-4.59	53	77.9

a, See legend to Table 7.

30 only (i.e. outside this range the potentials are zero). The truncated potentials produce a score $\bar{z}$ of -3.05 and the truncated field is still able to recognize 25 of 68 folds (37%). This result is only possible if the distance range of 15-30 contains specific although quite incomplete information on the relationship between amino acid sequences and their native folds.

On the other hand the predictive power deteriorates if the interaction energies for distances greater than 30 are included in the total energy. One reason is that the mean force potentials for large distances are dominated by a few large proteins in the data base. Small proteins have only few if any distances greater than 30. In addition the number of pair interactions increases with the square of the sequence length. Hence, the few large proteins in the data base contribute the overwhelming majority of measurements and the potentials are neccessarily biased in this range. Presently the small number of large proteins in the data base prohibits an appropriate analysis in the far distance range. The effective range of mean force potentials may extend beyond 30, but presently this is the upper limit beyond which the data become unreliable.

A second parameter which affects the quality of the results is the interval width used to sample mean force potentials. The results assembled in Table 8 show that this is an extremely stable parameter (the distance cut off used is 30). An interval width of 1 produces the optimum score but for practical purposes values between 0.5 and 2.0 yield comparable

Table 9: **Dependency of predictive power on data base size**[a]

number of proteins	$\bar{z}$	identified (%)	
70	-4.37	57	83.8
140	-5.50	61	89.7
199	-5.72	64	94.1

a, See legend to Table 7.

results. Even in the case of a grid size of 5 the score is $\bar{z} = -5.3$ with 60 folds successfully identified.

Table 9 shows the dependency of predictive power on the data base size used to compile mean force potentials. The results were obtained using a grid size of 1 and a cut off distance of 30. The predictive power $\bar{z}$ as well as the number of recognized native folds rises with the number of proteins in the mean force data base. The difference between the results obtained for a mean force data base of 140 proteins and a data base of 199 proteins is rather small. The result indicates that the current data base provides a representative set of globular proteins and that a large part of the information on the energetic features of protein architectures is already contained in the current data base.

The mean force potentials depend on the separation k of amino acid residues along the sequence [8]. Our current version of the force field contains individual potentials for $k = 1, \ldots, 6$. The remaining levels are condensed to two ranges of $k = 7, 8, 9$ and $k = 10, \ldots, \infty$. The question remains whether the separation of levels increases the predictive power as opposed to a single type of potentials which is independent of the separation of two residues along the sequence. If the energy is calculated for the short range contributions $k = 1, \ldots, 9$ only (i.e contributions for $k > 9$ are neglected), using the set of individual potentials the average score is $\bar{z} = -4.62$ and 57 proteins (from a set of 68) are successfully identified. For k-independent potentials the respective values are $\bar{z} = -4.28$ and 47 and hence, some information is lost in comparison to k-dependent potentials.

8 Conclusion

In the present study we concentrated on the development of tools for judging the predictive power of knowledge based mean fields. We used the polyprotein technique to derive an enlarged set of decoys. Since a substantial fraction of conformations obtained in this way contain linker-regions a most important question is whether or not the additional conformations are realistic. As demonstrated in Figure 4, the energy distribution obtained from the polyprotein is gaussian. This is only possible if the linker-containing and linker-free conformations are energetically indistinguishable. But we have an additional point concerning the quality of linker-containing fragments.

If the sequence of 1HMQ is combined with the conformation of 1MCP then this misfolded variant ranks at position 43,948, i.e. almost all conformations derived from the polyprotein are more favourable. In other words, almost all linker-containing conformations derived from the polyprotein are more favourable for the sequence of 1HMQ compared to the 1MCP fold, a linker-free conformation. This clearly demonstrates that the polyprotein constructed in this study is an efficient source for genuine decoys in the hide and seek test.

The hide and seek technique was applied to a mean field constructed from C^{β}-C^{β} pair potentials. It was clear at the outset that this is necessarily an incomplete energy model of protein solvent systems. Nevertheless, the incomplete mean field is able to recognize the native folds of most proteins among a set of approximately $45,000$ decoys. The exceptions are assembled in Table 5. Most of these proteins have unusual features, like 1CDV containing four heme groups or the high potential iron proteins 1HIP and 2HIP-A which contain iron-sulfur clusters. These groups should have a strong influence on the stability of the respective native folds and it is clear that C^β-C^β interactions alone are insufficient to account for the special features of such proteins. Nevertheless, most of the native folds of these proteins are among the most favourable structures.

There are however, other cases where the force field failed to recognize the native fold although the structure does not have large cofactors. One such example is calmodulin 1TRC-B (Table 5). The structure was determined to a very low resolution of 3.6 . When subjected to the hide and seek test the structure appears at position 259 in the energy sorted list. The most favourable structure encountered along the polyprotein produces a remarkable score of -6.19. The respective fragment corresponds to a section of 5TNC (troponin C) whose sequence is related to 1TRC-B. In terms of the mean field the fragment of troponin-C is a better model for the native fold of 1TRC-B compared to the low resolution model obtained from X-ray analysis.

In summary the knowledge based mean field investigated in this study seems to provide a reasonable energy model for the overwhelming majority of experimentally determined folds of proteins and the results obtained are rather insensitive to the global parameters. The model based on C^β interactions is quite incomplete so that the prospects for further improvement are rather exciting. The model based on C^β interactions lacks an explicit treatment of detailed side-chain interactions, for example, as well as the explicit consideration of protein solvent interactions. If solvent interactions are included the average score is raised from $\bar{z} = -6.72$ for C^β pair interactions alone (present study), to a value of $\bar{z} = -9.25$ (in preparation).

Recently research on knowledge based potentials has attracted some interest in the structural biology community. Several variants of pair potentials and their applications have been reported [22-28]. It is clear that the quality of results obtained depends on the particular implementation of mean force potentials. The present study reveals that the predictive power of the resulting mean field is rather insensitive to the global parameters. On the other hand small differences in predictive power may have a profound effect on the success or failure in particular applications. This applies in particular to fold recognition studies, where a slightly suboptimal implementation often prevents the recognition of distantly related folds (in preparation).

References

[1] C.B. Anfinsen, Principles that govern the folding of protein chains *Science.*, **181**, 223-230 (1973).

[2] F.A. Momany, R.F. McGuire, A.W. Burgess and H.A. Scheraga, Energy parameters in polypeptides. VII. Geometric parameters, partial atomic charges, nonbonded interactions, hydrogen bond interactions, and intrinsic torsional potentials for naturally occuring amino acids. *J.Phys.Chem.*, **79**, 2361-2381 (1975).

[3] P.K. Weiner and P.A. Kollman, AMBER: Assisted model building with energy refinement. A general program for modeling molecules and their interactions, *J.Comp.Chem.*, **2**, 287-299 (1981).

[4] U. Burkert and N.L. Allinger, *Molecular Mechanics*, American Chemical Society, Washington D.C., (1982).

[5] B.R. Brooks, R.E. Bruccoleri, B.D. Olafson, D.J. States, S. Swaminathan and M. Karplus, CHARMM: A program for macromolecular energy, minimization, and dynamics calculations., *J.Comp.Chem.*, **4**, 187 (1983).

[6] W.F. van Gunsteren, H.J.C. Berendsen, J. Hermans, W.G.J. Hol and J.P.M. Postma, Computer simulation of the dynamics of hydrated protein cristals and its comparision with x-ray data. *Proc.Natl.Acad.Sci.U.S.A.*, **80**, 4315-4319 (1983).

[7] M. Karplus and G.A. Petsko, Molecular dynamics simulations in biology. *Nature*, **347**, 631-639 (1990).

[8] M.J. Sippl, Calculation of conformational ensembles from potentials of mean force. An approach to the knowledge-based prediction of local structures in globular proteins. *J.Mol.Biol.*, **213**, 859-883 (1990).

[9] M.J. Sippl, Boltzmann's principle, knowledge based mean fields and protein folding. *J.Comput.Aided.Mol.Design*, **7**, 473-501 (1993).

[10] M.J. Sippl, Recognition of errors in three dimensional structures of proteins. *Proteins*, **17**, 355-362 (1993).

[11] M. Hendlich, P. Lackner, S. Weitckus, H. Floeckner, R. Froschauer, K. Gottsbacher, G. Casari and M.J. Sippl, Identification of native protein folds amongst a large number of incorrect models. *J.Mol.Biol.*, **216**, 167-180 (1990).

[12] S.J. Wodak and M.J. Rooman, Generating and testing protein folds. *Curr. Opp. Struct. Biol.*, **3**, 247-259 (1993).

[13] J.U. Bowie and D. Eisenberg, Inverted protein structure prediction. *Curr. Opp. Struct. Biol.*, **3**, 437-444 (1993).

[14] J. Novotny, R.E. Brucceroli and M. Karplus, An analysis of incorrectly folded protein models. Implications for structure predictions. *J.Mol.Biol.*, **177**, 787-818 (1984).

[15] J. Novotny, A.A. Rashin and R.E. Bruccoleri, Criteria that discriminate between native proteins and incorrectly folded models. *Proteins*, **4**, 19-30 (1988).

[16] F.C. Bernstein, T.F. Koetzle, G.J.B. Williams, E.F. Meyer Jr., M.D. Brice, J.R. Rodgers, O. Kennard, T. Shimanouchi and M. Tasumi, The protein data bank: A computer based archival file macromolecular structures. *J.Mol.Biol.*, **112**, 535-542 (1977).

[17] M.J. Sippl and H. Stegbuchner, Superposition of three-dimensional objects: a fast and numerically stable algorithm for the calculation of the matrix of optimal rotation. *Computers Chem.*, **15**, 73-78 (1991).

[18] W. Bode and R. Huber, Ligand-binding: Proteinase and proteinase inhibitor interactions. *Curr.Opp.Struct.Biol.*, **1**, 45-52 (1991).

[19] W. Bode and R. Huber, Natural protein proteinase inhibitors and their interaction with proteinases. *Eur.J.Biochem.*, **204**, 433-451 (1992).

[20] P.J. Folkers, P.M. van Duynhoven, A.J. Jonker, B.J. Harmsen, R.N. Konings, and C.W. Hilbers, Sequence-specific ^{1}H-NMR assignment and secondary structure of the Tyr41->His mutant of the single-stranded DNA binding protein, gene V protein, encoded by the filamentous bacteriophage M13. *Eur.J.Biochem*, **202**, 349-360 (1991).

[21] C.M. Ogata, P.F. Gordon, A.M. de Vos and S.H. Kim, Crystal structure of a sweet tasting protein Thaumatin I, at 1.65 A resolution. *J.Mol.Biol.*, **228**, 893-908 (1992).

[22] D.T. Jones, W.R. Taylor and J.M. Thornton, A new approach to protein fold recognition. *Nature*, **358**, 86-89 (1992).

[23] V.N. Maiorov and G.M. Crippen, A contact potential that recognizes the correct folding of globular proteins. *J.Mol.Biol.*, **227**, 876-888 (1992).

[24] M.J. Rooman, J.-P. Kocher and S.J. Wodak, Prediction of protein backbone conformation based on seven structure assignments: Influence of local interactions. *J.Mol.Biol.*, **221**, 961-979 (1991).

[25] M.J. Sippl and S. Weitckus, Detection of native-like models for amino acid sequences of unknown three-dimensional structure in a data base of known protein conformations. *Proteins*, **13**, 258-271 (1992).

[26] M.J. Sippl, M. Hendlich and P. Lackner Assembly of polypeptide and protein backbone conformations from low energy ensembles of short fragments: Development of strategies and construction of models for myoglobin, lysozyme, and thymosin β_4. *Protein Science*, **1**, 625-640 (1992).

[27] M. Wilmanns and D. Eisenberg, 3D profiles from residue-pair preferences: identification of sequences with β/α-barrel fold. *Proc.Natl.Acad.Sci.*, (1993 – in press).

[28] S.H. Bryant and C.E. Lawrence, An empirical energy function for threading protein sequence through the folding motif. *Proteins*, **16**, 92-112 (1993).

Optimized Energy Functions for Tertiary Structure Prediction and Recognition

Richard A. Goldstein, Zan A. Luthey-Schulten[†‡] and Peter G. Wolynes[†§]

Chemistry Department, Program in Protein Structure and Design, University of Michigan, USA and

[†] School of Chemical Sciences/ [‡] National Center for Supercomputing Applications, [§]Beckman Institute, University of Illinois, USA

Abstract

A theoretical basis for the alignment of a protein sequence to a set of protein structure templates is presented, based on a Bayesian statistical analysis. The optimal Hamiltonian for this threading is closely related to the Hamiltonian optimized for molecular dynamics based on spin-glass theory. The Bayesian theory provides the optimal penalty functions for insertions and deletions in the alignment, which can be put in the form of a chemical potential. In contrast to standard methods for determining gap penalties, these penalties involve the logarithm of the probability distribution of gaps in alignments against correct templates as compared to the probability distribution of gaps in alignments against random templates, as determined self-consistently. Sequences of unknown proteins can be aligned to known protein structures, identifying similar structural motifs and generating reasonably correct alignments.

1   Introduction

The ability to successfully predict the three-dimensional structure of globular proteins based only on knowledge of their amino acid sequences has long been one of the central pursuits of computational biophysics. This pursuit has gained added importance with the rapid burgeoning in the number of known DNA sequences, a knowledge base with the potential to revolutionize many fields of biology, biochemistry, and biotechnology. Unfortunately, for most applications, knowledge of the sequence is not enough, and the methodology to analyze these resulting sequences has not expanded at the same pace. Research into the specific functions and mechanisms of the encoded proteins generally starts with knowledge of the proteins three-dimensional structure. When the new sequence has high similarity to that of a protein of known structure, standard techniques for sequence comparisons can be used to infer a low-resolution structure, that can be further refined. Below a certain degree of sequence similarity, a "twilight zone" exists were it is difficult to infer structural similarity, although it is possible that such similarity does exist. While recent work has expanded our ability to predict protein structures [1–10], the only way of determining the structure of most of the remaining proteins are through laborious and time-consuming experimental methods, such as X-ray Crystallography or multi-dimensional NMR.

The reasons why this problem is difficult are well documented. Firstly, the number of possible conformations for even a protein of moderate size is astronomical, even ignoring side-chain degrees of freedom. With even a foolproof assay for distinguishing between correct and incorrect conformations, testing even a minuscule fraction of the number of possible conformations is currently impossible. Secondly, the roughness of the energy landscape characteristic of heteropolymers makes it difficult to arrive at an efficient search strategy for exploring these conformations. Lastly, we still lack a good method of distinguishing correct from incorrect conformations. Even assuming that the protein's folded structure represents a state of minimum free energy, the energy function for proteins, especially involving interactions between the protein and the solvent, are still matters of great uncertainty and debate.

One approach to this problem has been to take advantage of the large databases of protein sequences and structures now available, in order to make statistical statements about what is true for proteins in general. Initially this involved attempts at predicting secondary structure, but more recently has expanded in the direction of predicting tertiary structure. Often this has involved generating statistics about protein structures in order to ascertain an appropriate energy function, generally through the use of "potentials of mean force" [1–6]. These potentials, while having allowed important advances in protein structure prediction, suffer from a number of limitations. For instance, the potentials of mean force generally assume statistical independence of the various interactions. This is somewhat problematic, both for trivial reasons and deeper reasons having to do with the dynamics of folding [11]. It is not even clear that trying to model the "true" energy function in a simplified way is necessarily the best approach. Not only are proteins only marginally stable, but it is also easy to genetically alter proteins to increase their stability. In a similar way, it might be possible to find an optimal energy function that is better at distinguishing correct and incorrect conformations with better resolution than a realistic one. This is especially true if other information is available about either the protein of interest or proteins in general.

Another observation made possible by the growing protein data bank is that the number of possible protein structural motifs is limited, and there is may be many proteins of biological interest of unknown structure whose conformation is similar to that of a known protein. This suggests an intermediate goal towards protein tertiary structure prediction: identifying possible structural similarity in the absence of significant sequence similarity. Reva and Finkelstein developed a mean-field approach to self-consistently align a sequence onto a family of idealized beta-barrel structures [12]. Later, Bowie *et al.* introduced an approach for threading sequences onto known protein structural motifs, using an energy scheme that encoded how each amino acid residue of the protein was compatible with the location of that residue in the structure [1]. Because their energy function involved only single-body terms, they could use standard dynamic programming algorithms without further modification. Other approaches to energy functions have combined the simple one-body and multi-body potentials with mean-field alignment algorithms in the attempt to recognize protein structures [2,3,5,6,13].

In previous work, we have taken a different approach to the problem–trying to find an energy function optimized for the problem we were trying to solve, namely protein structure prediction [7–9]. These energy functions, designed to facilitate structure prediction using molecular dynamics, were based on minimizing the multiple-minima problem. Using methodologies derived from the study of spin glasses, we can consider two phase transitions possible from the ensemble of liquid-like states; if the random interactions in the protein dominate, a spin-glass transition can occur at a temperature T_g. If, however, the stabilizing interactions of the folded state dominate, folding can occur at a temperature T_f. In the thermodynamic limit, the spin-glass transition reflects a transition to non-ergodicity, where there are not enough states thermally accessible to allow the protein to find its native state. This then represents the situation where the protein dynamics is dominated by the local minima. Even near the spin-glass transition, the protein folding dynamics become slow and non-Arrhenius. This immediately suggests a measure of merit for energy functions for molecular dynamics–that the folding temperature should be high relative to the glass transition temperature, or since both temperatures can be scaled by scaling the Hamiltonian, that the ratio of T_f to T_g should be maximized. We showed that the thermodynamics of these two competing phase transitions can be treated by the random energy model, which postulates that the salient features of the transition can be captured by looking only at the distribution of energies of the random and correct conformations. If the energy of the correctly folded conformation is E_c, and the random conformation is modeled as a Gaussian distribution centered at energy E_r with width Γ_r, then T_f/T_g is maximized when $(E_r - E_c)/\Gamma_r$ is maximized. We further showed that as long as the Hamiltonian was linear with respect to a set of undetermined parameters, $\{\gamma_i\}$, the optimal values of these parameters could be solved for in closed form, simply by computing the energy of a set of random conformations and the correct conformation as a function of these parameters, and solving a system of linear equations. To briefly review this, we start by noting that $E_r - E_c$ is linear in γ_i while Γ_r^2 is quadratic in γ_i. For instance, if we can express the energy of protein ν in its native state as $E_T^\nu = \sum_i \xi_{Ti}^\nu \gamma_i$, and in random conformation ζ as $E_\zeta^\nu = \sum_i \xi_{\zeta i}^\nu \gamma_i$ for some coefficients ξ_{Ti}^ν and $\xi_{\zeta i}^\nu$, $E_r - E_c$ and Γ_l can be written in the form

$$E_r - E_c = A\gamma \tag{1}$$

$$\Gamma_r^2 = \gamma B \gamma \tag{2}$$

for vector A and matrix B, where

$$A_i = \xi_{Ti}^\nu - \left\langle \xi_{\zeta i}^\nu \right\rangle_\zeta \tag{3}$$

$$B_{ij} = \left\langle \xi_{\zeta i}^\nu \xi_{\zeta j}^\nu \right\rangle_\zeta - \left\langle \xi_{\zeta i}^\nu \right\rangle_\zeta \left\langle \xi_{\zeta j}^\nu \right\rangle_\zeta \tag{4}$$

and $\langle \rangle_\zeta$ indicates an average over random conformations. The value of γ that maximizes T_f/T_g is given by [7–9]

$$\gamma = B^{-1} A \tag{5}$$

The observation that the number of possible protein tertiary structures is limited, and as mentioned above, that seemingly non-homologous proteins share the same topologies, prompted us to consider the reduced problem–how to select the correct conformation from a set of possible conformations, and how to consider the set of alignments of a sequence on to a set of possible structural motifs as the set of possible conformations. We developed a Bayesian approach to the construction of the optimal energy function for structure recognition [10]. This Bayesian analysis also allows us to optimize gap penalties for sequence-structure alignment, which has previously been one of the more problematic features. Gap penalties, which can be put in the form of a chemical potential, can be constructed using both the probability distribution of gaps in correct alignments as well as the probability distribution of gaps in alignments between random pairs, as determined self-consistently. The Bayesian approach offers a general approach towards energy function optimization, naturally encorporating a wide variety of possible physicochemical interactions and non-physicochemical knowledge, taking into account the correlations that might exist between the various contributions. Optimization of the gap penalty is just one example of the incorporation of a non-physicochemical interaction into the energy function.

In this paper, we develop this Bayesian approach, and show that the function that results is related to that obtained previously for molecular dynamics, discussed above. We then demonstrate its applicability for a set of test cases for proteins of known structure. This method has recently been applied to the prediction of the tertiary structure of considerable biological interest, the hormone-binding domain of the superfamily of nuclear steroid receptor proteins [14].

2   Bayesian analysis of Hamiltonian

Let us start with a simpler problem, distinguishing the correct structure of a target protein sequence from a set of possible conformations, $\{k\}$, each described by the value of an "attribute" E_k, which we can consider an energy function or fitness function. We are interested in two questions:

1. What is the conditional probability that a structure with a given value of E_k is correct?

2. What is the optimal attribute for distinguishing correct from incorrect conformations?

We can write the conditional probability that a conformation with energy E_k is the correct one, relative to the probability that it is incorrect or random, as $\frac{P(\mathcal{C}|E_k)}{P(\mathcal{R}|E_k)}$, where $\mathcal{C}$ represents that the conformation is correct, and $\mathcal{R}$ represents that the conformation is incorrect or random. Bayes' theorem allows us to express this probability as the product of $P(E_k \,|\, \mathcal{C})$, the conditional probability that the correct conformation has energy E_k, times $P(\mathcal{C})$, the probability of any particular conformation being correct, divided by $P(E_k \,|\, \mathcal{R})$, the conditional probability that a random conformation has energy E_k, times $P(\mathcal{R})$, the probability of any conformation being incorrect.

$$\frac{P(\mathcal{C}\,|\,E_k)}{P(\mathcal{R}\,|\,E_k)} = \frac{P(E_k\,|\,\mathcal{C})}{P(E_k\,|\,\mathcal{R})}\frac{P(\mathcal{C})}{P(\mathcal{R})} \tag{6}$$

If we approximate the distribution of energies of the random conformations as a Gaussian, centered at E_r, with standard deviation Γ_r. and likewise approximate the distribution of energies of correct conformations also as a Gaussian centered at energy E_c, with a width Γ_c dividing the probability of a correct conformation having energy E_k by the corresponding quantity for random conformations yields

$$\frac{P(E_k\,|\,\mathcal{C})}{P(E_k\,|\,\mathcal{R})} \propto \exp\left\{ \frac{\Gamma_c^2 - \Gamma_r^2}{2\Gamma_r^2\Gamma_c^2}(E_k - E_c)^2 - \frac{E_r - E_c}{\Gamma_r^2}E_k \right\} \tag{7}$$

We are interested in evaluating probabilities in the energy regime characteristic of the correct conformations. We can then neglect the argument of the exponential quadratic in $E_k - E_c$, yielding

$$\frac{P(\mathcal{C}\,|\,E_k)}{P(\mathcal{R}\,|\,E_k)} = \mathcal{K}e^{-E_k/T} \tag{8}$$

where

$$\mathcal{K} = \frac{\Gamma_r}{\Gamma_c}\frac{P(\mathcal{C})}{P(\mathcal{R})}e^{(E_c+E_r)/2T} \tag{9}$$

and

$$T = \frac{\Gamma_r^2}{E_r - E_c} \tag{10}$$

This indicates that the probability of correctness of any conformation, relative to the probability that the conformation is incorrect, is given by a Boltzmann function of its energy, at a temperature T given by Eq. (10).

It is especially interesting to note that, as long as the ratio of Γ_r to Γ_c is approximately constant, the probability that the structure of lowest energy is the correct one is a monotonically increasing function of $(E_r - E_c)/\Gamma_r$. The optimal Hamiltonian for this Bayesian analysis corresponds exactly to the optimal Hamiltonian developed using ideas derived from the theory of spin-glasses discussed above [7–9], and can be computed using Eq. (5).

There is a connection between the optimization procedure based on spin-glass theory and the Bayesian statistical approach and the potential of mean force approach of, for instance, Sippl [4, 15]. With his approach, the energy contributions of particular interactions, the adjustable parameters γ_i^{mf}, are given by:

$$\gamma_i^{\mathrm{mf}} = -T \log \left(\frac{\xi_{Ti}^{\mu}}{\left\langle \xi_{\zeta i}^{\mu} \right\rangle_\zeta} \right) \tag{11}$$

This assumes that there are no correlations between different interactions. Making the same assumption in the optimization approach, and also assuming no correlation exists between the interactions in the liquid-like states, allows us to write B as a diagonal matrix, simplifying Eq. (5) to:

$$\gamma_i = \frac{\xi_{Ti}^{\mu} - \left\langle \xi_{\zeta i}^{\mu} \right\rangle_\zeta}{\left\langle (\xi_{\zeta i}^{\mu})^2 \right\rangle_\zeta - \left\langle \xi_{\zeta i}^{\mu} \right\rangle_\zeta^2} \tag{12}$$

Assuming a Poisson distribution of interactions of each type, $\left(\left\langle (\xi_{\zeta i}^{\mu})^2 \right\rangle_\zeta - \left\langle \xi_{\zeta i}^{\mu} \right\rangle_\zeta^2 \right)$ is proportional to $\left\langle \xi_{\zeta i}^{\mu} \right\rangle_\zeta$, leading to

$$\gamma_i \propto \frac{\xi_{Ti}^{\mu}}{\left\langle \xi_{\zeta i}^{\mu} \right\rangle_\zeta} - 1 \tag{13}$$

which is proportional to γ_i^{mf} in Eq. (11) in the limit of $\xi_{Ti}^{\mu}/\left\langle \xi_{\zeta i}^{\mu} \right\rangle_\zeta$ close to 1. So the spin-glass picture and the Bayesian statistical picture reduce to the potential of mean force result in the limit of small, uncorrelated interactions. This can be considered a justification of the potential of mean force, as well as pointing to one of its limitations, its inability to deal with correlated interactions.

Let us now consider the harder problem, of considering the set of possible alignments of the target protein to a set of possible scaffolds of known structure. We are interested, then, in $P(\mathcal{C} \mid E_k, \{G\}_k)$ the conditional probability that a particular alignment, represented by the choice of a structural motif and a set of insertions and deletions, notated by $\{G\}_k$, with energy E_k, is correct, relative to the conditional probability that it is incorrect or random, $P(\mathcal{R} \mid E_k, \{G\}_k)$. We can again use Bayes' theorem, and write this ratio as:

$$\frac{P(\mathcal{C} \mid E_k, \{G\}_k)}{P(\mathcal{R} \mid E_k, \{G\}_k)} = \frac{P(E_k \mid \{G\}_k, \mathcal{C})}{P(E_k \mid \{G\}_k, \mathcal{R})} \frac{P(\{G\}_k \mid \mathcal{C})}{P(\{G\}_k \mid \mathcal{R})} \frac{P(\mathcal{C})}{P(\mathcal{R})} \tag{14}$$

where $P(E_k \mid \{G\}_k, \mathcal{C})$ and $P(E_k \mid \{G\}_k, \mathcal{R})$ are the conditional probabilities that correct or random alignments with gaps $\{G\}_k$ will have energy E_k, respectively, and $P(\{G\}_k \mid \mathcal{C})$ and $P(\{G\}_k \mid \mathcal{R})$ are the respective conditional probabilities that correct or incorrect alignments will have gaps $\{G\}_k$.

We can categorize insertions and deletions by any parameters we consider relevant, such as the number of residues and the distance spanned in in insertion, or the distance spanned by a deletion. Assuming that the insertions and deletions are uncorrelated, we can calculate the probability of a given set of gaps by considering the probability of a particular set of gaps times the number of different ways of arranging the gaps in the alignment:

$$P(\{G\}|\mathcal{C}) = \frac{(N - \sum_l \lambda_l n_l)!}{(N - \sum_l (\lambda_l + 1)n_l)! \prod_l n_l!} \, p_0^{(N - \sum_m (\lambda_m + 1)n_m)} \prod_l p_l^{n_l} \tag{15}$$

where p_l is the probability of a gap of type l occurring at any location in the sequence in a correct alignment, λ_l is the number of residues in the inserted segment, and $p_0 = 1 - \sum_l p_l$ is the probability of no gap occurring. We can write the corresponding expression for random conformations,

$$P(\{G\}|\mathcal{R}) = \frac{(N - \sum_l \lambda_l n_l)!}{(N - \sum_l (\lambda_l + 1)n_l)! \prod_l n_l!} \, r_0^{(N - \sum_m (k_m + 1)n_m)} \prod_l r_l^{n_l} \tag{16}$$

where r_l is the probability of a gap of type l beginning at any location in the sequence in an alignment between random pairs of proteins, and $r_0 = 1 - \sum_l r_l$. Substitution yields

$$\frac{P(\{G\}_k|\mathcal{C})}{P(\{G\}_k|\mathcal{R})} = \left[\frac{p_0}{r_0}\right]^N \prod_l \left[\frac{p_l r_0^{\lambda_l+1}}{p_0^{\lambda_l+1} r_l}\right]^{n_l} \tag{17}$$

We again calculate $\frac{P(E_k|\{G\}_k,\mathcal{C})}{P(E_k|\{G\}_k,\mathcal{R})}$, by approximating the distribution of energies of the random conformations as a Gaussian, centered at $E = E_r(\{G\})$, with standard deviation $\Gamma_r(\{G\})$, with the distribution of energies of the correct conformation of an ensemble of target proteins as a similar Gaussian, centered at some energy $E_c(\{G\})$, with a width $\Gamma_c(\{G\})$. Both the center and width of these distributions can be a function of the number of gaps. Continuing to assume that we are interested in evaluating these quantities in the neighborhood of E_r so we can neglect terms quadratic in $E_k - E_r$, an analysis similar to that preceding Eq. (8) yields

$$\frac{P(\mathcal{C}|E_k,\{G\}_k)}{P(\mathcal{R}|E_k,\{G\}_k)} = \mathcal{K}' e^{-F(E_k,\{G\}_k)/T} \tag{18}$$

where

$$F(E_k,\{G\}_k) = E_k + \sum_l \mu_l n_l \tag{19}$$

$$\mu_l = \left(\frac{\Gamma_r^2}{E_r - E_c}\right) \log \left[\frac{r_l p_0^{\lambda_l+1}}{r_0^{\lambda_l+1} p_l}\right] \tag{20}$$

and

$$\mathcal{K}' = \frac{\Gamma_r}{\Gamma_c} \left[\frac{p_0}{r_0}\right]^N \frac{P(\mathcal{C})}{P(\mathcal{R})} e^{(E_c + E_r)/2T} \tag{21}$$

where T is still given by Eq. (10). Once again the probability that a conformation is correct is a simple Boltzmann-like function of the energy, where we now include a chemical potential μ_l for an insertion of a gap of type l.

As mentioned above, E_c, E_r, Γ_c, and Γ_r are dependent on the gap structure. We are interested, however, in again evaluating these parameters in the neighborhood of correct alignments. We are specifically interested in the tail of the distribution of random conformations where it most likely overlaps the distribution of correct conformations. The best approximation to this neighborhood is to consider the minima-energy alignments between random pairs of proteins. We approximate that for these minima, the number of gaps and the energy distributions are uncorrelated.

Since the concept of minima-energy alignments imply some energy function, yet the alignments are needed to calculate this energy function, this approach naturally leads into a notion of self-consistency: the free energy function used to find the minima should be the same as that calculated using those minima. In general, an iterative approach can be used, where random pairs of proteins are aligned in order to generate the next approximation to the Bayesian gap penalty. Sequence alignments can then be done using a mean-field alignment procedure described below.

3   Methods and results

A Hamiltonian based on local interactions was used:

$$\mathcal{H} = \sum_i \gamma_p(A_i, C_i) - \langle \gamma_p(A_i, C_i) \rangle_\zeta + \sum_\rho \sum_{i<j} \gamma_c^\rho(A_i, A_j) \left\{ u(r_c^\rho - r_{ij}) - \langle u(r_c^\rho - r_{ij}) \rangle_\zeta \right\}$$

(22)

where $\gamma_p(A_i, C_i)$ is the energy contribution for amino acid A_i to be in context C_i, where context is described as the combination of secondary structure and surface accessibility, and $\gamma_c^\rho(A_i, A_j)$ is the energy term for amino acids A_i and A_j to be in contact, where contact is defined by the C^β distance between the residues being closer than distance r_c^ρ, equal to 5 Å for $\rho = 1$ and 12 Å for $\rho = 2$. $\langle \gamma_p(A_i, C_i) \rangle_\zeta$ and $\langle u(r_c^\rho - r_{ij}) \rangle_\zeta$ represent the average of $\gamma_p(A_i, C_i)$ and $u(r_c^\rho - r_{ij})$ respectively over a set of random conformations $\{\zeta\}$.

This Hamiltonian was optimized using Eq. (5) [7–9] for a set of 42 example proteins selected from the PDB data base [16, 17]. To prevent excessive stabilization of structures with multiple cysteines the interaction between cysteines for $r_c^1 = 5$ Å was set to zero.

Gaps were divided into three different categories: insertions ($j - i > 1$ with $r_{i'j'} = 3.8$ Å, the distance between successive C_α atoms), deletions ($j - i = 1$ with $r_{i'j'} > 3.8$ Å), and more generalized gaps ($j - i > 1$ and $r_{i'j'} > 3.8$ Å). These three categories were treated independently, with simple functional forms used to describe their respective statistics. Insertions were modeled as having a certain probability of being initiated, combined with a different probability of being lengthened. The distance spanned by a deletion was modeled as a Gaussian distribution centered at some distance d. The length of inserted regions for more generalized gaps was also modeled with a particular probability to initiate the gap and another to continue it, with the end-to-end distance modeled as a generalization of the end-to-end distance of a freely-jointed chain of $j - i$ links. Gaps were not allowed in the middle of α-helices or β-sheet strands. Unphysical gaps, where $r_{i'j'}$ exceeded $3.8(j - i)$ Å by more than 5 Å when $j - i > 1$, or exceeded 7.5 Å in the case of $j - i = 1$, were also prohibited.

Statistics based on the alignment of 182 pairs of homologous proteins were used to calculate the statistical descriptors for correct alignments. In order to calculate the corresponding descriptors for random alignments, 382 pairs of unrelated proteins were aligned, using the optimized local-interaction Hamiltonian and an estimated set of gap penalties. These values were used to calculate the next value for the gap penalties, using successive overrelaxation.

The alignment of the sequence to the set of possible structures was done using the approach developed by Finkelstein and Riva [12], later used by Skolnick and co-workers [3, 13], based on the Smith and Waterman dynamic programming algorithm [18]. An

Target	Self-recognition			Prediction				
	F	q	RMS	Scaffold	F	q	RMS	% I_h
1ALC	10.34	1.00	0.00	2LYZ	9.83	0.68	1.98	37
1FX1	13.53	1.00	0.00	3FXN	9.61	0.51	2.99	31
3DFR	10.70	0.95	0.73	8DFR	9.49	0.45	4.67	31
1MBA	11.07	1.00	0.00	5MBN	9.04	0.62	2.91	28
2LHB	12.18	1.00	0.00	5MBN	8.84	0.52	4.50	26
2LH4	9.84	1.00	0.00	5MBN	8.61	0.41	5.74	19
1FDH(G)	9.71	1.00	0.00	5MBN	8.12	0.65	2.44	23
3HFM(L)	9.78	0.85	1.56	3HFM(H)	8.09	0.27	8.54	26
4FAB(L)	11.01	0.83	2.62	3HFM(H)	7.93	0.30	5.89	21
4DFR(B)	11.48	0.95	0.66	8DFR	7.36	0.58	3.68	34
1HDS(B)	8.70	1.00	0.00	5MBN	7.20	0.72	1.82	25
1F19(L)	10.46	0.75	3.64	3HFM(H)	7.12	0.32	5.94	24
2RHE	7.50	1.00	0.00	3HFM(H)	6.86	0.57	3.62	29
2HHB(A)	7.72	1.00	0.00	5MBN	6.48	0.64	3.26	26
5CYT(R)	7.52	1.00	0.00	1RBB(A)	5.94	0.17	15.06	40
1REI(A)	8.63	1.00	0.00	1CD4	5.63	0.30	5.65	30

Table 1: Results of aligning 16 test proteins to a set of scaffold proteins including itself and the 42 example proteins used to optimize the Hamiltonian. The proteins are listed using their Protein Data Base designations, with subunits identified in parenthesis. In all cases, the alignment with lowest free-energy was self-alignment. In 15 of the test cases, the alignment with the next lowest free energy, was with a structural homolog, the one failure being tuna cytochrome C (5CYT), a protein largely stabilized by interactions with a heme protein not included in the current Hamiltonian. For both self-recognition and prediction, listed for the alignments are the value of the free energy F, the q-score for the alignment, defined as the proportion of the pairwise distances which are correct within some tolerance [7], and the RMS deviation of the structure form the structure determined by X-ray crystallography. The large RMS deviation for the prediction of 3HFM(L) represents a change in the orientation of the two domains. The RMS deviations for the two halfs of the protein considered independently were 5.79 and 3.82 Å respectively. Also listed is % I_h, the percentage identity of the target protein with the most similar example protein.

initial alignment is made where each residue interacts with the other residues of the scaffold protein. These residues are then replaced by the residues of the target protein aligned to that respective position, and the process is repeated in an iterative fashion.

As shown in Table 1, neglecting self-alignments, a correct assignment was made to a structural homolog in 15 out of 16 of the test cases, even in cases that often fail using standard sequence alignment methods (Leghemoglobin with Sperm whale myoglobin and Bence-Jones protein with a CD-4 fragment). The alignments were generally quite accurate. The one failure involved the prediction of the structure of Tuna Cytochrome C (5CYT). As no account is taken of the interactions of the protein with the heme prosthetic group, this suggests that including prosthetic groups explicitly in this model might be necessary for small proteins where interactions with these groups furnish much of the stabilization energy.

16 test proteins, each structurally similar to one of the scaffold proteins, but with limited sequence similarity (17 to 40 % sequence identity) were selected to test this method. As this method assumes that a structurally-similar protein is available in the set of scaffold proteins of known structure, it is therefore appropriate to test this method on a set of test proteins structurally similar to proteins in the example set. It would be interesting to consider optimizing the Hamiltonian and the gap penalties for specific scaffold proteins.

The test proteins were aligned to a set of possible scaffold proteins including the 42 proteins used to optimize the Hamiltonian plus itself. In all of the test cases, the structure with the lowest free energy was the alignment to itself, demonstrating correct self-recognition. There were at most minor shifts in the alignment.

4   Conclusion

Sequence-structure alignment algorithms have been hampered by the determination of an appropriate cost function, and especially an appropriate penalty function for insertions and deletions. In this paper, we have discussed a way to use a Bayesian analysis to generate an energy function for sequence-structure alignment, including the optimal gap penalties which enter as a chemical potential. In contrast to previous work, where these gap penalties have been determined in an *ad-hoc* manner [19, 20], or by looking at the statistical distribution of gaps in correct alignments only [21, 22], we find that the optimal choice involves the statistical distribution of gaps in alignments of random pairs of proteins. These results can be applied to other forms of alignment methodologies.

Acknowledgments

This work was supported by grant NIH 1 RO1 GM44557 to P. G. Wolynes. Computational support was provided by the National Center for Supercomputing Applications and the FMC Corporation. Helpful assistance by David Evensky is gratefully acknowledged. The application of this methodology to the prediction of the hormone binding sites of the nuclear receptor proteins was done in collaboration with John Katzenellenbogen and Donald Seielstad.

References

[1] J. U. Bowie, R. Luthy and D. Eisenberg, *Science* , **25**, 164–170 (1991).

[2] D. T. Jones, W. R. Taylor and J. M. Thornton, *Nature (London)*, **358**, 86–89 (1992).

[3] A. Godzik, A. Kolinski and J. Skolnick, *J. Mol. Biol.*, **227**, 227–238 (1992).

[4] M. J. Sippl and S. Weitckus, *Proteins*, **13**, 258–271 (1992).

[5] S. H. Bryant and C. E. Lawrence, *Proteins: Struct., Funct., and Genetics*, **16**, 92–112 (1993).

[6] C. Ouzounis, C. Sander, M. Scharf and R. Schneider, *J. Mol. Biol.*, **232**, 805–825 (1993).

[7] R. A. Goldstein, Z. A. Luthey-Schulten and P. G. Wolynes, *Proc. Nat. Acad. Sci., U.S.A.*, **89**, 4918–4922 (1992a).

[8] R. A. Goldstein, Z. A. Luthey-Schulten and P. G. Wolynes, *Proc. Nat. Acad. Sci., U.S.A.*, **89**, 9029–9033 (1992b).

[9] R. A. Goldstein, Z. A. Luthey-Schulten and P. G. Wolynes, In: *Proc. 26th Annual Hawaii International Conference on System Sciences*, (T. N. Mudge, V. Milutinovic, and L. Hunter, eds.), **1**, 699–707, IEEE Computer Society Press (1993a).

[10] R. A. Goldstein, Z. A. Luthey-Schulten and P. G. Wolynes, *In: Proc. 27th Annual Hawaii International Conference on System Sciences*, IEEE Computer Society Press (1994). In the press.

[11] N. Gō, *Ann. Rev. Biophys. Bioeng.*, **12**, 183–210 (1983).

[12] A. V. Finkelstein and B. Reva, *Nature (London)*, **351**, 497–499 (1991).

[13] A. Godzik and J. Skolnick, *Proc. Nat. Acad. Sci., U.S.A.*, **89**, 12098–12102 (1992).

[14] R. A. Goldstein, J. Katzenellenbogen, Z. A. Luthey-Schulten, D. Seielstad and P. G. Wolynes, *Proc. Nat. Acad. Sci., U.S.A.*, **90**, 9949–9953 (1993b).

[15] M. J. Sippl, *J. Mol. Biol.*, **213**, 859–883 (1990).

[16] F. C. Bernstein, T. F. Koetzle, G. J. B. Williams, E. F. Meyer, Jr., M. D. Brice, J. R. Rodgers, O. Kennard, T. Shimanouchi and M. Tasumi, *J. Mol. Biol.*, **112**, 535–542 (1977).

[17] E. E. Abola, F. C. Bernstein, S. H. Bryant, T. F. Koetzle and J. Weng, *In: Crystallographic Databases—Information Content, Software Systems, Scientific Applications*, (F. H. Allen, G. Bergerhoff and R. Sievers, eds) pp. 107–132. Data Commission of the International Union of Crystallography, Bonn, (1987).

[18] T. F. Smith and M. S. Waterman, *J. Mol. Biol.*, **147**, 195–197 (1981).

[19] W. M. Fitch and T. F. Smith, *Proc. Nat. Acad. Sci., U.S.A.*, **80**, 1382–1368 (1983).

[20] D. F. Feng, M. S. Johnson and R. F. Doolittle, *J. Mol. Evol.*, **21**, 112–125 (1985).

[21] E. J. Demchuk, G. N. Esipova and V. G. Tumanyan, *Studia Biophysica*, **2–3**, 193–199 (1989).

[22] S. A. Benner, M. A. Cohen and G. H. Gonnet, *J. Mol. Biol.*, **229**, 1065–1082 (1993).

Prediction of 3D Structures of Globular Proteins based on Self–consistent Molecular Field Theory

Alexei V. Finkelstein, Rumen A. Dimitrov, Azat Ya. Badretdinov and Boris A. Reva[*]

Institute of Protein Research and [*] Institute of Mathematical Problems of Biology, Russian Academy of Sciences, Pushchino, Moscow Region, Russian Federation

Abstract

A novel approach to protein 3D structure prediction is proposed. A strategy of this approach can be formulated as follows: (1) A set of possible folding patterns is generated. (2) Protein chain is "inscribed" into each of these patterns, and its free energy is minimized by threading using molecular field theory to describe the long-range interactions and one-dimensional statistical mechanics to find out the chain pathway in the field. (3) Free energy minimum indicates the most favorable folding pattern and the stable fold within it. The application of this approach to β-domains is encouraging: the similarity of calculated and observed structures is rather close.

1 Introduction

Determination of protein structure from its amino acid sequence is a long-standing objective of molecular biophysics. The key point is a prediction of protein fold which gives in outline three major aspects of the protein architecture: the secondary structures of which it is composed, their relative arrangement in space, and the path taken through the structure by the polypeptide chain [1, 2, 3].

The real difficulty for protein fold prediction is caused by the interactions between many remote chain regions [4]. When these interactions are absent (as in Zimm-Bragg's model [5]), strict and fast search for the stable structure is possible regardless of the multitude of competing folds. This search is based on matrix formalism of statistical mechanics of one-dimensional systems [6] which is very similar to dynamic programming [7] in essence [8]. The problems caused by multiparticle interactions are common in statistical physics, and are usually overcome using a molecular field method [9]. The main idea of this approach is to replace (approximate) the actual forces acting at a particle from the side of other ones by a "molecular field" (Fig.1) acting at this particle (= chain link in our case). The field can be different for different links and different points of three-dimensional space.

In this study molecular field theory is used to find the stable protein fold. Practical applicability of this approach is based on that (i) chain pathway in any 3D force field can be strictly and easily calculated [4, 8], (ii) the main features of a molecular field acting at the chain are determined by protein folding pattern [2, 4], and (iii) a set of potentially stable folding patterns (i.e. of the patterns which only can be stable for a vast majority of

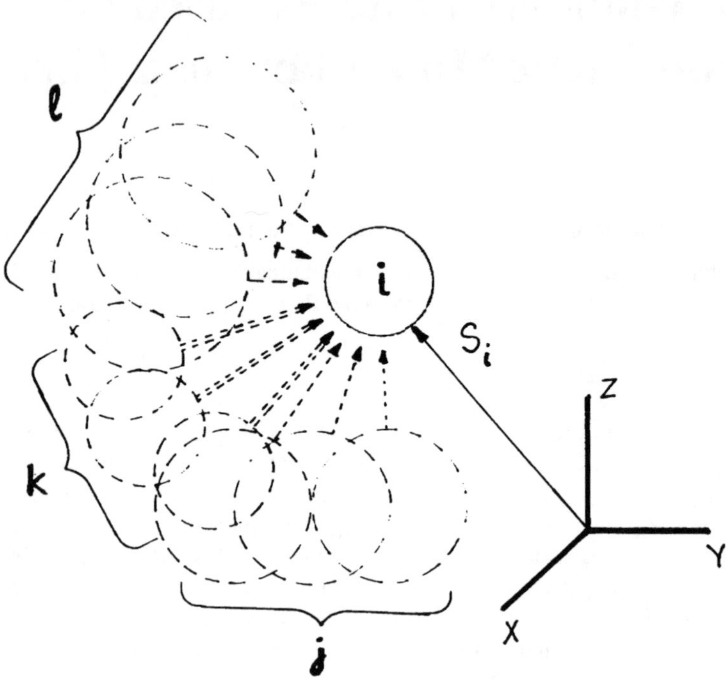

Figure 1: Mean force field acting at the link "i" in the given point of 3D space is a result of averaging of the momentary force field acting at this link over the fluctuations of the other ones.

sequences [10]) is rather limited [2] and can be obtained without reference to a concrete amino acid sequence. A strategy of our approach [2] can be formulated as follows: (1) A set of the "potentially stable" folding patterns is generated. In practice, it is enough to generate that part of this set which corresponds to the given chain length, hydrophobicity, and the expected secondary structure type. (2) Protein chain is inscribed into (or treaded through, as people say now) each of these patterns; the free energy is minimized using molecular field theory to describe the long-range interactions and one-dimensional statistical mechanics to calculate the chain pathway in the field. Thus one obtains the most stable chain fold corresponding to the given folding pattern. (3) Free energies of all these stable folds are compared. The minimal free energy indicates the most stable of them.

2 Molecular field theory for chain molecules

2.1 Energy function and the chain free energy functional

We consider a chain of N links assuming that its energy (including free energy of chain-solvent interactions) has a form

$$E(S_1, \ldots, S_N) = \sum_{i=1}^{N} h_i(S_i) + \sum_{i=2}^{N} u_{i-1,i}(S_{i-1}, S_i) + \frac{1}{2} \sum_{i=1, i \neq j}^{N} \sum_{j=1}^{N} \varepsilon_{ij}(S_i, S_j) \qquad (1)$$

Here h, u and ε are the elementary energy functions which depend on the links (i.e. on their chemical sorts) and on their coordinates S : $h_i(S_i)$ describes interactions within link "i"; $u_{i-1,i}(S_{i-1}, S_i)$ describes interactions between neighbor links, in particular their covalent bonding; $\varepsilon_{ij}(S_i, S_j) \equiv \varepsilon_{ji}(S_j, S_i)$ describes non-covalent interactions between links "i" and "j".

The problem is to find out link coordinates corresponding to thermodynamic equilibrium.

We look for the equilibrium state rather than for energy minimum (though they coincide when protein structure is stable) because, when the energy functions are known approximately (which is a real case!), the search for the equilibrium structure held at room temperature (strictly [11]: at a temperature just below the temperature of protein melting) gives a more robust result than energy minimization [11].

Thermodynamic equilibrium corresponds to free energy minimum or, in other words, to the minimum of free energy functional [9]

$$F\{W\} = \sum_{S_1} \ldots \sum_{S_N} [RT \cdot \ln W(S_1, \ldots, S_N) + E(S_1, \ldots, S_N)] \cdot W(S_1, \ldots, S_N) \qquad (2)$$

over all the probability functions W which obey the normalization

$$W(S_1, \ldots, S_N) \geq 0 \text{ for any values of } S_1, \ldots, S_N; \qquad (3)$$

$$\sum_{S_1} \ldots \sum_{S_N} W(S_1, \ldots, S_N) = 1$$

$W(S_1, \ldots, S_N)$ is probability of the fold with link "1" in position $S_1, \ldots$, link "N" in position S_N; T is the temperature, R the gas constant.

A strict result of free energy minimization is very well known [9]: there is only one equilibrium state; its free energy is

$$F = -RT \cdot \ln[\sum_{S_1} \ldots \sum_{S_N} \exp\left(-E(S_1, \ldots, S_N)/RT\right)]; \qquad (4)$$

the equilibrium value of the coordinate $S_i(i = 1, \ldots, N)$ is

$$< S_i > = \sum_{S_i} S_i W_i(S_i) \qquad (5)$$

where

$$W_i(S_i) = \sum_{S_i} \ldots \sum_{S_{i-1}} \sum_{S_{i+1}} \ldots \sum_{S_N} \exp((F - E(S_i, \ldots, S_N))/RT) \qquad (6)$$

is probability to find link "i" in position S_i (the last sum is taken over all the coordinates but for S_i).

The above solution is easy to write, but the values of F and W_i's cannot be computed in practice, unless the terms ε_{ij} in Eq.(1) are absent, which is not a case for protein chains.

2.2 Molecular field approximation

The advantage of Eq.(2) is that it is applicable to *any* probability function; this allows to find an approximate solution of free energy minimization problem. This solution is accurate when one may neglect a contribution of correlated link fluctuations to the energy of long-range interactions [9], i.e. as far as

$$\sum_{S_i} \sum_{S_j} \varepsilon_{ij}(S_i, S_j) \cdot W_{ij}(S_i, S_j) \approx \sum_{S_i} \sum_{S_j} \varepsilon_{ij}(S_i, S_j) \cdot W_i(S_i) \cdot W_j(S_j) \tag{7}$$

Here

$$W_{ij}(S_i, S_j) \equiv \sum_{S_1} \cdots \sum_{S_{i-1}} \sum_{S_{i+1}} \cdots \sum_{S_{j-1}} \sum_{S_{j+1}} \cdots \sum_{S_N} W(S_1, \ldots, S_N)$$

(the sum is taken over all the coordinates but for S_i and S_j) is a probability of that link "i" is in position S_i and, simultaneously, link "j" is in S_j) , while $W_i(S_i)$, $W_j(S_j)$ are the probabilities (see Eq.(6)) of that given link is in given position, irrespectively of position of another link.

It is noteworthy that, strictly speaking, Eq.(7) must hold only for that probability function which corresponds to thermodynamic equilibrium. This condition is valid for non-covalent interactions in native proteins (because a range of thermal fluctuations of their atoms is 0.3–0.5 Å [12], smaller than the potential wells of non-covalent forces which are ~ 1Å wide [13]), but never for covalent bonds (their potential wells are less than 0.1Å wide [13]). Thus, covalent bonding of links (terms u in Eq.(1)) must be considered explicitly [14], while the non-covalent interactions (terms ε in Eq.(1)) can be considered under a molecular field approximation.

2.3 Minimization of free energy functional

Under this approximation (see Eqs.(1),(2),(6),(7)), the free energy functional can be represented as

$$F\{W\} = \sum_{S_1} \cdots \sum_{S_N} [RT \ln W + \sum_{i=1}^{N} h_i + \sum_{i=2}^{N} u_{i-1,i}]W + \frac{1}{2} \sum_{i=1,i\neq j}^{N} \sum_{j=1}^{N} \sum_{S_i} \sum_{S_j} \varepsilon_{ij} W_i W_j \tag{8}$$

To find a minimum of $F\{W\}$, one has to consider a variation of $F\{W\}$ with W. Let $W \to W' = W + \delta W$, and both W and W' obey the normalization given by Eq.(3), i.e. $\sum_{S_1} \cdots \sum_{S_N} \delta W(S_1, \ldots, S_N) = 0$. Then

$$\delta F\{W, \delta W\} \cong \sum_{S_1} \cdots \sum_{S_N} \delta W[RT \cdot \ln W + \sum_{i=1}^{N} h_i + \sum_{i=2}^{N} u_{i-1,i} + \sum_{i=1}^{N} \psi_i^W] \tag{9}$$

where

$$\psi_i^W(S_i) = \sum_{j=1_{j\neq i}} \sum_{S_j} \varepsilon_{ij}(S_i, S_j) \cdot W_j(S_j) \tag{10}$$

is the molecular field potential (Fig.1) acting at link "i" in point S_i from the side of the other links when the distribution of these links j ($j \neq i$) is given by the probability functions $W_j(S_j)$.

Molecular field potentials approximate the action of long-range interactions. They help (cf. [9]) to split the otherwise very complicated search for a stable state of a multi-particle system into two easier ones: 1) calculation of a field from the chain state, and 2) calculation of a chain state from the field.

Suppose that a probability function $\tilde{W}(S_1, \ldots, S_N)$ corresponds to a minimum of $F\{W\}$. Then $\delta F(\tilde{W}, \delta W)$ is zero for any small variation δW (if $\sum_{S_1} \ldots \sum_{S_N} \delta W = 0$, of course). This means that $\tilde{W}$ satisfies the equation

$$RT \cdot \ln \tilde{W}(S_1, \ldots, S_N) = \tilde{F}\Psi - \sum_{i=1}^{N}(\psi_i(S_i) + h_i(S_i)) + \sum_{i=2}^{N} u_{i-1,i}(S_{i-1}, S_i) \qquad (11)$$

where $\psi_1, ldots, \psi_N$ are some field potentials, and the $\tilde{F}_\Psi$ value corresponds to normalization of $\tilde{W}$, see Eq.(3). Thus, one can solve the free energy minimization problem taking into account only those probability functions $\tilde{W}^\Psi$ which have a form given by Eq.(11).

This form of $\tilde{W}^\Psi$ function corresponds to the equilibrium state of a chain having the energy $\tilde{E}^\Psi = \sum(\psi_i(S_i) + h_i(S_i)) + \sum u_{i-1,i}(S_{i-1}, S_i)$ which includes the same short-range interactions (terms h and u) as Eq.(1), but (explicitly) no long-range ones; the latter are replaced now by a force field $\Psi = \sum_i \psi_i(S_i)$. The absence of explicit interactions between remote chain links allows to find, *strictly* and *quickly* the value of $\tilde{F}^\Psi$ (chain free energy in the field Ψ), the probabilities $\tilde{W}_i^\Psi(S_i)$ of all the positions of all links of the chain in field Ψ, the minimum of $\tilde{E}^\Psi$, etc. (for the algorithms and mathematical details, see [8] and the references therein).

2.4 Iterative free energy minimization: search for a self–consistent solution

Now one has to find the potentials $\psi_i(S_i)$ of force field Ψ which minimize the free energy functional. It is easy to show that, for a probability function $\tilde{W}^\Psi$ given by Eq.(11), this functional (see Eq.(8)) has a form

$$F\{\tilde{W}^\Psi\} = \tilde{F}^\Psi + \frac{1}{2}\sum_{i=1}^{N}\sum_{S_i} \tilde{W}_i^\Psi \psi_i^{\tilde{W}^\Psi} - \sum_{i=1}^{N}\sum_{S_i} \tilde{W}_i^\Psi \psi_i \qquad (12)$$

(where the potentials $\psi_i^{\tilde{W}^\Psi}(S_i)$ are calculated from the probabilities $\tilde{W}^\Psi$ after Eq.(10)), and that the variation of free energy is

$$\delta F\{\tilde{W}^\Psi, \delta\psi_1, \ldots, \delta\psi_N\} \cong -\sum_{S_1}\ldots\sum_{S_N}\left[\tilde{W}^\Psi \cdot \sum_{i=1}^{N}(\psi_i^{\tilde{W}^\Psi} - \psi_i) \cdot \sum_{j=1}^{N}(\delta\psi_j - \sum_{S_j}\tilde{W}_j^\Psi \delta\psi_i)\right]/RT \qquad (13)$$

when the variations of potentials $\delta\psi$ are small.

When the chain is in equilibrium, the "new" potentials $\psi_i^{\tilde{W}^\Psi}(S_i)$ *formed* by the chain fold (i.e. determined by $\tilde{W}^\Psi$ functions) coincide with the "old" potentials $\psi_i(S_i)$ which *form* this fold. This means that the field in equilibrium is a self-consistent one [9].

When the difference between "new" and "old" potentials is not zero (or constant), a variation of "old" potentials

$$\delta\psi_i(S_i) = \lambda \cdot (\psi_i^{\tilde{W}^\Psi}(S_i) - \psi_i(S_i)) \ (i = 1, \ldots, N; \lambda > 0 \text{ and small}) \qquad (14)$$

decreases the $F\{\tilde{W}^\Psi\}$ value; it is easy to show that λ must be not greater than a unity, and we used $\lambda = 0.5$ in all our calculations.

Thus, an iterative modification of field potentials leads to free energy minimization and converges (when the main approximation of the theory, Eq.(7), is valid) to a self-consistent solution. The iteration begins with some "starting" field Ψ (i.g. $\Psi \equiv 0$) and continues until a self-consistent solution is obtained.

3   Folding patterns classify the self-consistent solutions

It is well known that, depending on starting conditions, molecular field method leads not only to the stable, but also to metastable states of a system [9]. Thus, one has to look through all the obtained self-consistent solutions and to pick up that of them which has the lowest free energy.

Fortunately, in a case of protein globules, one can classify a set of possible self-consistent solutions before any calculations.

The analysis of Eq.(10) shows that repulsive and H-bonding constituents of a molecular field produce *some compartment* for any secondary structure region, and this compartment is stretched along the chain. The same compartments are typical for protein folding patterns (see [1, 2, 3, 15]); thus, one can use these patterns for an *a priori* classification of self-consistent solutions.

In principle, the variety of folding patterns is great, but it is known that this variety is strongly limited by some structural rules (see [1, 3, 15, 16] which can be rationalized as prohibition of defects (such as overcrossed loops) which, for "random" amino acid sequences, increase the protein fold free energy [2, 10]. A set of "stable" patterns is very limited: there are, for example, only 60 such patterns among $\sim 10^5$ different ones corresponding to a sandwich with 4 antiparallel β-strands in each sheet [2].

Moreover, folding patterns determine the coarse features of a molecular field [2, 4]: they show which compartments are dipped in globule and which are exposed to water; which are involved in H-bond network and which are not (Fig.2).

3.1   Lattice model of protein globule

The folding patterns allow to make 3D lattice models of protein folds [4]. The lattices are usual counterparts of molecular field theories [9]: they bear the most rigid forces which are hardly approximated by Eq.(7). In proteins, lattice models (Fig.2) allow for excluded (due to steric repulsion) volumes and regular (due to H-bonding) arrangement of secondary structures [2].

Within a given folding pattern, a fold is determined by the coordinates of N- and C-ends of secondary structure regions. Thus, a fold which contains K secondary structures is described by 3D *lattice* of K bars, by *topology* giving the chain pathway and saying what half a bar contains a given end of the secondary structure segment (Fig.2), and by 2K *composite coordinates* $S_i = (b_i, s_i)$ of these ends (i=1,2;...;2K-1,2K) which give their positions on the bars (b_i) and in the sequence (s_i).

Therefore, it is convenient to search for a stable fold using the ends of K secondary structure segments as 2K "links" with coordinates $S_i = (b_i, s_i)$.

4   Chain fold energy

4.1   Energetic parameters

The present study is based on residue-to-residue interactions rather than on atom-to-atom ones; we take into account only the main long-range interactions (steric repulsion as excluded volume; H-bonds in backbone; hydrophobic forces) and ignore the side chain degrees of freedom. This means that we describe actually the molten globule [17, 18] rather than a "solid" native protein.

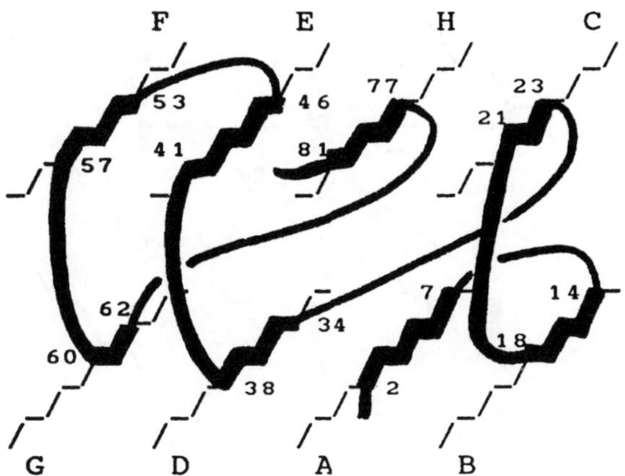

Figure 2: Three-dimensional lattice model for a β-sandwich domain of eight β-strands. The lattice allows for excluded volume of β-strands (bars show their possible room in space) and for hydrogen bonding between them. "In" and "out" pleating of the lattice corresponds to "in" and "out" positions of side chains. For simplicity, the β-sheets are not twisted. The first domain of γ-crystallin[22] is shown as a chain pathway (heavy line) through this lattice. The lattice model shows the native location of the β-strands in space; their location in the sequence is given by Arabic numerals. β-Strands of γ-crystallin are numbered by capitals. The loops are presented schematically. One can see that coarse features of molecular field arise from the folding pattern: the field is stronger for the residues of the inner β-strands (A,D,E,G) and weaker for the edge ones (B,C,F,H).

In this study we consider only β-proteins and, as in [4], take into account the following energetic terms:

1. H-bonds (and connected interactions) in β-sheets (ξ^H in Fig.3);

2. Inter-sheet hydrophobic interactions (ξ^{hp} in Fig.3)

3. Intrinsic energy of a β-strand residue: $\xi^{\text{ext}}(A) = F_\beta(A) + 2\Delta F_e(A)$, F_β being a free energy difference between interior of β-sheet and a coil, and ΔF_e an additional energy of residue A at a β-sheet edge.

4. Free energies of loops and their minimal lengths are estimated as in [4]. Initiation of any allowed loop costs 4 kcal/mol. The parameters of loop elongation and immersion (the last only for the internal loops, like C-D, D-E, G-H in Fig.2) are $\Delta u_0 = 0.2$ kcal/mol, $p_0 = 0.05$; they served as adjustable parameters.

The only sequence-dependent parameters of our computations are thermodynamic parameters of hydrophobicity, G_{hp}, and those of β-sheet stability, F_β and ΔF_e.

The G_{hp} values are taken directly from experiments on amino acid transfer from water to organic solvent [19].

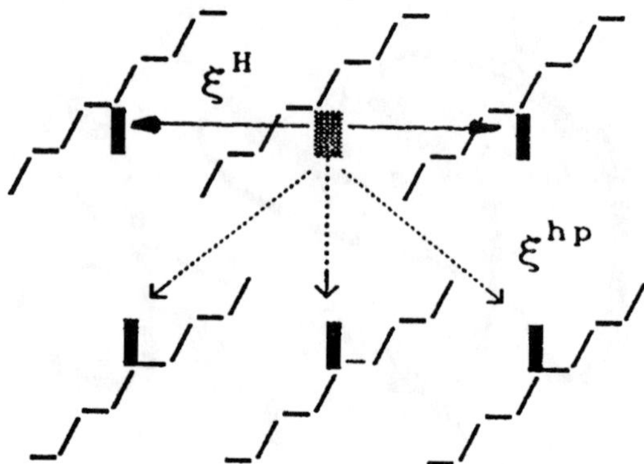

Figure 3: Long-range interactions within a β-sandwich. For one residue (striped), all the possible long-range H-bonds within the β-sheet (thin arrows; criterion: $C_\alpha - C_\alpha$ distance less than 6Å), and hydrophobic contacts with the other β-sheet (dotted arrows; criterion: $C_\beta - C_\beta$ distance less than 7Å) are indicated; side chains are shown only for the partner residues. The energy ξ of each interaction depends on its type (intra-sheet: $\xi = -\Delta F_e(A_p) - \Delta F_e(A_q)$; inter-sheet: $\xi^{hp} = \Delta G_{hp}(A_p) + \Delta G_{hp}(A_q)$) and on chemical sort ($A$) of the interacting residues "p" and "q". $\xi = 0$ when the partner residue is absent. $\Delta F_e(A)$ is an additional energy of residue A at a β-sheet edge [4]; $\Delta G_{hp}(A) = \frac{1}{6}(G_{hp}(A) - G_{hp}(Ala))$, $G_{hp}(A)$ being water $\rightarrow$ organic solvent transfer free energy for amino acid A [19].

The parameters F_β and ΔF_e are listed in [4]. For some residues they had been estimated from thermodynamics F_β and kinetics (ΔF_e) of β folding in polypeptides, and for a majority of amino acids they have been calculated from the conformational analysis [20] because the experimental date where insufficient and we did not want to base our thermodynamic theory on protein statistics.

At present, one can compare many of these calculated β-propensities with the recent [21] experimental data and see a good correlation between them — better than between any of them and protein statistics (A.Finkelstein, manuscript in preparation).

4.2   Calculation of molecular field potentials

Equation (10) gives a general algorithm to calculate molecular field potentials from the distribution of particles in space and the energies of interactions between them. When the interactions are connected with hydrophobic H-bonded contacts (Fig.3), it is easy to find the corresponding molecular field potential acting at chain residues (its chemical sort is A_s) in lattice point b:

$$\Delta\psi(b,s) = [-\Delta F_e(A_s) - <\Delta F_e>_b] <n_H>_b + [\Delta G_{hp}(A_s) + <\Delta G_{hp}>_b] <n_{hp}>_b$$

$$(15)$$

Table 1: **Choice of the most stable folding pattern for a given amino acid sequence.** Here each amino acid sequence has to choose between three β-structural folding patterns, one of which is the native pattern for this sequence. In the schemes of folding patterns, β-strands are numbered by capitals according to their positions in the chains (cf. Fig.2). The strands are shown by dots (when they come to the reader) and crosses (when they go from the reader). For each sequence and each pattern, we calculate and present $F_{\mathrm{P}}^{\mathrm{AA}}$ (kcal/mol), free energy of the amino acid sequence AA in the folding pattern P, and then the resulting thermodynamic probability (%) of the pattern P for the sequence AA: $W_{\mathrm{P}}^{\mathrm{AA}} = 100\%W \cdot \exp(-F_{\mathrm{P}}^{\mathrm{AA}}/RT)/\sum_{\mathrm{P}'=1}^{3} \exp(-F_{\mathrm{P}'}^{\mathrm{AA}}/RT)$.

Folding pattern

Amino acid sequence	FCHA $\cdot + \cdot +$ $+ \cdot + \cdot$ EDGB	$\left(\beta - \text{domain}\right)$ CAP,	FEHC $\cdot + \cdot +$ $+ \cdot + \cdot$ GDAB	$\left(\gamma\text{--CR. domain}\right)$	DAGH $\cdot + + \cdot$ $+ \cdot + \cdot$ CBEF	(Prealb.)	
CAP (β-domain)	-5.5	98.7%	-2.9	1.2%	-1.6	0.1%	= 100%
γ-CR.(domain 1)	-6.4	10.5%	-7.7	89%	-4.6	0.5%	= 100%
γ-CR.(domain 2)	-6.7	30%	-7.1	69%	-4.5	1%	= 100%
Prealbumin	-14.5	21%	-14.3	16%	-15.2	63%	= 100%

Here $< n_H >_b$ is the average number of neighbors which can form H-bonds with a residue occupying the point b; $< \Delta F_e >_b$ is their average β-edge energy; and $< n_{hp} >$, $< \Delta G_{hp} >$ are the same for the residues which can form inter-sheet hydrophobic contacts with a residue in the point b. These values are easily calculated from the distribution of secondary structure ends in space (at the lattice bars) and in the chain.

For any given coordinate of a "link" (= end of secondary structure segment) in the chain and lattice, one can find the potential acting at this "link" as a sum of corresponding $\Delta\psi$ values.

5 Results and discussion

First, it was checked whether we could find the positions of strands in space and sequence, provided the native folding pattern is given.

When the program starts to work, it "knows" (besides energetic parameters and sequence) only the folding pattern, which says that the chain first comes (somewhere) to the bar A (see Fig.2), goes along it in a definite direction, then (somewhere) leaves this bar, comes to the bar B, goes along it in the opposite direction, etc. The program has to find out the equilibrium positions of the secondary structure ends in chain and lattice, as well as the fluctuations of these positions.

The number of particular folds (i.e. of different positions of secondary structure ends)

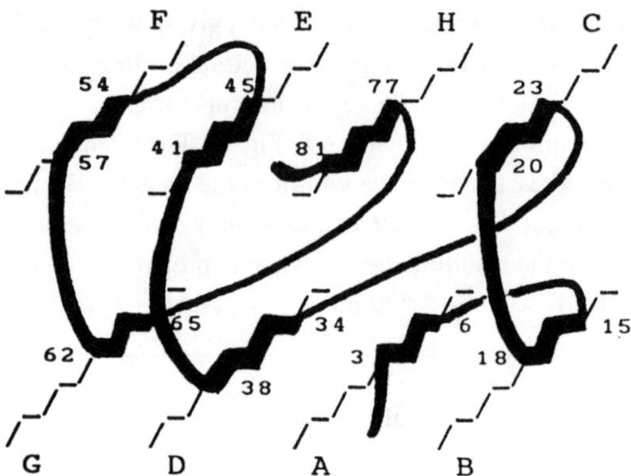

Figure 4: Lattice model of γ-crystallin fold computed from its amino acid sequence. The equilibrium location of β-strands in space and in the sequence is shown. The computed range of fluctuations (at room temperature) is about ± 1 lattice point for the positions of the β-strand ends in space, and ± 2 residues for their positions in the chain. These fluctuations are due partly to the movement of a strand as a whole but mainly to the independent vibrations of β-strand ends.

within a folding pattern is great: $\sim 10^{25}$ [4]. Nevertheless, the iterative search of a self-consistent solution within a pattern converges rather fast (~ 10 steps $=\sim 10$ minutes on IBM PC 286), and the solution does not depend on the starting field. This happens mainly because the molecular field is rather simple: it looks like a hydrophobic drop pierced by the lattice, so that only the shape of this drop is refined in course of iterations.

After some adjustment of the non-specific loop parameters Δu_0 and p_0 , the calculated structures of γ-crystallin (Fig.4) and of many other proteins became very close to the native ones. Substantial deviations were found mostly for the edge β-strands.

Second, we took several sequences forming different β-folding patterns and examined whether a sequence could choose its native one. Table 1 presents the results for both γ-crystallin domains [22], β-domain of catabolite gene activator protein (CAP) [23], and prealbumin [24]. It shows that a chain attains the lowest free energy in its own folding pattern and, though a free energy difference from the closest competitor is rather small, the dif- ference in probabilities of these patterns is sufficiently large. The results remain essentially the same when a usual twist of β-sheets is taken into account.

In conclusion, the search for the most stable fold was done among a large set of possible folding patterns. We took the chains with different antiparallel β-folds of 7 and 8 strands and, *assuming* that they can form only β-sandwiches of a similar kind, threaded each of the chains onto all 96 "potentially stable" (according to the known structural rules [2]) patterns of 7 and 8 antiparallel β-strands.

We obtained the following result: for each sequence with *short* loops the native folding pattern is among the few most stable ones calculated (Tab.2), despite the coarse parameters of the model, and nearly all other patterns are strictly rejected (their probability is much less than 1%). However, the native fold rating is much worse for the sequences with *long* loops. The last result is trivial: when the loops are long, we do not take into account a great part of interactions (actually, it means that we have to incorporate the long loops in our lattice

Table 2: **Calculated rating of native folding patterns.** The table includes 15 sequences of antiparallel β-proteins (domains) with comparatively short loops. The results for sequences with long and/or structured loops (i.g. for superoxide dismutase [30] and actinoxantin [31]) are discussed in the text. Each sequence has to choose between 96 possible folding patterns, one of which is the native one for theis sequence. The list of 96 possible folding patterns consists of different topologies of antiparallel aligned β-sandwiches. It includes 36 patterns for a sandwich of 7 β-strands, plus 60 patterns for a sandwich of 8 β-strands. These topologies satisfy to the following restrictions: (a) parallel arrangement of the strands adjacent along the chain is prohibited; (b) loops must not cross each other or screen one another from water; (c) left superhelices are prohibited. For explanations see [2, 3, 10].

Sequence	Rating of the native fold in the list of 96 possible ones
γ-Crystallins, 8 different domains [22, 25]	I-V
Catabolite activator protein, β-domain [23]	V
Satellite tobacco necrosis virus, β-domain [26]	I
Human pancreatic lipase, β-domain [27]	II
Immunoglobulin, 4 domains (CL,CH1,CH2,CH3) [28, 29]	I-V

model).

Thus, the present study verifies the results of our earlier paper [4]. A 3D fold is stable when it is "consistent" with the amino acid sequence: (1) the order of inner and edge β-strands matches the order of the more and the less hydrophobic regions in the chain; (2) strands which are close in space have hydrophobic surfaces similar in size; (3) the loops required by a folding pattern correspond to the lengths of hydrophilic chain regions.

Acknowledgements

We are grateful to O. B. Ptitsyn, E. I. Shakhnovich and A. M. Gutin for valuable discussions. This work was supported by Russian Foundation for Fundamental Researches (grant 93-04-6636).

References

[1] J. S. Richardson, The anatomy and taxonomy of protein structure, *Adv. Prot. Chem.*, **34**, 167–339 (1981).

[2] A. V. Finkelstein and O. B. Ptitsyn, Why do globular proteins fit the limited set of folding patterns?, *Progr. Biophys. and Mol. Biol.*, **50**, 171–190 (1987).

[3] C. Chothia and A. V. Finkelstein, The classification and origins of protein folding patterns, *Ann. Rev. Biochem.*, **59**, 1007–1039 (1990).

[4] A. V. Finkelstein and B. A. Reva, Search for the most stable folds of protein chains, *Nature*, **351**, 497–499 (1991).

[5] B. H. Zimm and J. R. Bragg, Theory of the phase transition between helix and random coil in polypeptide chains, *J. Chem. Phys.*, **31**, 526–535 (1959).

[6] T. M. Birstein and O. B. Ptitsyn, Conformations of Macromolecules, New York: Interscience Publ. (1966).

[7] A. V. Aho, J. E. Hopcroft and J. D. Ullman, The Design and Analysis of Computer Algorithms, Reading: Addison-Wesley Publ. Comp. (1976).

[8] A. V. Finkelstein and B. A. Reva, Search for a stable state of a short chain in a molecular field, *Protein Engineering*, **5**, 617–624 (1992).

[9] R. Kubo, Statistical Mechanics Amsterdam: North-Holland Publishing Company (1965).

[10] A. V. Finkelstein, A. M. Gutin and A. Ya. Badretdinov, Why are the same protein folds used to perform different functions?, *FEBS Letters*, **325**, 23–28 (1993).

[11] A. V. Finkelstein, A. V. Gutin and A. Ya Badretdinov, A good temperature for protein structure prediction and folding, Abstr. Meet. on Predict. and Recogn. Antigen. Determ., p. L7 (1992).

[12] J. Doucet and J. P. Benoit, Molecular Dynamics Studied by Analysis of X-Ray Diffuse Scattering from Lysozyme Crystal, *Nature*, **325**, 643–646 (1987).

[13] H. A. Scheraga, Calculation of Conformations of Polypeptides, *Adv. Phys. Org. Chem.*, **6**, 103–183 (1968).

[14] I. M. Lifshitz, A. Yu. Grosberg and A. R. Khokhlov, Some problems of the statistical physics of polymer chains with volume interaction, *Rev. Mod. Phys.*, **50**, 683–713 (1979).

[15] M. Levitt and C. Chothia, Structural patterns in globular proteins, *Nature*, **261**, 552–557 (1976).

[16] C. Branden and J. Tooze, Introduction to Protein Structure, Garland Publ., New York — London (1991).

[17] E. I. Shakhnovich and A. V. Finkelstein, Theory of cooperative transitions in protein molecules. I. Why denaturation of globular protein is the first order phase transition, *Biopolymers*, **28**, 1667–1680 (1989).

[18] O. B. Ptitsyn, The molten globule state, In: Protein Folding (T.E.Creighton, ed), Freeman & Co, New York, 243–300 (1992).

[19] I. I. Fauchere and V. Pliska, Hydrophobic parameters π of amino-acid side chains from partitioning of N-acetyl-amino- acid amides, *Eur. J. Med. Chem. Chim. Ther.*, **18**, 369–375 (1983).

[20] A. V. Finkelstein and O. B. Ptitsyn, Theory of Protein Molecule Self-Organization. I. Thermo-dynamic parameters of local secondary structures in the unfolded protein chains, *Biopolymers*, **16**, 469–495 (1977).

[21] A. Kim and J. M. Berg, Thermodynamic á-sheet propensities using a zinc-finger host protein, *Nature*, **362**, 267–270 (1993).

[22] G. Wistow, B. Turnell, L. Summers, C. Slingsby, D. Moss, L. Miller, P. Lindley and T. Blundell, X-ray analysis of eye lens protein g-crystallin at 1.9 A resolution, *J. Mol. Biol.*, **170**, 175–202 (1983).

[23] D. B. McKay, I. T. Weber and T. A. Steitz, Structure of catabolite gene activator protein at 2.9 A resolution, *J. Biol. Chem.*, **257**, 9518–9524 (1982).

[24] C. C. F. Blake, M. J. Geisow, S. J. Oatley, B. Rerat, and C. Rerat, Structure of prealbumin: secondary, tertiary and quaternary interactions determined by Fourier refinement at 1.8 A, *J. Mol. Biol.*, **121**, 339–356 (1978).

[25] Yu. N. Chirgadze, N. A. Nevskaya, Yu. V. Sergeev and N. P. Fomenkova, Evolutionary conservatism of molecular structure of vertebrate g-crystallins, *Mol. Biol. (USSR)*, **21**, 368–376 (1987)

[26] L. Lilias, T. Unge, A. Jones, K. Fridborg, S. Lovgren, U. Skoglund and B. Strandberg, Structure of Satellite tobacco necrosis virus at 3.0 A resolution, *J. Mol. Biol.*, **159**, 93–108 (1982).

[27] F. K. Winkler, A. D'Arcy and W. Hunziker, Structure of human pancreatic lipase, *Nature*, **343**, 771–774 (1990)

[28] F. A. Saul, L. M. Amzel and R. J. Poljak, Preliminary refinement and structural analysis of the Fab fragment from human immunoglobulin NEW at 2 A resolution, *J. Biol. Chem.*, **253**, 585–597 (1978).

[29] J. Deisenhofer, Crystallographic refinement and atomic models of a human Fc fragment and its complex with fragment B of protein A from Staphylococcus Aureus at 2.9 and 2.8 A resolution, *Biochemistry*, **20**, 2361–2370 (1981).

[30] J. A. Tainer, E. D. Getzoff, K. M. Beem, J. S. Richardson and D. C. Richardson, Determination and analysis of the 2 A structure of Cu, Zn superoxide dismutase, *J. Mol. Biol.*, **160**, 181–217 (1982).

[31] V. Z. Pletnev, A. P. Kuzin, S. D. Trakhanov and P. V. Kostetsky, Three-dimensional structure of actinoxantin. IV. A 2.5 A resolution, *Biopolymers*, **21**, 287–300 (1982).

A Potential Function that Identifies Correct Protein Folds

Gordon M. Crippen and Vladimir N. Maiorov

College of Pharmacy, University of Michigan, Ann Arbor, Michigan 48109, USA

Abstract

The thermodynamic approach to calculating protein folding has attracted many researchers over many years, but it has been hard to carry out for two main reasons. First one must devise a function of polypeptide conformation that resembles the free energy of a solvated polypeptide chain at least as far as favoring the native conformation over all others; and second, one must search over this conformation space for the global minimum of the function in order to predict the folding of a given amino acid sequence. Building accurate molecular mechanics potential functions has consumed countless man-years of effort, and the global search problem is in the general case infeasible for molecules as large as proteins. We have recently avoided both problems by building most of the information about polypeptides into the conformation space, rather than the potential function. Then our very simple potential function favors the native over essentially all alternatives for practically every compact, globular, water soluble protein in the Brookhaven Protein Data Bank. We anticipate this will be a generally useful tool for protein structure prediction, the inverse folding problem, homology modeling, and the early stages of solving protein crystal and NMR structures.

1   Introduction

In spite of recent attention to chaperonins and the problems of *in vivo* protein synthesis, export, and folding, the experimental fact remains that many proteins fold reversibly *in vitro* to a unique native conformation, apparently determined solely by their amino acid sequence[48]. To be able to predict the conformation from the sequence is possibly the most important current problem in molecular biology, especially now that so many genes are being sequenced.

In order to eliminate confusion at the outset, let us make it perfectly clear that the ultimate objective of our research is the classical protein folding problem: given an amino acid sequence, calculate its native, unique, globular conformation under physiological conditions, if it indeed has one. There has been a great deal of confusion in the recent literature about the degree of success in such calculations, because there is no clear consensus on what constitutes a correct prediction, and everyone wants to present his work in the best possible light. We hold ourselves to a much higher standard than "generally correct topology" or "strong resemblance of interresidue contact matrix". As explained in more detail below, we demand that the calculated conformation must be unmistakably close to the experimentally

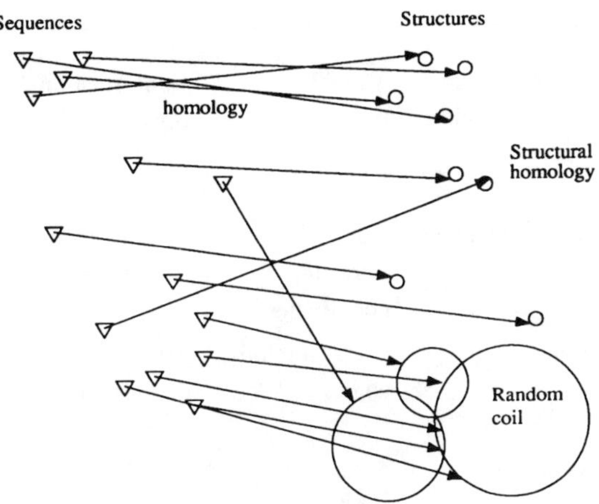

Figure 1: Our current experimental knowledge about the correspondence between amino acid sequence and protein conformation. Triangles represent sequences, and their proximity corresponds to sequence similarity. Small circles represent precise globular folds, large circles are more disordered states, and proximity again implies similarity.

determined native structure, when available, e.g. $D(native, calcd) < 3$ Å for a 100-residue protein, where D is the root-mean-square C^α coordinate deviation after optimal rigid body translation and rotation of one structure onto the other.

There are many different motivations for protein structure calculations. One might attempt to calculate protein conformation from first principles, quantum mechanics, or all-atom molecular dynamics in an effort to understand the relative importance of different physical factors in determining the conformation. We, on the other hand, am willing to be very empirical, if that is a useful expedient to reach the goal, and we believe it is. Alternatively, one may strive to simulate the folding *pathway* as an object of scientific interest itself and/or as a guide to reaching the native structure. Our view is that there is very little experimental evidence — and that of low accuracy — concerning the folding mechanism with which to compare the calculations. Further, the means that real proteins use to find their native structures probably does not map into an efficient computer calculation. Finally there is the inverse folding problem: given a globular protein conformer, find any and all amino acid sequences that would have this structure as their native one. This is of great interest to molecular biologists looking through protein sequence databases for sequences distantly homologous to a given protein crystal structure. The computational chemist is relieved of the burden of producing atomic coordinates, and the criteria for success are relatively unclear. Our main reservation is that while we have detailed experimental evidence on the mapping of a few sequences into globular structures, we have little information about the set of sequences matching a given structure (see figure 1).

Figure 1 schematically summarizes our experimental knowledge about protein folding. Nature has sampled only an infinitesimal portion of all possible amino acid sequences and has tended to keep either fibrous or globular proteins. While it is not at all clear what proportion of all sequences will fold up to unique globular structures[49], many naturally

Table 1: Different versions of the optimization approach for BPTI.

Method	Protein Representation	Variables	Conformation Space	Size of Conformation Space
molecular mechanics	10^3 point atoms	atomic Cartesian coordinates	box in 3×10^3 dimensions	10^{3528}
rigid valence geometry	same	> 250 torsion angles	250 dimensional torus	10^{389}
rotational isomeric	same	5 states per residue	combinations of states	10^{41}
lattice	2 points per residue	5 lattice positions per residue	combinations of lattice positions	10^{41}
Crippen & Maiorov	5 atoms per residue	choice of whole protein conformation	10^4 alternative conformations	10^4

occurring ones do (small circles in the figure) and even some carefully designed sequences don't (large circles). Substantially different sequences tend to fold up rather differently, with some remarkable exceptions[36], but very similar sequences generally fold up to similar structures. This is the basis for all the activity in protein homology modelling. We are unaware of any example of two very similar sequences folding to very different, unique, globular structures. In any case, all we really know is in the direction from sequence to structure: given one of a relatively small number of sequences, we know whether it does or does not fold to a unique globular structure, and if so, what the structure is. Going in the opposite direction, we can start with a given X-ray crystal structure and know that at least that protein's sequence folds to it, but we have no assurances about how many different sequences may assume that fold.

While there are many approaches to the protein folding problem, as we will outline below, we choose to pursue the "optimization" attack. The physical basis is simply that a system at equilibrium at constant temperature and pressure is in the thermodynamic state of globally minimal Gibbs' free energy, compared to all other kinetically accessible states. In a more general sense, we seek any scalar function of the system (i.e. protein and solvent described at some appropriate level of detail and realism) such that this function is at its global minimum for the native conformation compared to all available alternative conformations. This must hold for essentially all proteins having experimentally determined globular native conformations. Of course, this also entails devising some method to locate the global minimum of the function.

It is illuminating at this point to make some back-of-the-envelope estimations of the magnitude of the optimization task for the favorite test protein, pancreatic trypsin inhibitor (BPTI). Keep in mind these estimates are not even to the nearest order of magnitude, but they illustrate some important general trends. Of course, the purist would prefer evaluating the free energy of a conformational state by some ensemble average over the ab initio quantum mechanical energy, but even semi-empirical quantum mechanical methods are infeasible for even a single protein conformer. The most detailed feasible calculations we have listed as "molecular mechanics", which tacitly includes molecular dynamics, using a standard empirical, all-atom potential, such as Amber, Charmm, or MM2. Obviously, atoms are not really isotropic points with fixed partial charges, but these potential functions agree with many experiments by building in knowledge of bond lengths, bond angles, etc. as special terms in the potential. As a global search problem, the atoms are each free to move independently in three dimensions, say anywhere in a 15 Å cube. If we wanted to position each atom within one Angstrom, there are an astronomical number of conformations to examine. Early in the history of conformational analysis it was realized that the physically significant states lay near local minima of the potential function, so one need only seek out all local minima. From an arbitrary starting configuration, one can efficiently move to a nearby local minimum by some unconstrained *local optimization* algorithm, but there may be substantially better minima elsewhere: the famed "multiple minima problem." This requires *global optimization*, which is a fundamentally different process — an idea still resisted by many chemists. It is possible to prove that a thorough global search may require a computing effort on the order of an exhaustive search, corresponding roughly to the numbers in the "conformation space size" column[14, 37]. One way to improve the situation is to restrict the search to conformers having reasonable bond lengths and bond angles, the "rigid valence geometry" method in the table. The corresponding potential, such as ECEPP, now needs to know less about the structure of amino acids because that information is built into the conformation space. An exhaustive search over all combinations of torsion angles in 10° steps still corresponds to an unacceptably large conformation space. Once again, one can focus on the numerous local energy minima by local optimization with respect to all torsion angles. Roughly equivalently, the "rotational isomeric" model amounts to trying combinations of conformations built out of local minima of single residues. Another favorite approach from polymer chemistry is the "lattice" approximation, where the amino acid residues are substantially simplified, and conformations are walks restricted to the points on some sort of regular three-dimensional lattice[11]. This makes it easier to avoid the steric overlaps between residues distant in sequence that one might encounter in the rotational isomeric model, but the combinatorial difficulties are comparable. Once again, a special but even simpler potential is required to match the simplification of the residue representation.

Finally, we come to the "Crippen & Maiorov" entry in the table, the subject of this proposal. As will be explained below, the conformation space knows yet more about protein structure, while the potential functions knows less. Choosing a conformation is no longer a matter of picking dihedral angles or atomic coordinates, but simply selecting an entire protein conformer out of a set that has been constructed from the Brookhaven Protein Data Bank (PDB). The very simple contact function guides the search toward those conformers having the lowest function values, and the conformation space is so small that it is at last feasible to conduct a global search. Note the trends in the table from top to bottom: simpler potential functions, more complicated conformation spaces, but smaller searches. Although our potential is not perfect, and our conformation space is not as large as it could be, this is still the first time in the history of the protein folding problem that anyone has succeeded in making the general optimization approach work for such a wide variety of proteins.

Of course, optimization is not the only approach to the protein folding problem. One protocol, long advocated by Scheraga and explored by people too numerous to list, is the "secondary then tertiary structure" approach. First predict the secondary structural features of the sequence, such as runs of helix, extended strands, and bends, and then pack these preformed pieces together. The first part is usually treated as a rule-based, or knowledge engineering problem of deducing rules by studying known protein crystal structures. We now have from X-ray crystallography and NMR more than 700 examples of the native conformations of various water soluble globular proteins under roughly physiological conditions, along with their sequences. In spite of a vast effort on the part of many investigators, the repeatedly occurring structural patterns or motifs[9] do not correlate clearly enough with sequence to be able to deduce a set of rules for going reliably from sequence to local or secondary structure[44, 43, 41], much less to the overall tertiary conformation. In fact, a five-residue segment of a given sequence may occur in quite different conformations in the context of different protein crystal structures, and the percentage of sequence identity between two segments required to give a high likelihood of conformational similarity decreases only gradually with increasing chain length[46]. The problem is that, just as a fragment of the polypeptide chain comprising a protein domain will not fold up, neither can one calculate the conformation of a protein by looking at only a piece of it at a time. Certainly there are tendencies, and secondary structure prediction is about 70% correct, but the overall tertiary folding modifies the tendencies in the other 30% of the residues.

Perhaps a better understanding of the pathway of protein folding would lead to methods that calculate the final native conformation by simulating the folding process. A direct simulation by molecular dynamics is impractical for the foreseeable future[27] because a 1 nanosecond trajectory of even a small protein with the essential solvent molecules is at the current limit of supercomputers, whereas real proteins fold on a time scale of seconds. Even less detailed calculations of the folding pathway[31, 7, 29] by statistical mechanical and thermodynamic methods are hindered by two problems: the lack of detailed experimental evidence for many proteins concerning their folding mechanisms[1], and the need to account for pathways from the many different conformations of the unfolded state to the native conformation[60].

This leaves the optimization approach, where there has been a flurry of activity on constructing potentials or scoring schemes that distinguish between the native and incorrect folds. Of course, every published article demonstrates "satisfactory" performance, but careful reading reveals important differences.

The most popular approach for devising potentials is to survey the distributions of amino acid spatial separations[2, 57] in PDB files broken down by residue types and sequence separation along the polypeptide chain, and attribute these histograms to a Boltzmann distribution due to the potential of mean force[39, 26, 27]. For example, if $f(d)$ is the frequency of occurrence over the whole database of Leu-Arg residues separated by 5 residues in sequence and d Å in space, then the potential must have a term proportional to $-RT \ln f(d)$ applied to every Leu-Arg pair in the protein, if they are 5 residues apart in the sequence. Tabulating such histograms as $f(d)$ for every residue pair type and several different ranges of sequence separation is quite voluminous, but the resulting potential function is a direct, unadjusted consequence of the database, not of some fitting procedure. The theoretical weakness is that all these observed pairwise interactions are not independent geometric events, due to packing constraints in three dimensions and the connectivity of the fixed sequences of polypeptide chain. Worse yet, such potentials do not always favor the native over the alternative conformations. For example, the Miyazawa and Jernigan potential[35] ranks the

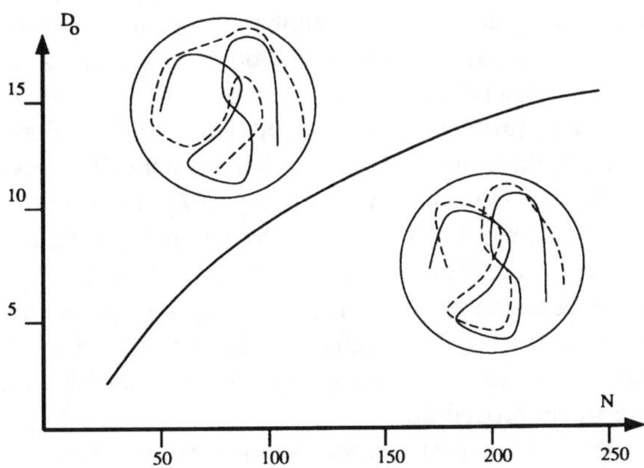

Figure 2: A plot of D_o (in Å) as a function of polypeptide chain length, N. The circles represent examples of two different globular protein structures, solid and dashed lines, superimposed and constrained to be compact. When $D < D_o$, only minor changes in conformation are possible, but when $D > D_o$, substantial rearrangements may occur.

native in the best 2% of conformations found in a restricted lattice search, but that means many alternative structures were better[12]. Skolnick and coworkers find it necessary to add to this same potential extra local interactions tailored to the particular protein being studied in order to favor a native-like conformation[53]. The potential of mean force used by Hinds and Levitt preferred an alternative conformation of BPTI (bovine pancreatic trypsin inhibitor) that was a distant 5.65 Å from the native in coordinate rms deviation[27]. Sippl and coworkers in a series of publications have derived many such potentials, depending on which protein's structure is to be predicted, but there are at least 20 proteins for which such a potential fails to favor the native over some alternatives[50, 26, 6, 52]. Wolynes and coworkers build potential functions of protein C^α coordinates by a method derived from spin-glass theory, but in order to favor the native conformation of a given protein, there can only be a few proteins used to derive the potential, and some of these must be homologous with the protein in question[21, 47, 3, 25]. The common feature of all these studies has been a focus on correct, native protein conformations. By looking at many examples of the right way to fold proteins, one tries to encode this information into a potential function of some form. The expectation is that this potential will then guide proteins away from incorrect structures toward correct ones, even though the potential has not seen incorrect folds during the training phase.

The next most popular way to derive a suitable potential is based on the early observations that the native fold buries more hydrophobic groups or surface area, while exposing more ionizable groups than other globular, compact, alternative conformations[5, 38]. One can discriminate between a few alternatives by calculating the solvation free energy relatively carefully[8], but a more rapid method is preferable when there are many alternatives. Holm and Sander[28] have a simplified assessment of solvation of all sidechain heavy atoms that correctly prefers the native over one globular alternative for 14 different proteins;

Eisenberg and coworkers' profile[4, 59], which includes solvation and secondary structural contributions, is able to favor the native over one globular alternative for seven different proteins, even in some cases when the conformations were quite similar[32]. Whereas the Sippl potentials have been tested on large numbers of proteins and larger numbers of alternative conformations, it is not known how the Sander or Eisenberg potentials perform in such a test. In any case, the intent has been to apply these and other evaluation methods[22, 23, 40] to the inverse folding problem. In our approach to the standard protein folding problem, we must score many different conformers taken from the PDB, given one particular sequence. In effect, we change the sidechains of many different protein crystal structures into the sequence of interest. However, any crystal structure is adjusted for optimal packing of its original sidechains, not this new sequence. In order to get a fair estimate of the solvation, one has to carry out a lengthy readjustment of packing, once the sidechains have been changed[38]. For our purposes, it is much more efficient to have a potential that is not so dependent on the fine details of the conformation. In this way, our results are also not dependent on the readjustment procedure.

The second part of the optimization approach to protein structure is to find the conformation corresponding to the global minimum of the potential. In our earlier work, the protein was represented as one or two points per residue, each having continuously variable Cartesian coordinates. Energy embedding was an expensive but successful algorithm for locating very low energy minima in such a continuous conformation space[16], but the conformation space of even a small protein is so large that one is never sure of locating the global minimum. The calculations can be speeded up by going to a rotational isomeric model of protein conformations, or representing the polypeptide chain by a walk on some sort of lattice. One can then either exhaustively enumerate all walks on a sufficiently restrictive lattice[27, 12] or proceed along a sequence of conformations from some arbitrary starting point by a Monte Carlo or simulated annealing procedure[54]. One can also generate conformations as combinations of overlapped short segments, each segment having possible conformations seen in known protein structures[51]. We have preferred to generate alternative conformations by selecting all possible contiguous chain segments of the desired length drawn from PDB files of all larger proteins[15, 26]. The advantage is that global optimization amounts to simply evaluating the potential for each of the many alternatives, but the drawback is that the Protein Data Bank probably does not contain all possible folding patterns.

2   Methods

The first question is: what is the goal? Given the N-residue sequence, we want to calculate a conformation such that the root-mean-square deviation in C^α coordinates after optimal rigid-body superposition ("coordinate RMSD", as opposed to the rms deviation in interresidue distances) $D(calc, nat) < D_o$, where

$$D_o = -10.82 + 4.31 N^{1/3} \qquad\qquad (1)$$

Qualitatively speaking, if two globular protein structures have $D < D_o$, they may differ in terms of small, local displacements, but their overall folding motif must be the same. (Some would use the word "topology" here, but that is mathematically inappropriate because all polypeptide chains without disulfide bridges have the same covalent topology[13, 17]). This sort of behavior is observed in cubic lattice walks and fragments of PDB entries where the chain is self-avoiding and so densely packed that only small conformational changes can

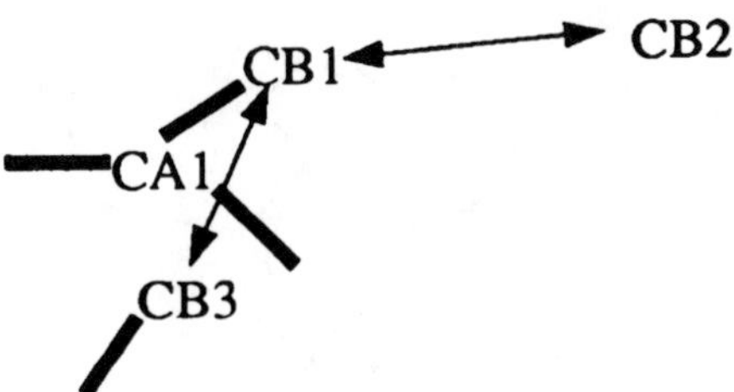

Figure 3: The screening effect in the contact definition. Atoms CB1 and CB2 are in contact, but CB1 and CB3 are not because CA1 lies too close to the line segment joining CB1 and CB3.

be made without substantial internal rearrangements. When comparing a native protein structure with alternatives consisting of N-residue contiguous segments taken from larger protein crystal structures, $D < D_o$ occurs only when the two proteins are clearly homologous structures[34]. Eq. (1) is the very accurate result of a least-squares fit of the minimal observed D for N-residue segments taken from non-homologous protein pairs as a function of $N = 36, \ldots, 256$ (see Fig. 2).

Given that we are trying to calculate the conformation of proteins by some sort of global optimization of a potential function, the next step is to define the potential (which need not correspond to a real thermodynamic function like Gibbs' free energy). Basically, the idea is to calculate the potential E as a sum of interresidue contacts, the contributions depending on how close they are in space, their separation in sequence, the residue types involved, and whether the sidechains or backbones are interacting. In order to keep the number of adjustable parameters, ϵ_{ij}, down to a manageable 112, we grouped sequence separations into short- and long-range categories, and residue types into groups of helix formers, hydrophobic, etc. See [33] for details.

$$E = \sum_{\substack{contacts \\ i,j}} \epsilon_{ij} V\left(d_{ij}, U, L\right) \prod_k \left[1 - V\left(d_{ijk}, U', L'\right)\right] \tag{2}$$

where d_{ij} is the distance between atoms i and j, d_{ijk} is the distance of a third atom k from the line segment joining i and j, and V is the smooth sigmoidal function

$$V\left(d, U, L\right) = \begin{cases} \frac{(d-U)^2(2d-3L+U)}{(U-L)^3} & \text{if } L \leq d \leq U \\ 1 & \text{if } d < L \\ 0 & \text{if } d > U \end{cases} \tag{3}$$

The upper and lower bounds on the transition between contact and non-contact (U, U', L, and L') depend on whether it is a screening distance, a contact involving sidechains, or one involving backbone atoms. The point is that a very small change in atomic coordinates makes a small change in E, so that homologous crystal structures have similar potential values. By "homologous" we really mean structures where the same protein, or a point mutant of it, is crystallized with different ligands or in a different crystal form.

While we have not optimized the functional form of the potential extensively, we believe it has important features that facilitate the proposed work. The key is to count contacts over

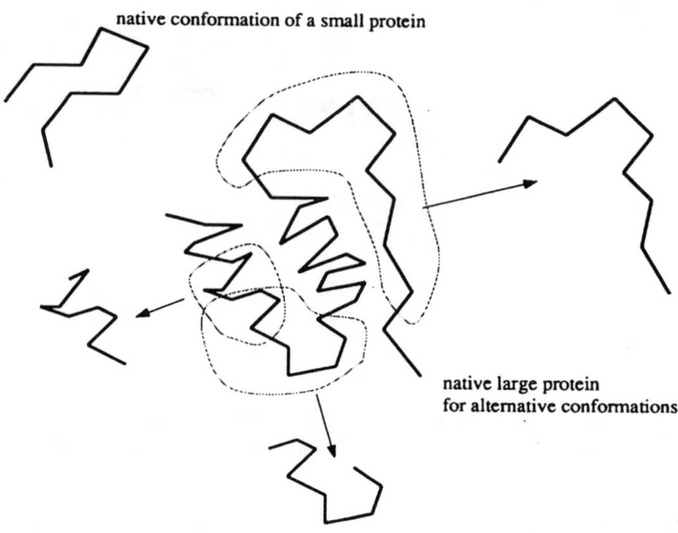

Figure 4: The "threading" procedure for generating alternative conformations of a small protein by taking contiguous chain segments out of a larger protein's crystal structure.

generously long distances as long as there are no interfering atoms between. Realizing that the alternatives are taken from other PDB files corresponding to different sequences than that of the chosen native protein, backbone-backbone, backbone-sidechain, and sidechain-sidechain contacts depend only on the coordinates of each residue's N, C', O, and C^β atoms. Then for example, C^β atoms as far apart as 9 Å are said to be in contact on the grounds that long sidechains would have room to touch each other, and short sidechains still might interact, given a minor adjustment of the backbone[56]. Because of the possible screening effect of extraneous atoms between two that are otherwise close enough to pass the contact definition, the potential is actually something of a sum of three-body interactions (see Fig. 3).

Assuming for the moment that Eq. (2) is an adequate potential function, we need to construct a conformation space over which the global optimization is to occur. As Table 1 suggests, one should try constructing a cleverer conformation space that "knows" more about polypeptide structures. The idea is to build a conformation space for a given protein consisting of a large but finite set of alternative conformations that each incorporate all we know about the invariants of peptide geometry, so that the potential can be simple. Yet the set of alternatives should cover the whole range of possible conformations. Our solution[15] was to extract from the entire PDB every contiguous piece of chain of the correct length. Thus, one can generate 21 alternatives to a 100 residue protein from the crystal structure of a 120 residue protein by using the coordinates of its residues 1-100, 2-101, ..., 21-120 (see Fig. 4). Sippl and coworkers[26] independently hit upon this same trick at about the same time. As noted in Table 1, this procedure produces over 12,000 alternatives for a 58-residue protein, such as BPTI. For multimeric proteins, such as a pair of chains having lengths N_1 and N_2, we extract all combinations of non-overlapping N_1 and N_2 residue segments from any PDB chain longer than $N_1 + N_2$ residues. The only shortcoming is that novel folds not yet observed experimentally, or even standard motifs but with unusual arrangements of insertions and deletions, are not represented in this universe of conformations. Although essentially all the PDB files are for globular proteins, many (but not all) alternatives taken

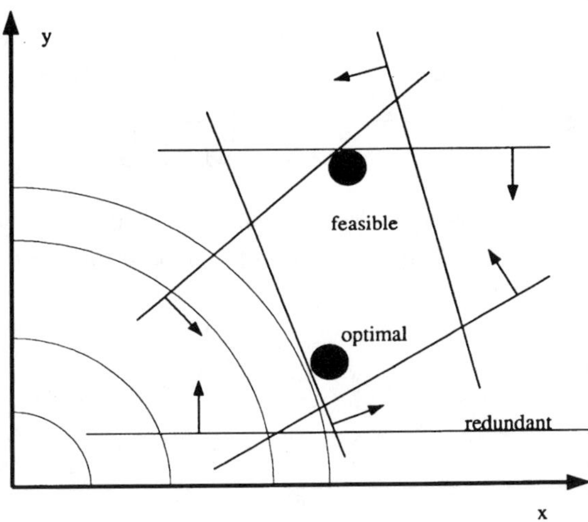

Figure 5: Example of redundant and active linear inequalities in two variables, x and y, defining a finite feasible region. All inequalities not marked as redundant are nonredundant. Each inequality permits the half-plane indicated by the boundary line and arrow. Within the irregular, five-sided feasible region, one solution to the inequalities is marked by a dot, corresponding to two active inequalities. Another solution, labelled as optimal, satisfies all the inequalities, has only one active constraint, and otherwise minimizes the objective function, $x^2 + y^2$, which is indicated by circular level lines.

from fragments of these are not compact. In this respect, our conformation space goes beyond the set of all globular protein folds. Note that although the conformation space consists of a finite, discrete set of alternatives, each one of these is a well-constructed piece of polypeptide chain having good local geometry, reasonable packing, proper right-handed α-helices, etc. They are not distorted by forcing them onto some regular three-dimensional lattice.

At the moment, our global search technique is stunningly simple: evaluate the potential for the native and every alternative conformation; choose the structure with the lowest value. For roughly 300 residue and smaller proteins having many tens of thousands of alternatives, this is not a very lengthy calculation.

In order for a global search over all these alternative conformations to succeed, we must adjust the ϵ_{ij}'s of Eq. (2) such that for all native structures and all their alternatives

$$E\left(alt\right) - E\left(nat\right) > M \ \forall \ alt \tag{4}$$

where we arbitrarily set the margin $M = 3.0 \times D(nat, alt)$. (In contrast, the spin-glass approach seeks to maximize the mean potential difference between alternatives and the native[21, 47, 3, 25]). By our choice of the functional form of E, this amounts to solving a set linear inequalities in the ϵ_{ij}'s. Now solving a set of linear inequalities is not like linear regression on a set of observations. Linear regression always gives a fit, sometimes a good one and sometimes bad, but a given set of inequalities may have no solution, called the

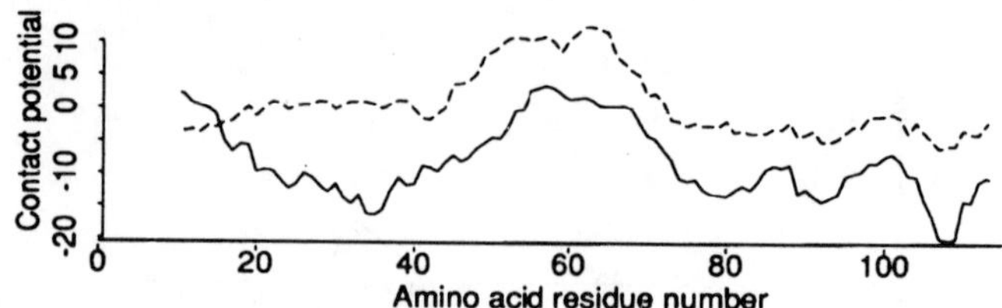

Figure 6: RuBisCO contact potential vs. residue number, smoothed. Solid line: correct sequence on correct structure, $E = -894$. Dashed line: correct sequence on incorrect structure, $E = +22$.

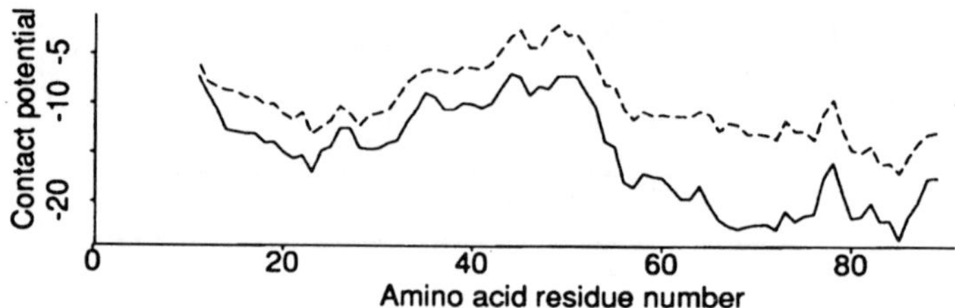

Figure 7: HIV protease contact potential as a function of residue number, smoothed over a 21-residue window. Solid line: 5hvp sequence on 5hvp structure, $E = -1384$. Dashed line: 5hvp sequence on 1hvp structure, $E = -922$.

"feasible region", or the feasible region may be large or small. It is trivial to write down just two inequalities in any number of variables such that there is no solution, so the existence of a feasible region can by no means be taken for granted. As illustrated in Fig. 5, some inequalities may be "redundant" (have no bearing on the feasible region), and there may be more so-called "nonredundant" inequalities bounding the feasible region than there are adjustable parameters. When we solve Eq. (4), we locate one, not necessarily unique point in the feasible region, whereas a linear least-squares fit is always unique. At such a solution, there may be between 0 and n "active" inequalities that are barely satisfied, where n is the number of variables. In an early example of adjusting such a contact potential in terms of 4 parameters, all 802,074 inequalities were satisfied at a feasible point that was defined by four active inequalities, corresponding to four important alternatives that all happened to be globular conformations[15]. We have never yet imposed such strong inequalities that there is no feasible region, so we use the remaining degrees of freedom to minimize

$$\sum_{i,j} \epsilon_{ij}^2. \tag{5}$$

If there is any feasible solution to the inequalities, there is a unique solution to this constrained optimization problem, called a "quadratic program".

While our quadratic program consists of on the order of 10^6 linear inequalities in 112 variables, experts estimate only some 200 of these are actually nonredundant. In practice, we regularly adjust the parameters in a couple hours on a workstation using the mathematical programming package MINOS and a subset of 6000 inequalities containing the nonredundant ones. If some day we should choose a set of mutually inconsistent inequalities, the program will unambiguously identify the event. Minimizing the sum of the squares of the parameters gives relatively small values so that the inequalities are not satisfied by the cancellation of large interaction terms of opposite sign. Furthermore, introduction of

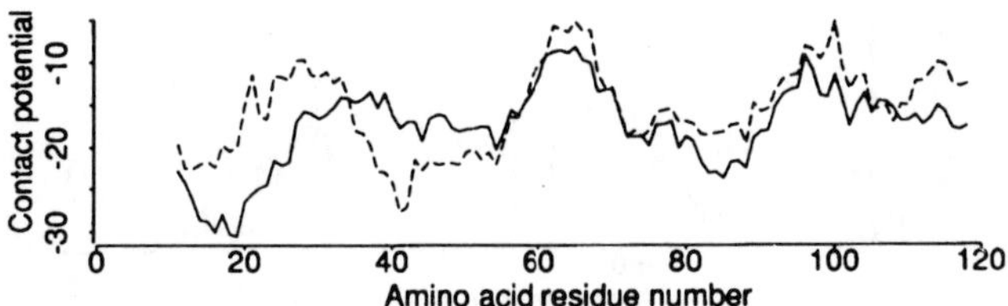

Figure 8: Interleukin 4 contact potential vs. residue number, smoothed. Solid line: NMR sequence on NMR structure, $E = -2216$. Dashed line: NMR sequence on model structure, $E = -1996$.

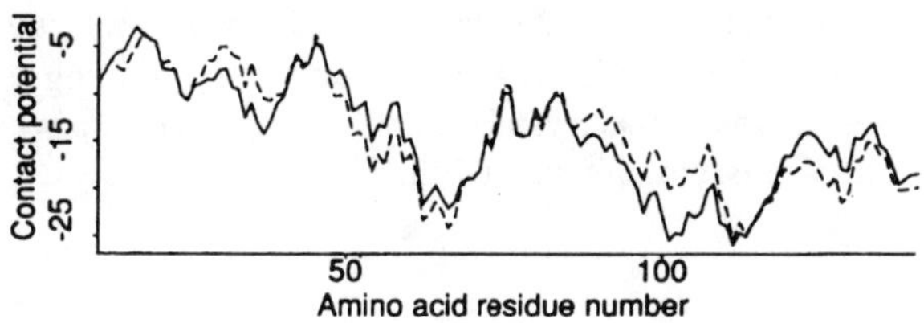

Figure 9: Solid line: 1mle sequence on 1mle structure, $E = -2144$. Dashed line: 1mle sequence on 1mns structure, $E = -2177$.

new inequalities (e.g. derived from new proteins) may either contract the feasible region or leave it unchanged, and even in the former case, the optimal solution may or may not move. This is in contrast to any averaging approach, such as potentials of mean force, where the training set must not only be extensive over all possible protein folds, but also uniformly representative of all different folds, and addition of new proteins to the training set always changes the resulting force field, at least a little.

One of the special features of adjusting parameters by solving sets of inequalities rather than by averaging, is that introducing a single bad constraint may result in an inconsistent set of inequalities and no solution at all, rather than a somewhat poorer solution. Consequently, we have had to carefully select the native structures used in Eq. (4), according to objective criteria we have developed. For example, very small peptides that form many *inter* molecular contacts in the crystal have a native conformation that cannot be explained in terms of the *intra* molecular contacts. Looking at all N-residue native and alternative structures generated from the PDB as a function of N, we found that the minimal observed radius of gyration (in Å) very accurately fits

$$r_{g,min} = -1.26 + 2.79 N^{1/3} \tag{6}$$

and the maximum number of contacts observed (according to the above definition of inter-residue contacts) fits

$$n_{c,max} = -53.17 + 4.25 N. \tag{7}$$

Then for every native structure having actual radius of gyration r_g and number of contacts n_c satisfying

$$\frac{r_g}{r_{g,min}} < 1.3 \text{ and } \frac{n_{c,max}}{n_c} < 1.5, \tag{8}$$

we demand that Eq. (4) be satisfied for all alternative structures. One can quibble about the exact cutoffs used in this definition, but it is certainly objective, and both criteria must be

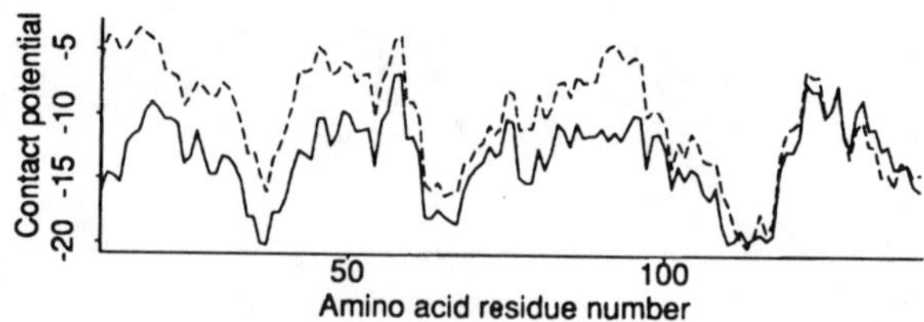

Figure 10: Solid line: 1mns sequence on 1mns structure, $E = -2055$. Dashed line: 1mns sequence on 1mle structure, $E = -1482$.

examined[33]. If a single chain in a crystal structure fails this compactness test, we have difficulty adjusting the potential function if it is included in the training set. However, if a small set of neighboring chains in the crystal can be selected that as a whole pass the test, then we have no difficulty satisfying its inequalities.

3   Results

We are delighted by our current ability to distinguish between correctly and incorrectly folded proteins. In developing the potential we considered a subset of PDB consisting of 212 proteins (176 X-ray + 17 NMR + 19 models) and 477 homologues of these. For these we generated a grand total of 1,627,714 alternatives. Of the 212, only 161 are compact (totalling 1,242,182 alternatives). Out of the compact native structures we selected a training set of 58 proteins (50 X-ray + 8 NMR), and a few of their alternatives, totalling 6,566 constraints. Solving these constraints (Eq. (4)) determined the potential function. Applying it to the whole set of 212 proteins, we found only one alternative in violation of Eq. (4) for each of two compact, experimentally determined structures. Model or non-compact proteins had more violations. Explicit knowledge of disulfide bridges, prosthetic groups, or ligands is not required. It prefers the correct structure over the intentionally incorrect fold in all 12 cases we could check[38, 28]; the correct X-ray structure of RuBisCO[18] over the earlier incorrect chain tracing[19]; the X-ray structure of HIV protease[20] and the earlier model structure[58]; and the NMR structure[55] of interleukin 4 over the model structure[10]. In the case of muconate lactonizing enzyme[42, 24, 45] vs. mandelate racemase[30], which have very similar structures but little sequence similarity, each sequence prefers the more highly refined 1mns crystal structure, but by a small margin.

In addition, the potential works for multimeric proteins as well as single chains, as long as the aggregate is compact. We have succeeded with 10 cases of two-chain proteins from PDB, ranging between 500,000 and 2,000 alternatives each. Melittin dimer is non-compact and has many violations; the tetramer (the stable form in solution and in the crystal) is compact and has no violations in the 1.28 million alternatives checked. The potential doesn't know which cysteines are linked in disulfide bridges, so the A and B chains of insulin have many alternative structures they prefer over the crystal structure. This is in agreement with the observation that once the disulfides of insulin are reduced, reoxidation does not reconstitute the correct pairing. Biologically, the single chain proinsulin folds up, not insulin itself.

Our assumption all along has been that the native conformation is stabilized by the sum of virtually all the interactions along the whole polypeptide chain, and that any alternative is less favorable because the sum of all its interactions is not quite so good. This is in accord with the many experiments showing that nearly the entire protein chain is required in order for a domain to refold, and the observation that short segments of chain having a given sequence can adopt a variety of conformations when that same sequence occurs in different proteins[46]. Indeed, the sum of the interactions between a five-residue segment and all other residues in the native is not necessarily more favorable than the sum of interactions experienced by those same five residues in some alternative conformation. However, we find that all 21-residue segments in the native have a net more favorable interaction (within the segment and between residues of the segment and residues outside it) than the corresponding segment does in any alternative. Thus, like the Eisenberg group's profile[32], our potential can locate portions of the chain that are misfolded.

This work was supported by grants from the National Institutes of Health (GM37123) and the National Institute on Drug Abuse (DA06746). We are indebted to all the crystallographers who deposited their protein structural data in the Protein Data Bank.

References

[1] R. L. Baldwin and H. Roder. Characterizing protein folding intermediates. *Curr. Biol.*, 1:218–20, 1991.

[2] G. Boehm and R. Jaenicke. Correlation functions as a tool for protein modeling and structure analysis. *Protein Sci.*, 1:1269–78, 1992.

[3] H. G. Bohr and P. G. Wolynes. Initial events of protein folding from an information-processing viewpoint. *Phys. Rev. A*, 46:5242–8, 1992.

[4] J. U. Bowie, R. Luthy, and D. Eisenberg. A method to identify protein sequences that fold into a known three-dimensional structure. *Science*, 253:164–70, 1991.

[5] S. H. Bryant and L. M. Amzel. Correctly folded proteins make twice as many hydrophobic contacts. *Int. J. Peptide Prot. Res.*, 29:46–52, 1987.

[6] G. Casari and M. J. Sippl. Structure-derived hydrophobic potential. hydrophobic potential derived from x-ray structures of globular proteins is able to identify native folds. *J. Mol. Biol.*, 224:725–32, 1992.

[7] G. Chelvanayagam, Z. Reich, R. Bringas, and P. Argos. Prediction of protein folding pathways. *J. Mol. Biol.*, 227:901–16, 1992.

[8] L. Chiche, L. M. Gregoret, F. E. Cohen, and P. A. Kollman. Protein model structure evaluation using the solvation free energy of folding. *Proc. Nat. Acad. Sci. U.S.A.*, 87:3240–3243, 1990.

[9] C. Chothia and A. V. Finkelstein. The classification and origins of protein folding patterns. *Ann. Rev. Biochem.*, 59:1007–39, 1990.

[10] F. Cohen and S. Presnell. personal communication.

[11] D. G. Covell. Folding protein α-carbon chains into compact forms by monte carlo methods. *Proteins: Struct., Funct., Genet.*, 14:409–20, 1992.

[12] D. G. Covell and R. L. Jernigan. Conformations of folded proteins in restricted spaces. *Biochemistry*, 29:3287–3294, 1990.

[13] G. M. Crippen. Topology of globular proteins. *J. Theor. Biol.*, 45:327–338, 1974.

[14] G. M. Crippen. Global optimization and polypeptide conformation. *J. Comp. Phys.*, 18:224–231, 1975.

[15] G. M. Crippen. Prediction of protein folding from amino acid sequence over discrete conformation spaces. *Biochemistry*, 30:4232–4237, 1991.

[16] G. M. Crippen and T. F. Havel. Global energy minimization by rotational energy embedding. *J. Chem. Inf. Comp. Sci.*, 30:222–227, 1990.

[17] G.M. Crippen. Topology of globular proteins. ii. *J. Theor. Biol.*, 51:495–500, 1975.

[18] P.M.G. Curmi, D. Cascio, R.M. Sweet, D. Eisenberg, and H. Schreuder. Crystal structure of the unactivated form of ribulose-1,5- bisphosphate carboxylase/oxygenase from tobacco refined at 2.0-a resolution. *J. Biol. Chem.*, 267:16980–16989, 1992.

[19] D. Eisenberg. personal communication.

[20] P.M.D. Fitzgerald, B.M. McKeever, J.F. van Middlesworth, J.P. Springer, J.C. Heimbac, C.-T. Leu, W.K.Herber, R.A.F. Dixon, and P.L. Darke. Crystallographic analysis of a complex between human immunodeficiency virus type 1 proteins and acetylpepstatin at 2.0-angstroms resolution. *J. Biol. Chem.*, 265:14209–14219, 1990.

[21] M. S. Friedrichs, R. A. Goldstein, and P. G. Wolynes. Generalized protein tertiary structure recognition using associative memory hamiltonians. *J. Mol. Biol.*, 222:1013–1034, 1991.

[22] A. Godzik, A. Kolinski, and J. Skolnick. Topology fingerprint approach to the inverse protein folding problem. *J. Mol. Biol.*, 227:227–38, 1992.

[23] A. Godzik and J. Skolnick. Sequence-structure matching in globular proteins: Application to supersecondary and tertiary structure determination. *Proc. Natl. Acad. Sci. U. S. A.*, 89:12098–102, 1992.

[24] A. Goldman, D. L. Ollis, and T. A. Steitz. Crystal structure of muconate lactonizing enzyme at 3 angstroms resolution. *J. Mol. Biol.*, 194:143–153, 1987.

[25] R. A. Goldstein, Z. A. Luthey-Schulten, and P. G. Wolynes. Protein tertiary structure recognition using optimized hamiltonians with local interactions. *Proc. Natl. Acad. Sci. U. S. A.*, 89:9029–33, 1992.

[26] M. Hendlich, P. Lackner, S. Weitckus, H. Floeckner, R. Froschauer, K. Gottsbacher, G. Casari, and M. J. Sippl. Identification of native protein folds amongst a large number of incorrect models. the calculation of low energy conformations from potentials of mean force. *J. Mol. Biol.*, 216:167–180, 1990.

[27] D. A. Hinds and M. Levitt. A lattice model for protein structure prediction at low resolution. *Proc. Natl. Acad. Sci. U. S. A.*, 89:2536–40, 1992.

[28] L. Holm and C. Sander. Evaluation of protein models by atomic solvation preference. *J. Mol. Biol.*, 225:93–105, 1992.

[29] R. S. Judson. Teaching polymers to fold. *J. Phys. Chem.*, 96:10102–4, 1992.

[30] J. A. Landro, J. A. Gerlt, J. W. Kozarich, C. W. Kod, V. J. Shah, G. L. Kenyon, D. J. Neidhart, S. Fujita, J. Clifton, and G. A. Petsko. On the role of lysine 166 in the mechanism of mandelate racemase from pseudomonas putida: Mechanistic and crystallographic evidence for stereospecific alkylation by (r)-alpha-phenylglycidate, 1993. To be published.

[31] P. E. Leopold, M. Montal, and J. N. Onuchic. Protein folding funnels: a kinetic approach to the sequence-structure relationship. *Proc. Natl. Acad. Sci. U. S. A.*, 89:8721–5, 1992.

[32] R. Luethy, J. U. Bowie, and D. Eisenberg. Assessment of protein models with three-dimensional profiles. *Nature*, 356:83–85, 1992.

[33] V. N. Maiorov and G. M. Crippen. Contact potential that recognizes the correct folding of globular proteins. *J. Mol. Biol.*, 227:876–888, 1992.

[34] V. N. Maiorov and G. M. Crippen. Significance of root-mean-square deviation in comparing three-dimensional structures of globular proteins. *J. Mol. Biol.*, in press, 1993.

[35] S. Miyazawa and R. L. Jernigan. Estimation of interresidue contact energies from protein crystal structures: Quasi-chemical approximation. *Macromolecules*, 18:535–552, 1985.

[36] D. J. Neidhart, G. L. Kenyon, J. A. Gerlt, and G. A. Petsko. Mandelate recemase and muconate lactonizing enzyme are mechanistically distinct and structurally homologous. *Nature*, 347:692–694, 1990.

[37] J. T. Ngo and J. Marks. Computational complexity of a problem in molecular structure prediction. *Protein Eng.*, 5:313–21, 1992.

[38] J. Novotny, A. A. Rashin, and R. E. Bruccoleri. Criteria that discriminate between native proteins and incorrectly folded models. *Proteins: Struct., Funct., Genet.*, 4:19–30, 1988.

[39] M. Oobatake and G. M. Crippen. Residue-residue potential function for conformational analysis of proteins. *J. Phys. Chem.*, 85:1187–1197, 1981.

[40] J. Overington, D. Donnelly, M. S. Johnson, A. Sali, and T. L. Blundell. Environment-specific amino acid substitution tables: tertiary templates and prediction of protein folds. *Protein Sci.*, 1:216–26, 1992.

[41] S. Rackovsky. On the nature of the protein folding code. *Proc. Natl. Acad. Sci. U. S. A.*, 90:644–8, 1993.

[42] P. A. Rice, A. Goldman, and T.A. Steitz. A helix-turn-strand structural motif common in alpha-beta proteins. *Proteins*, 8:334–340, 1990.

[43] M. J. Rooman, J. P. A. Kocher, and S. J. Wodak. Extracting information on folding from the amino acid sequence: Accurate predictions for protein regions with preferred conformation in the absence of tertiary interactions. *Biochemistry*, 31:10226–38, 1992.

[44] M. J. Rooman, J. Rodriguez, and S. J. Wodak. Relations between protein sequence and structure and their significance. *J. Mol. Biol.*, 213:337–50, 1990.

[45] C. Sander and L. Holm. Muconate lactonizing enzyme heavy atom coordinates calculated from available x-ray c-alpha coordinates, 1993. personal communication.

[46] C. Sander and R. Schneider. Database of homology-derived protein structures and the structural meaning of sequence alignment. *Proteins*, 9:56–68, 1991.

[47] M. Sasai and P. G. Wolynes. Unified theory of collapse, folding, and glass transitions in associative-memory hamiltonian models of proteins. *Phys. Rev. A*, 46:7979–97, 1992.

[48] R. Seckler and R. Jaenicke. Protein folding and protein refolding. *FASEB J.*, 6:2545–52, 1992.

[49] E. Shakhnovich, G. Farzitdinov, A. M. Gutin, and M. Karplus. Protein folding bottlenecks: a lattice monte carlo simulation. *Phys. Rev. Lett.*, 67:1665–1668, 1991.

[50] M. J. Sippl. Calculation of conformational ensembles from potentials of mean force. *J. Mol. Biol.*, 213:859–883, 1990.

[51] M. J. Sippl, M. Hendlich, and P. Lackner. Assembly of polypeptide and protein backbone conformations from low energy ensembles of short fragments: Development of strategies and construction of models for myoglobin, lysozyme, and thymosin beta-4. *Protein Science*, 1:625–640, 1992.

[52] M. J. Sippl and S. Weitckus. Detection of native-like models for amino acid sequences of unknown three-dimensional structure in a database of known protein conformations. *Proteins: Struct., Funct., Genet.*, 13:258–71, 1992.

[53] J. Skolnick and A. Kolinski. Dynamic monte carlo simulations of a new lattice model of globular protein folding, structure and dynamics. *J. Mol. Biol.*, 221:499–531, 1991.

[54] Jeffrey Skolnick and Andrzej Kolinski. Computer simulations of globular protein folding and tertiary structure. *Annu. Rev. Phys. Chem.*, 40, 207-35, 1989.

[55] L.J. Smith, C. Redfield, J. Boyd, G.M.P Lawrence, R. G. Edwards, R. A. G. Smith, and G. M. Dobson. Human interleukin 4: The solution structure of a four-helix-bundle protein. *J. Mol. Biol.*, 224:899–904, 1992.

[56] R. Varadarajan and F. M. Richards. Crystallographic structures of ribonuclease s variants with nonpolar substitution at position 13: packing and cavities. *Biochemistry*, 31:12315–27, 1992.

[57] G. Vriend and C. Sander. Quality control of protein models: directional atomic contact analysis. *J. Appl. Crystallogr.*, 26:47–60, 1993.

[58] I.T. Weber, M. Miller, M. Jaskolski, J. Leis, A.M. Skalka, and A. Wlodawer. Molecular modeling of the hiv-1 protease and its substrate binding site. *Science*, 243:928–931, 1989.

[59] M. Wilmanns and D. Eisenberg. Three-dimensional profiles from residue-pair preferences: Identification of sequences with β/α-barrel fold. *Proc. Natl. Acad. Sci. U. S. A.*, 90:1379–83, 1993.

[60] R. Zwanzig, A. Szabo, and B. Bagchi. Levinthal's paradox. *Proc. Natl. Acad. Sci. U. S. A.*, 89:20–2, 1992.

Genetic Algorithm Codings used in Protein Structure Prediction by Energy Minimization

Frank Herrmann

Department of Molecular Biophysics, German Cancer Research Center, Im Neuenheimer Feld 280, D-69120 Heidelberg, Germany

Abstract

One method to predict protein tertiary structure is to minimize empirical energy functions. Genetic Algorithms (GAs) are optimization techniques that employ a model of natural adaptation. They are used to adapt a protein conformation to a minimal energy. The conformations are represented by dihedral angles of rotatable bonds. The GA uses binary strings coding for the angles during the optimization. An analysis of the standardly used binary representation is presented that reveals properties of the coding that potentially cause convergence to suboptimal solutions.

1   Introduction

Different optimization techniques are used to predict protein tertiary structure by energy minimization. GAs as global optimization techniques are used to locate the global energy minimum. The second section gives a brief introduction to GAs and how they are used to predict protein tertiary structure.

A key word in GAs is coding: The objects to be optimized are represented by strings over a certain alphabet. The coding plays an important role in the convergence theory of GAs, the so called schema theorem. Codings not suited for the optimization task are called deceptive since the optimization will not converge to the optimum. Results of optimizations of the pentapeptide Met-Enkephalin suggest that the standardly used binary coding might be deceptive when minimizing empirical force fields. The third section introduces the schema theorem and the concept of deception that provide the means for an analysis of a coding.

The fourth section contains an analysis of the deception of the standard binary coding. The search space of Met-Enkephalin is too huge to decide analytically whether the coding is deceptive. All energy values of the finite search space are needed for that analysis. The analysis is therefore done by using a di- and a tripeptide where it can be carried out in reasonable time and disk space. The results are presented as bar charts that show the deception of the standard binary coding.

2 Protein structure prediction using GAs

The evolution of GAs started with Hollands book [1]. Goldbergs book [2] is the standard reference and a good introduction to the field. Genetic Algorithms are optimization methods, that employ inheritance mechanisms similar to those occuring in nature. They investigate a whole population of search points and sample the search space at many points at a time. In this way, information is obtained that directs the search to promising regions of the search space. Search points that contain good partial solutions are combined to form possibly better solutions. Similarities in the representation of the best points are used to identify subspaces with a higher frequency of good solutions. The information about previously reached good points is compactly stored in the composition of the population of search points.

The candidate solutions contained in the population are encoded as binary strings. A sequence of generations is generated by building new populations from the previous ones. The members of the new population are selected from the old one according to their objective values. The better the value the greater the probability that a member occurs in the next population. After this stage of selection the genetic operators are applied. The genetic operators introduce the mechanisms of inheritance. The most important operators are crossover and mutation. The crossover combines parts of two parent solutions to a new solution. It swaps a part of the bitstring that is determined by a randomly chosen crossover point. The mutation operator randomly changes the binary representation of a solution. The solutions are then decoded and the objective values are computed for the next generation. This main loop of a GA is iterated until a termination criterion is fulfilled.

Genetic Algorithms can be applied to many problems. Different applications have different objective functions and different representations of the solutions. The user supplies the objective function and determines how the binary strings are mapped to solutions. The rest of the GA is problem independent since the only information used to generate new solutions are the objective values.

An empirical energy function can be used as objective function when predicting protein structure by energy minimization. The different conformations are represented by vectors of dihedral angles of rotatable bonds in the backbone and the side chains. The bond lengths and bond angles are fixed in this rigid geometry model. Dihedral angles instead of cartesian coordinates are used so that identical local foldings always have identical codings and can be combined to conformations.

The pentapeptide Met-Enkephalin (TYR GLY GLY PHE MET) was optimized using 24 dihedral angles ranging from 0 to 360 degrees, binary coded at a resolution of 12 bits each. The Tripos [3] empirical energy function is used to compute the conformational energies. The optimization was repeated, using different random number generator seeds. They all converged to the same minimum structure. This structure is not the same as the best so far known structure reported by Nayeem, Vila and Scheraga [4]. The GA converged to this best minimum if a large fraction of the initial population (1024 out of 16384) is seeded with this structure. The search was tracked away from the minimum if only 5 and 512 copies were used. The population drifts to the suboptimal solution, despite that all parts of the minimum are present in the population. There are several sources of failure for a GA to converge to a suboptimal solution [5]. One of these is a property of the objective function and the associated coding called 'Deception'.

Deception can be detected analytically if the objective values of the whole search space are given. The search space of Met-Enkephalin is too huge for such an analysis, so it is done with small peptides. Although one cannot conclude from the results for small peptides to all peptides, it shows a tendency of this coding to be deceptive. The schema theory provides the means for the analysis.

3 Schema Theory and Deception

Deceptive problems can be distinguished from non deceptive ones by investigating the schemata. They provide a convergence theory for non deceptive problems by explaining why Genetic Algorithms are no random search techniques. New search points are generated randomly but more frequently in promising regions of the search space. In non deceptive problems the promising regions are really the better ones. Schemata [6] are similarity templates of strings. They define subspaces of the search space that consist of strings with common substrings. Bits that differ are marked with don't care symbols.
Example:

Schema	Strings contained in schema
10##1:	$\{10001, 10011, 10101, 10111\}$

The strings in the population are reproduced according to their fitness. The Schemata are implicitly reproduced according to their average fitness. The better the average fitness of strings contained in a schema is, the more strings of this schema will appear in the population. The population shifts to these promising regions of the search space where again the better subspaces will gain increasing numbers of strings in the population.
Example:
The Argument of $x \rightarrow x^2, x \in [0, 32]$ is binary coded. Good strings have 1's at the higher valued positions that will converge first.

Schema	Fitness
1####	average(16..31)=573.5
0####	average(0..15)=77.5
11###	average(24..31)=761.5
10###	average(16..23)=385.5

The proliferation rate of a schema also depends on its defining length, the distance of the outermost bits in the schema. The smaller the defining length the bigger the probabilty not to be destroyed by crossover. Short schemata with high fitness are called building blocks [7] since they combine to build better solutions.

Schemata with the same fixed positions define a partition of the search space, a set of disjunct schemata. The Schemata compete for the allocation of trials and the one with the highest average fitness will win. Problems are deceptive [8], if the optimum is not contained in the winning schema. The search is guided into the false subspace and a suboptimal solution will be the result.
Example:

minimum:	10111111	
schema1:	##0#####	5.0
schema2:	##1#####	8.0

Table 1: Rotatable bonds of the peptides defining the conformational space.

Angle No.	Peptide No. 1	Peptide No. 2
1	ALA1.N-ALA1.CA	TRP1.C-TRP1.CA
2	ALA1.CB-ALA1.CA	TRP1.CB-TRP1.CA
3	ALA1.C-ALA1.CA	TRP1.CG-TRP1.CB
4	ASN2.N-ASN2.CA	PRO2.N-TRP1.C
5	ASN2.CB-ASN2.CA	PRO2.C-PRO2.CA
6	ASN2.C-ASN2.CA	PHE3.N-PHE3.CA
7	ASN2.CG-ASN2.CB	PHE3.CB-PHE3.CA
8	ASN2.ND2-ASN2.CG	PHE3.CG-PHE3.CB

```
minimum:     10111111
schema1:     #####00#     3.0
schema2:     #####01#     4.0
schema3:     #####10#     9.0
schema4:     #####11#     6.0
```

The schema averages can be computed directly or using partition coefficients [9] and walsh transforms [10, 11]. Partition coefficients allow to easily compute each of the 3^l schema averages once the 2^l partition coefficients are computed from the 2^l objective values. Since this requires to solve a linear equation system of 2^l variables, here the schema averages are computed directly.

4   Schema Analysis

To judge whether a given function coding combination is deceptive, one has to compute all function values of the search space. The schema averages have to be computed and compared for sets of competing schemata. This investigation is here made with short peptides for whom it is feasible to calculate all energy values. The peptides are chosen to have not too few dihedral angles and to allow for clashes to be comparable to larger peptides. Figure 1 shows the two peptides and table 1 gives the rotatable bonds. The search space is defined by the rotatable bonds and the employed resolution. The angles are sampled at equidistant values according to the number of bits of their binary representation. Here eight angles are used with a resolution of three bits corresponding to 45 degrees. The search space contains therefore 8^8 points.

The angles are then coded and matched against the competing Schemata of a partition to calculate the average energy values. Normal binary coding and gray coding are used. All Schemata that have one, two or three sucessive fixed positions are investigated, since they may be building blocks. The schema averages are then plotted as a separate barchart for one, two and three bit schemata. The figures 2 and 3 show the bargraphs for Peptide No. 1 in normal and gray coding. The bar groups are ordered according to the position of the fixed bits in the string from left (000) to right (111). The bars in the bar groups are sorted according to the values of the fixed positions. Each bar group is labelled by the values that

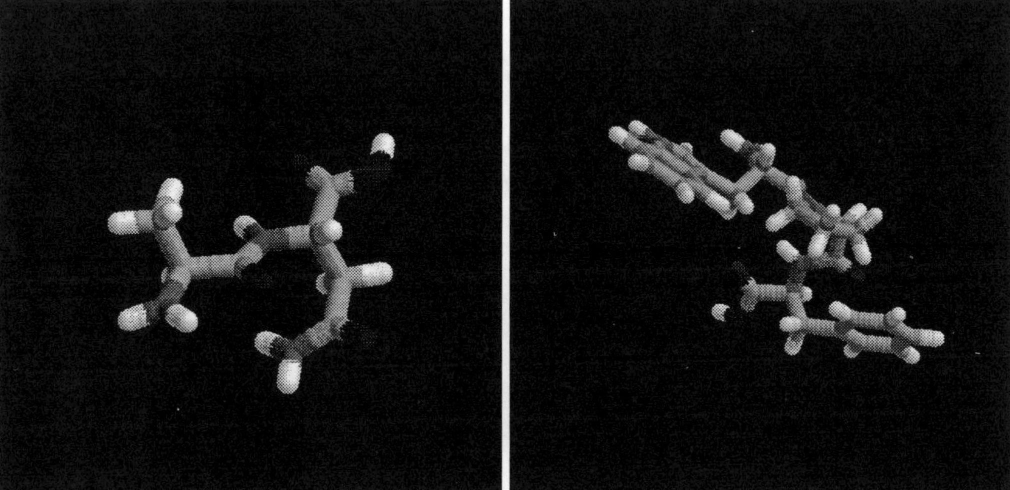

Figure 1: For color version see Color Plate 5, page xvii. Peptide No. 1: (ALA ASN) Peptide No. 2: (TRP PRO PHE). The conformational space of the peptides is sampled at a resolution of 45 degrees for 8 rotatable bonds. The associated energy values are used to compute average energy values of certain subspaces.

the fixed bits have in the minimum. Deceptive Schemata are those where the bar with the smallest value has not the fixed positions of the minimum. They are marked by a star.

The energy values are computed using the SEARCH command of the molecular modelling program Sybyl. The rotatable bonds were defined and the angle increment was adjusted according to the number of bits used in the coding. Peptide No. 1 required 15 hours and Peptide No. 2 required 2 days and 18 hours CPU time on an Evans and Sutherland Graphics Workstation consuming 107 Megabytes of disk space each. The resulting angle file contains the angle values of all conformations along with the corresponding energy values. The angle file is then translated into a binary format by coding the angle values of each conformation according to the coding used in the GA. The schema averages were computed in the next step. Each entry of the binary format file is matched against all schemata of a partition and contributes to the appropriate schema average. This is repeated for all partitions of length one, two and three. The schema averages are displayed using Mathematica.

Additional studies are done with subspaces of Met-Enkephalin, defined by the known minimum [4] and three out of 24 dihedral angles. It was the aim to see if even a subspace of the minimum contains deceptive partitions. Each dihedral angle is sampled at a resolution of 7 bits corresponding to about 3 degrees. This finer resolution is more like that used in the GA Applications. The energy values are computed at a resolution of 2 degrees since the SEARCH command allows only integral angle increments. The coded angles differ therefore from the angles contained in the angle file but are mapped to the closest value. Table 2 gives the three used sets of angles. Set No. 1 contains the first dihedral angle of each of the three side chains. Set No. 2 contains dihedral angles of the backbone at the glycines and Set No. 3 contains angles at the PHE=MET-End to allow for clashes of the sidechains. Figure 4 shows the barchart for the binary coded angle set No. 1.

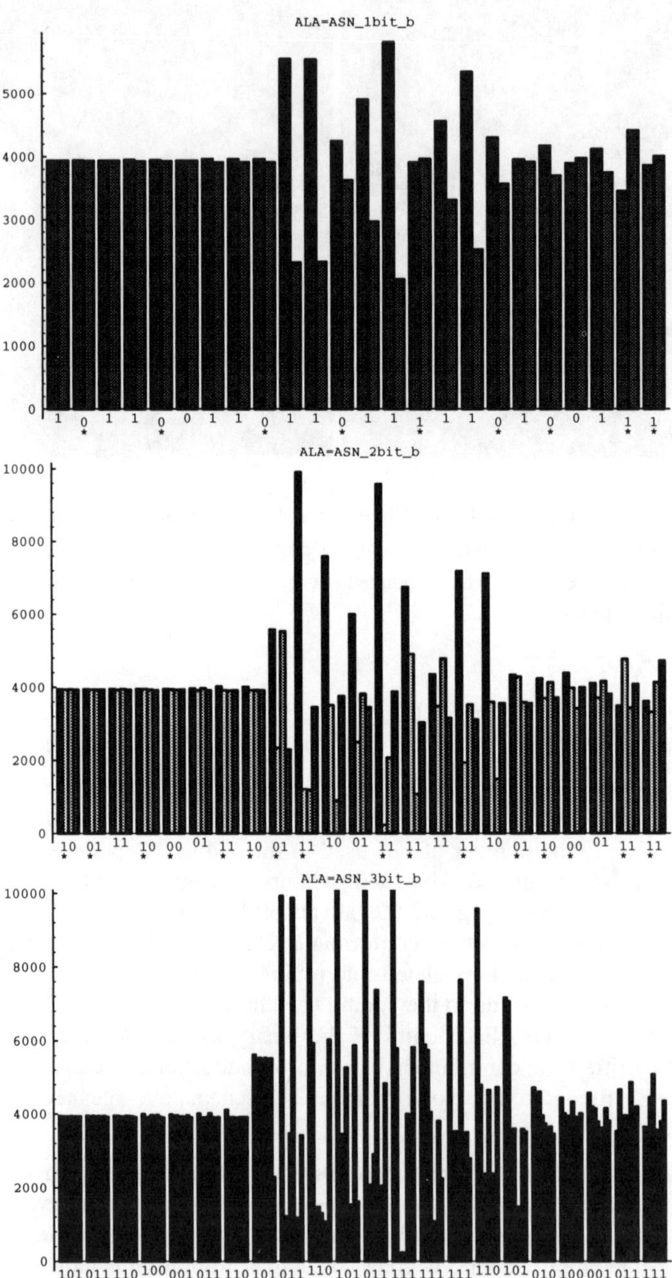

Figure 2: Peptide No. 1 binary coding. Each bar group corresponds to a partition of the conformational space. The bars show the average energy values of each subspace. The star marks partitions where the minimal energy value is not contained in the subspace with the minimal average energy value.

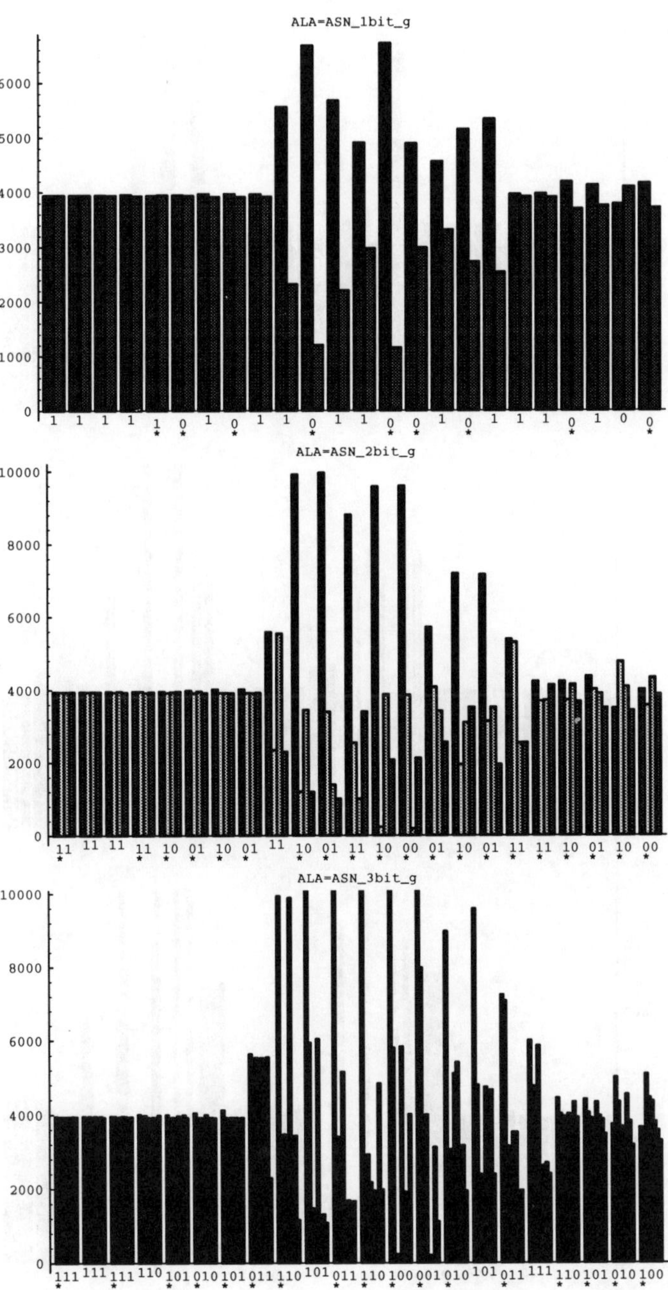

Figure 3: Peptide No. 1 gray coding

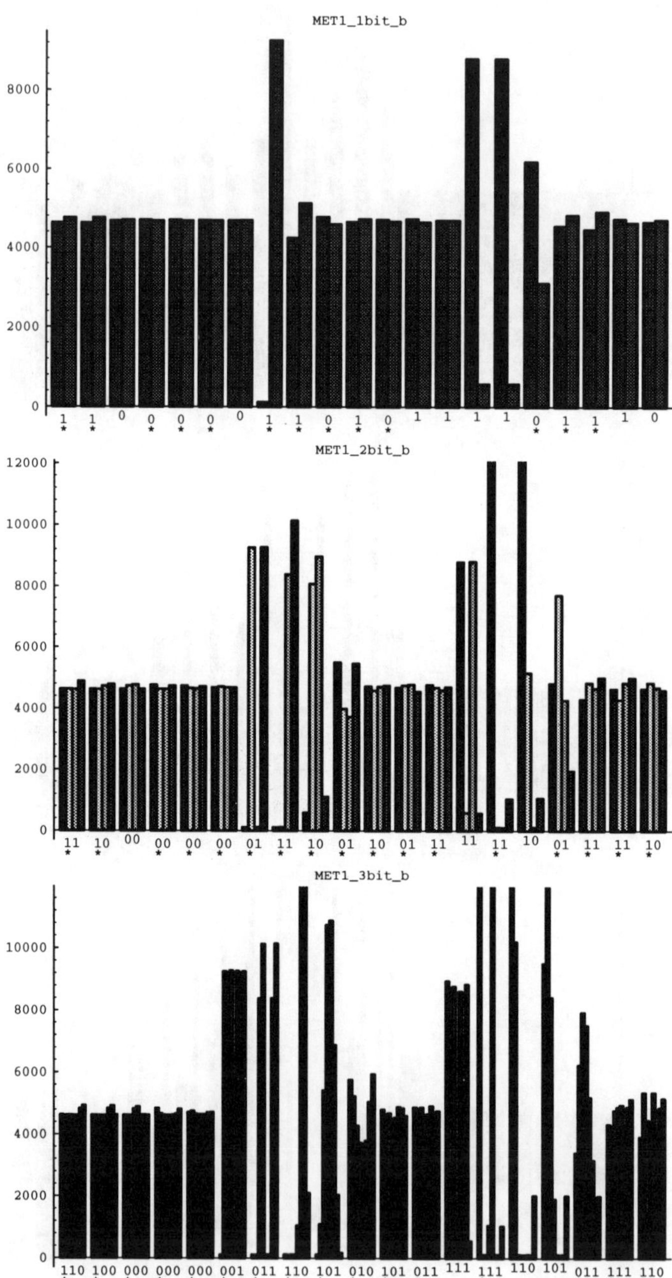

Figure 4: Met-Enkephalin Subspace No. 1 binary coding

Table 2: Rotatable bonds for Met-Enkephalin Subspaces

Angle No.	Angle Set No. 1	Angle Set No. 2	Angle Set No. 3
1	TYR1.CA-TYR1.CB	PHE4.CA-PHE4.CB	MET5.CA-MET5.CB
2	GLY2.N-GLY2.CA	GLY2.CA-GLY2.C	GLY3.N-GLY3.CA
3	PHE4.CA-PHE4.CB	MET5.N-MET5.CA	MET5.CA-MET5.CB

5   Discussion

It can be seen that almost all first to third order building blocks are deceptive. There is no significant difference in the deceptiveness of ordinary binary and gray coding. Some partitions have very different schema averages indicating that there are subspaces with mainly high energy values and those with mainly low ones, e.g. the schemata corresponding to the ϕ,ψ and χ_1 angles of ASN in Peptide No. 1. Many of them are deceptive with the minimum located in the subspace that contains mainly high energy values. The partitions with nearly equal schema averages, e.g. those corresponding to the ϕ and χ_1 angle in ALA of Peptide No. 1, show that these dihedral angles have almost no effect on the energy.

The fact that these small problems are GA-deceptive suggests that larger problems are GA-deceptive too. This reinforces the assumption that deception is the source of failure in structure prediction by energy minimization. Deception can be tackled by utilizing different representations and codings. These are the two main starting points for improvements in this field. Most GA applications to structure prediction published so far use dihedral angles and standard binary or gray coding so that there are no different approaches that can be compared.

The investigation of small problems can be extended. Empirical studies performing GA-optimizations of the investigated peptides will show if the predicted convergence to suboptimal solutions occurs. If the empirical studies show convergence to a suboptimal solution, then the schema analysis can be repeated using the suboptimal minimum instead of the global one. If it shows no deception, this is another hint that deception is the source of failure. Other sources of failure can be investigated. Crossover disruption occurs when highly fit low order schemata have large defining lengths and are therefore destroyed by crossover with a high probability. An analysis of these schemata will require a greater effort since there are more of them than there are building blocks. Sampling errors occur when high schema variances impede an appropriate estimation of schema averages using the samples provided by the population. An analysis have to compute schema distributions instead of schema averages.

To summarize, the small problems can be used as scaled down models of the larger problems to analytically and empirically investigate the properties of this representation coding combination. New representations and codings have to be developed to compare them and to improve the performance for larger problems.

References

[1] John H. Holland, Adaptation in Natural and Artificial Systems, The University of Michigan Press, 1975.

[2] David E. Goldberg, Genetic algorithms in search, optimization, and machine learning, Addison-Wesley Publishing Company, Inc., 1989.

[3] M. Clark, R. D. Cramer III, and N. Van Opdenbosch. Validation of the general purpose tripos 5.2 force field. *Journal of Computational Chemistry*, **10**, 982–1012 (1989).

[4] Akbar Nayeem, Jorge Vila, and Harold A. Scheraga. A comparative study of the simulated-annealing and monte carlo-with-minimization approaches to the minimum-energy structures of polypeptides: [met]-enkephalin. *Journal of Computational Chemistry*, **12**, 594–605 (1991).

[5] Gunar E. Liepins and Michael D. Vose. Deceptiveness and genetic algorithm dynamics. In Gregory J. E. Rawlins, editor, Foundations of Genetic, 36–50. Morgan Kaufmann Publishers, 1991.

[6] David E. Goldberg. Genetic algorithms in search, optimization, and machine learning, 18–20 and 28–33. Addison-Wesley Publishing Company, Inc., 1989.

[7] David E. Goldberg. Genetic algorithms in search, optimization, and machine learning, 41–45. Addison-Wesley Publishing Company, Inc., 1989.

[8] David E. Goldberg. Genetic algorithms in search, optimization, and machine learning, 46–52. Addison-Wesley Publishing Company, Inc., 1989.

[9] David E. Goldberg. Genetic algorithms in search, optimization, and machine learning, 373–379. Addison-Wesley Publishing Company, Inc., 1989.

[10] David E. Goldberg. Genetic algorithms and walsh functions: Part i, a gentle introduction. *Complex Systems*, **3**, 129–152, 1989.

[11] David E. Goldberg. Genetic algorithms and walsh functions: Part ii, deception and its analysis. *Complex Systems*, **3**, 153–171, 1989.

Secondary and Supersecondary Structures, Folds and Conformations in Proteins

Super–secondary Structures in Proteins

Alexander V. Efimov

Institute of Protein Research, Russian Academy of Sciences, Pushchino, Moscow Region,
Russia

Abstract

A number of super-secondary structures of different types is described in this
report. Super-secondary structures can be defined as commonly occurring
folding units consisting of two or more elements of secondary structure. As
a rule, they are formed by secondary structure segments adjacent along the
polypeptide chain. Super-secondary structures of each given type found in both
homologous and non-homologous proteins have a very similar overall fold and
arrangement of α–helices and/or β–strands, but the loops can differ in length
and conformation. A distinctive feature of super-secondary structures is that
many of them have a unique handedness. All this together shows that the
information about arrangement of super-secondary structures and their features
may be of particular value in protein modelling and prediction.

1 Introduction

The problem of protein structure prediction includes at least two major tasks, prediction of
all the secondary structure elements and packing of the elements into a three-dimensional
structure. The rules that govern the packing together of individual (not connected) α-
helices and/or β-sheets and that result primarily from the principle of close packing and
geometric constraints have been described previously [1, 7]. On the other hand, connecting
loop regions play as important a role in protein folding as α-helical and β-sheet regions.
Super-secondary structures are the main subject of the present study as they represent the
simplest models of the packing together of secondary structure elements connected by loops.
Other features in structural organization of proteins resulting from the connectivity of the
polypeptide chain will also be discussed in this paper.

2 α-Helical Super-secondary Structures

Pairs of connected α-helices packed so that to form α-α-hairpins represent the simplest
type of α-helical super-secondary structure. α-α-Hairpins can be both right-turned and left-
turned depending on whether the second α-helix runs on the right or left relative to the first
one when viewed from the same side (for example, from the hydrophobic core). In the cases
of short connections, it is possible to determine whether the α-helices are packed into a right-
or left-turned α-α-hairpin as they have different sequence patterns for the key hydrophobic,
hydrophilic and glycine residues [8]. α-α-Hairpins are widespread in globular proteins and
usually included into more complex structures, in part, into four-α-helical super-secondary

                                                                    Alexander V. Efimov

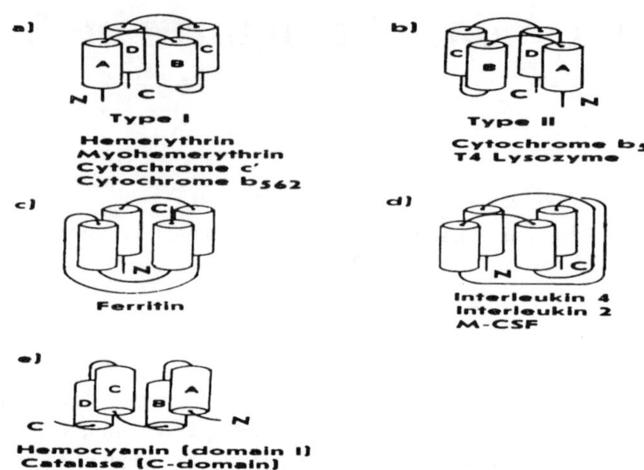

Figure 1: Four-α-helical super-secondary structures with the alinged packing of α-helices and some examples of known proteins where they have been found: hemerythrin [16], myohemerythrin [17], cytochrome c' [18], cytochrome b$_{562}$ [19], cytochrome b$_5$ [20], T4 phage lysozyme [21], ferritin [22], interleukin 4 [23], interleukin 2 [24], macrophage colony-stimulating factor, M-CSF [25], hemocyanin [26], catalase [27].

structures. The four-α-helical bundle with up-and-down topology was the first example of a large repeating folding unit found in α-helical proteins and domains [9]. There are two types (I and II) of packing of four α-helices into such bundles (Figs. 1a,b) that can be defined as the right- and left-handed bundles, respectively, Four consecutive α-helices can be packed into bundles of other types if they have one (Fig. 1c) or two (Fig. 1d) long overhand connections. The four-α-helix bundles with overhand connections can also be right- and left-handed (for details, see [10]). Fig. 1e shows one more super-secondary structure in which four consecutive α-helices are packed into a right-handed superhelix. Similar superhelices can also be formed by three (see, e.g., annexin, [11]), five (domain I of δ-endotoxin, [12]) and more α-helices (lipovitellin-phosvitin [13], cellulose [14], glucoamylase [15]). All the α-helical superhelices found so far are right-handed.

As seen, the four-α-helix bundles with overhand connections (Figs. 1c and d) can be formed if the corresponding connections are long enough. So four connected α-helices should be packed into the bundles of type I, type II or into the right-handed superhelix (Fig. 1e) if all the connections are short or if some of them are short and the others are not so long to form overhand connections. In theory, it is possible to determine whether the α-helices are packed into a right- or left-turned α-α-hairpin from their sequence patterns. So if, for example, both the hairpins, A-B and C-D, are encoded to be left-turned (when viewed from the hydrophobic core), they should be packed into a bundle of type I irrespective of the length and conformation of loop BC. In the case when both the hairpins, A-B and C-D, are encoded to be right-turned, a bundle of type II is formed. When hairpin A-B is left-turned and hairpin C-D is right-turned a superhelical structure (Fig. 1e) is formed.

In proteins with the predominantly orthogonal packing of α-helices, the right-handed α-α-corner is a highly recurrent folding unit. The α-α-corner is a super-secondary structure formed by two consecutive α-helices packed approximately crosswise and connected by an interhelical loop (Fig. 2). The α-α-corner with the short connection has the $\alpha_m\gamma\alpha_L\beta\beta\alpha_n$-

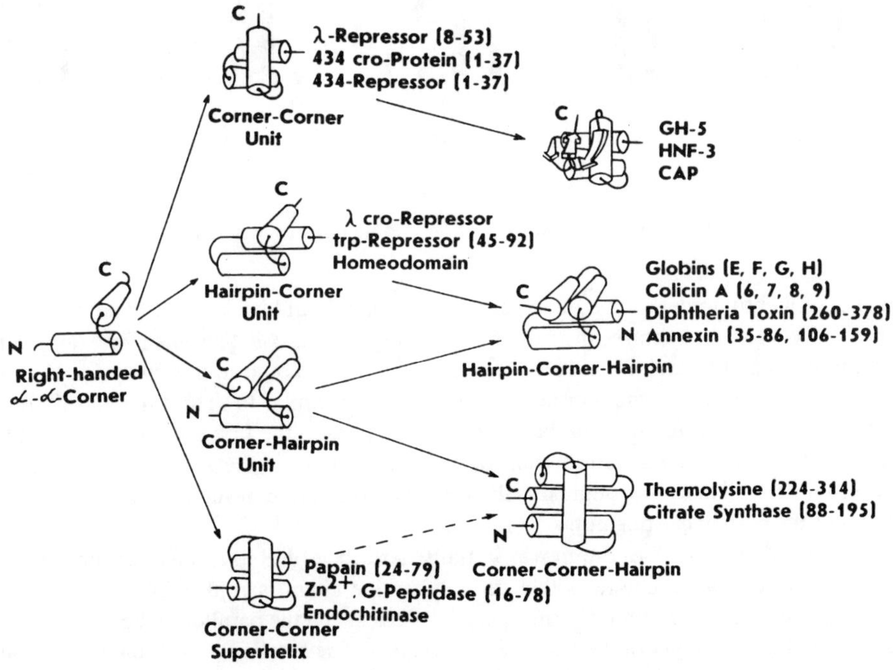

Figure 2: Commonly occurring folding units in proteins with the orthogonal packing of α-helices. Structural information has been taken from the following papers: l-repressor [30], phage 434 cro-protein [31], phage 434 repressor [32], cro-repressor from bacteriophage l [33], trp- repressor [34], engrailed homeodomain [35], papain [36], G-peptidase [37], endochitinase [38], GH5 [39], HNF-3 [40], CAP [41], colicin A [42], diphtheria toxin [43], thermolysin [44], citrate synthase [45], annexin [11].

conformation and a strictly definite sequence pattern for the key hydrophobic, hydrophilic and glycine residues (for details, see [28, 29]). α-α-Corners are widespread in known proteins and occur practically always in the right-handed form independent of the length and conformation of the interhelical connection. Other recurrent structural motifs in proteins with the orthogonal packing of α-helices are shown in Fig. 2. Each of these structures can be represented as a combination of righthanded α-α-corners and α-α-hairpins and can be obtained, for example, by stepwise addition of α-helices to the α-α-corner. Possible three-α-helix structures are presented in the left column. If both the pairs of α-helices are packed into α-α-corners a corner-corner unit or a left-handed corner-corner superhelix can be formed. Analysis of known proteins shows that the left-handed corner-corner superhelix is formed when both the loops are long. In contrast, the corner-corner units are formed if the connections are short (in the corresponding units of l-repressor, 434 cro-protein and 434-repressor both the α-α-corners have the $\alpha_\mathrm{m}\gamma\alpha_\mathrm{L}\beta\beta\alpha_\mathrm{n}$-conformations) or medium-sized and short (GH5, HNF-3 and CAP).

If one pair of α-helices is folded into an α-α-corner and the other into an α-α-hairpin the hairpin-corner or corner-hairpin units are formed. In the case of short connections, it is possible to differentiate which α-helices should form the α-α-hairpin and the α-α-corner

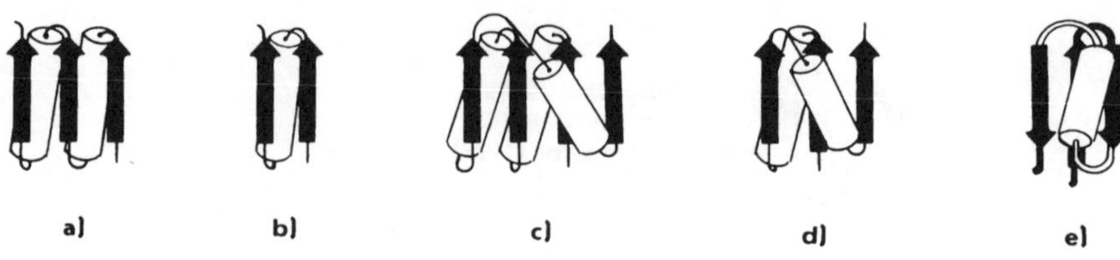

Figure 3: Commonly occuring folding units in α/β-proteins.

and consequently to determine which of the two units should be formed. In some proteins, there are V-shaped structures instead of the corresponding α-α-hairpins (for example E-F pairs in globins) but such differences are not taken into account in this analysis. Note that the overall fold of the corner-corner unit may be quite similar to that of the hairpin-corner unit especially when the Ω angle between the α-helices in the first α-α-corner is less than 90° and the α-α-corner is reminiscent of an α-α-hairpin. If the second α-α-corner of a corner-corner unit has the Ω angle less than 90° the overall fold of such a unit may be similar to that of the corner-hairpin unit.

The repetitive β-α-β-α-β-unit was initially described by Rao and Rossmann [49]. The main feature of this structure is that its polypeptide chain is folded into a right-handed superhelix so that the β-strands form a parallel β-sheet and the parallel packed α-helices are located in the other layer of the superhelix (Fig. 3a). The smaller β-α-β-unit is also folded into a similar right-handed superhelix (Fig. 3b) in the large majority of observed examples [50]. It is important to note that the handedness and the overall folds of these folding units are not directly dependent on the length and conformation of their loops and they can have both short and long loops.

Addition of an α-helix to the hairpin-corner or corner-hairpin unit results in formation of a hairpin-corner-hairpin unit that is the commonly occurring folding unit in globins, colicin A, diphtheria toxin, domains of annexin and other proteins. Note that the overall fold of globins is very similar to that of the colicin A region including the helices from 4 to 9 (helices, A, B, E, F, G and H, of globins correspond to helices, 4, 5, 6, 7, 8 and 9, of colicin A, respectively). The overall fold of region 225-378 of diphtheria toxin is also similar to that of globins. Addition of an α-helix to the corner-hairpin unit can also result in formation of a corner-corner-hairpin structure that occurs in thermolysine and citrate synthase. A search for other possible structures that can be obtained by stepwise addition of α-helices and β-strands to the α-α-corner is described in detail elsewhere [28].

3   Super-secondary Structures of α/β-Proteins

The majority of the α/β-proteins are composed of α-helices and β-strands that roughly alternate along the polypeptide chain. They can be grouped into three subclasses depending on their overall folds. Some of them are folded into α/β-barrels, the others into three-layer structures having a parallel β-sheet and α-helices packed upon its faces, and some miscellaneous α-β-proteins are folded into two-layer structures having both the parallel and antiparallel packing of β-strands (see, e.g., [46, 48]). Super-secondary structures found in these proteins are presented in Fig. 3.

In the three-layer α/β-proteins and domains, there are more complex super-secondary

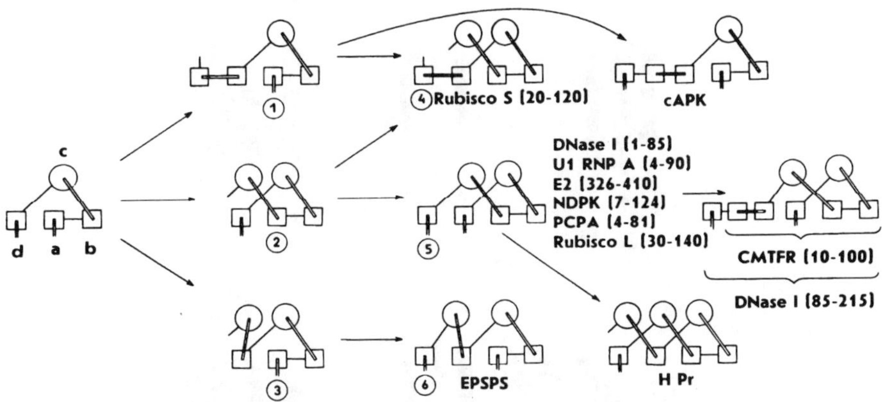

Figure 4: A scheme of search for possible structures of two-layer α/β-proteins using a stepwise addition of α-helices and β-strands to the abcd-unit. The structures are viewed end on with α-helices represented as circles and β-strands as rectangles. The near connections are shown with double lines and the far connections with thin lines. It is presumed that each structure can have both the directions of the polypeptide chain. Abbrevations used: Rubisco S and L, the small and large subunits of spinach Rubisco [54]; DNase I, bovine pancreatic deoxyribonuclease I [55]; U1 RNP A, U1 small nuclear ribonucleoprotein A [56]; E2, E2 DNA-binding domain [57]; NDPK, nucleoside diphosphate kinase [58]; PCPA, procarboxypeptidase A [59]; EPSPS, EPSP synthase [60]; cAPK, cAMP-dependent protein kinase [61]; CMTF R, aspartate carbamoyltransferase R-subunit [62]; HPr, histidine-containing phosphocarrier protein [63].

structures involving five or seven segments of regular secondary structure (Fig. 3c, d). They are widespread in this group of α/β-proteins, for example, the five-segment structure occurs in flavodoxin and other proteins having flavodoxin-like structures and the seven-segment structure is found in dehydrogenases. As a rule, these super-secondary structures are located on the edges of molecules and domains. In some proteins, there are structures analogous to those shown in Fig. 3c, d, which have the opposite directions of the polypeptide chains.

In two-layer α/β-proteins, the abcd-unit in which region c is presented as an α-helix [51] is a commonly occurring folding unit (Fig. 3e). This structure has an antiparallel β-sheet formed by strands b, a and d, and a right-handed β-α-β-superhelix formed by regions b, c and d. The unit having the opposite direction of the polypeptide chain (as compared to that of the unit shown in Fig. 3e) is also widespread in proteins. In known proteins, the abcd-units are located on the edges of molecules and domains and arrangements of other α-helices and β-strands can be obtained by stepwise addition of the segments to the unit using simple rules as follows:

1. An α-helix and β-strands can not be packed in one layer because of dehydration of the NH- and CO-groups of the β-strands [4]; so a subsequent α-helix should be packed into the α-helical layer and a β-strand into the β-layer (see also [52]).

2. Crossing of connections is prohibited in proteins [53].

3. β-α-β-units should be folded into right-handed superhelices [49, 50].

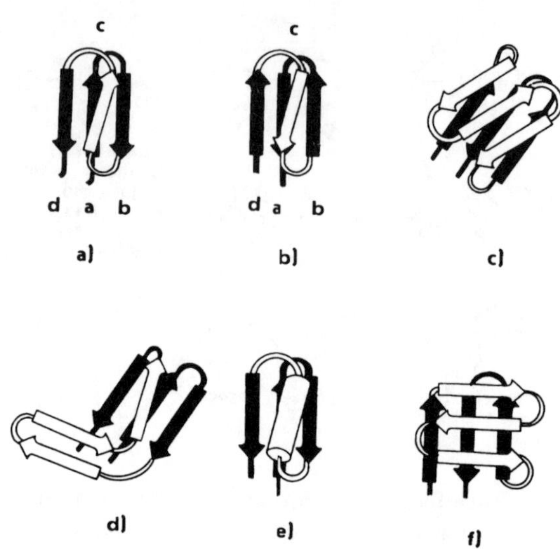

Figure 5: A schematic representation of the abcd-unit and its variants found in β-proteins. Strands a, b and d forming the antiparallel β-sheet are shown by black arrows and the strands of the c-regions by white arrows; the cylinder represents an α-helix.

If a molecule or a domain has an α-helix joined to strand a, it is packed next to helix c that results in formation of structure 2 (Fig. 4). If there is no an α-helix joined to strand a in a molecule but there is an α-helix joined to strand d it is packed next to helix c (see structures 3 and 6 in Fig. 4). β-Strands joined to strand d are packed into the β-sheet next to strand d (see structures 1 and 4 and the structure observed in cAPK). Addition to structure 2 of a β-strand joined to the edge α-helix results in structure 5 that is found in many proteins and domains. Addition to structure 5 of an α-helix joined to strand d results in the structure found in HPr. β-Strands joined to the edge β-strand of structure 5 are packed into the β-sheet as in CMTFR and DNase I. A more complete description of this analysis will be published elsewhere (Efimov, A., in preparation).

4   Super-secondary Structures of β-Proteins

β-Proteins can be grouped into two basic classes, depending on the β-sheet packing. One class includes the β-proteins with the aligned β-sheet packing and the other class includes β-proteins with the orthogonal β-sheet packing. They can also be grouped into several families, depending on the recurrent folding unit or the super-secondary structure that occurs in each family.

In the family of bilayer β-proteins with the aligned β-sheet packing, the abcd-unit is the commonly occurring folding unit [51]. The simplest variant of the abcd-unit consists of four consecutive β-strands three of which (a, b and d) lie in one layer forming an antiparallel β-sheet and one strand (c) is located in the other layer. Strands b, c and d form the righthanded

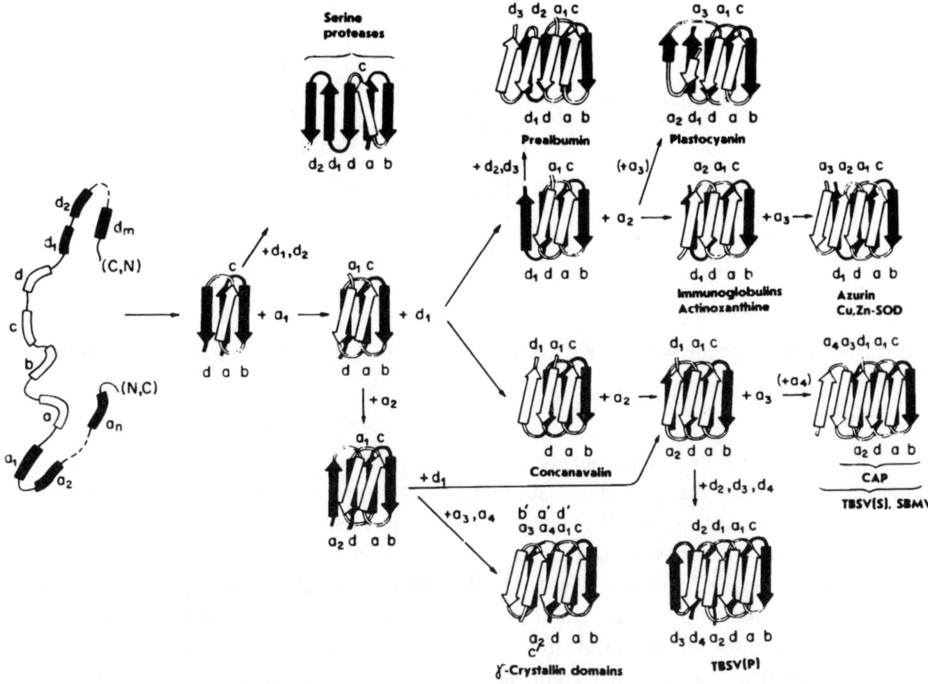

Figure 6: A schematic diagram showing possible ways of packing of β-strands relative to the abcd-unit in the known β-proteins if the strands are added to it step by step. Black arrow represent β-strands in the far b-layers and white arrows are β-strands in the near c-layers. The order of the strands in the b-layer is shown at the botton of each structure by letters with subscripts, and the order of the strands in the c-layer is shown at the top. For some proteins the direction of the arrows coincides with the chain direction from the N- to C-ends for others not. Structural information on the proteins is taken from the papers cited in [29, 51].

superhelix, bcd that is analogous to the superhelix formed by β-α-β-units. Strand a is always situated between strands b and d in a direction antiparallel to them. There are two variants of the simple abcd-unit which differ from one another in the direction of the polypeptide chain (Fig. 5a, b). A more complex variant of the abcd-unit where region c is presented as a triple-strand β-sheet (Fig. 5c) also occurs in the β-proteins with the aligned β-sheet packing (for example, in the V domain of human IgG New [64]. For comparison, Figs. 5d, e and f show other complex variants of the abcd-unit which are found in other proteins. As seen, the overall fold of the polypeptide chains and the three-dimensional arrangement of segments a, b, c and d are very similar in all the observed abcd-units.

Analysis of known proteins with the aligned β-sheet packing shows that the abcd-unit is always located at the edge of the double layer. The other strands of a molecule or a domain are situated on the side of the d-strand (Fig. 6). Arrangement of other β-strands relative to the abcd-unit can be obtained if the next strands are added to it step by step (see Fig. 6). In each step, the next segment can be packed into the b- or c-layer depending on the structure that has been formed in the preceding step (for details, see [51, 29]). In β-proteins with the orthogonal β-sheet packing there are other types of super-secondary structure, the

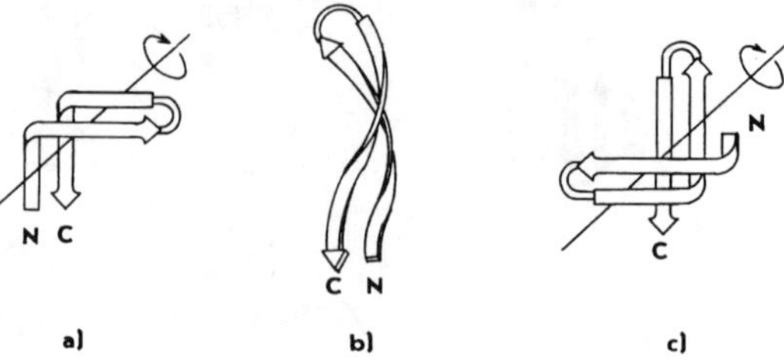

Figure 7: A schematic representation of the right-handed β-β-corner (a), coiled β-β-hairpin (b) and 3β-corner (c).

so-called β-β-corners and 3β-corners (Fig. 7). The β-β-corner can be represented as a long β-β-hairpin folded on itself so that β-strands of the two halves are packed approximately orthogonally in two different layers. β-β-Corners are right-handed in proteins of known structure. This means that the strands rotate about an imaginary axis in the right-handed direction when passing from one β-sheet to the other. The other important feature is that β-β-corners are formed by right-turned β-β-hairpins if they are viewed from the concave sides irrespective of the length and conformation of the loops. A β-β-hairpin can be strongly twisted and coiled so that to form a right-handed double-strand superhelix (Fig. 7b). When viewed from the concave side, the β-β-hairpin is always rightturned in this structure [65].

The triple-strand corner (or the 3β-corner) can be represented as an antiparallel triple-strand β-sheet folded on itself so that the two β-β-hairpins are packed approximately orthogonally in different layers (Fig. 7c). The 3β-corners are right-handed in proteins, i.e. the central strand rotates about an imaginary axis in the right-handed direction when passing from one layer to the other. In the right-handed 3β-corner the N-terminal β-β-hairpin is right-turned and the C-terminal β-β-hairpin is left-turned when viewed from the concave side. The 3β-corners are widespead in proteins and are situated at the edges of molecules or domains. The other β-strands of a molecule or a domain are arranged on the concave side of the 3β-corner to form a compact structure (for details, see [66]).

5   Super-secondary Structures involving Triple-strand β-sheets

Triple-strand β-sheets having up-and-down topology are widespread in proteins and occur in two forms denoted here as S-like and Z-like β-sheets. In the S-like β-sheet, the first and second β-strands form a left-turned β-β-hairpin but the second and third β-strands form a right-turned β-β-hairpin. The opposite arrangement is observed in the Z-like β-sheet. In the cases of long loops, it is rather difficult to find preference for one form of the β-sheet if they are considered as isolated flat structures. The preference is observed at the level of supersecondary structures of higher order that involve the β-sheets or if the triple-strand β-sheets fold into three-dimensional structures themselves.

All the 3β-corners observed in proteins can be represented as the Z-like β-sheets folded on themselves when viewed from the concave sides (see Fig. 7c). A Z-like β-sheet (when

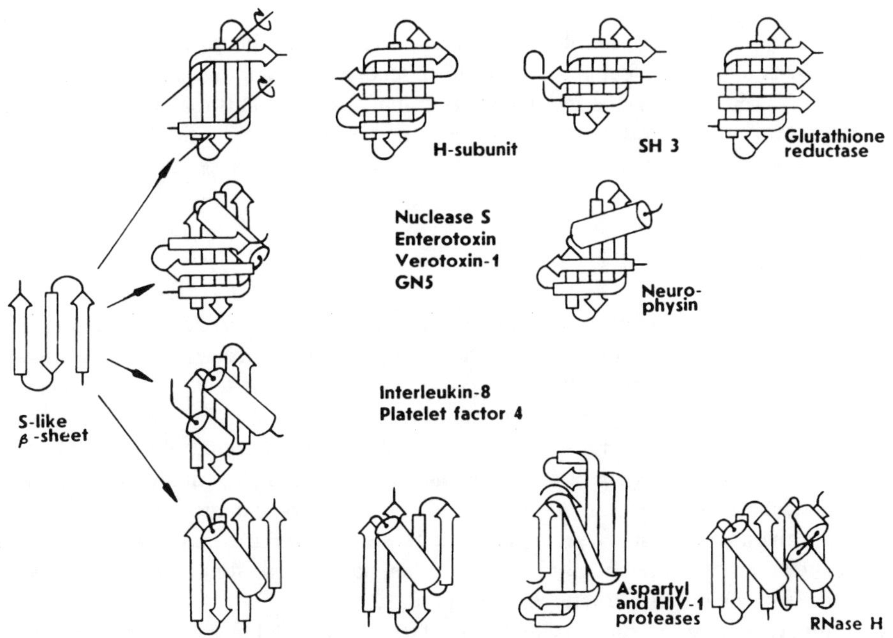

Figure 8: A schematic representation of structures involving S-like β-sheets when viewed from inside. β-Strands are shown as arrows directed from the N- to C-termini and α-helices as cylinders. The structural information has been taken from the following papers: H-subunit of the reaction centre [67]; SH3, the Src-homology 3 domain [68]; glutathione reductase [69]; nuclease S [70]; enterotoxin [71]; verotoxin-1 [72]; GN5, the gene 5 protein of bacteriophage fd [73]; bovine neurophysin [74]; interleukin 8 [75]; bovine platelet factor 4 [76]; HIV-1 protease [77]; aspartyl protease [78]; RNase H, ribonuclease H [79].

viewed from the hydrophobic core) can be involved in the abcd-unit with the aligned β-sheet packing (Fig. 5c). The structures involving S-like β-sheets are shown in Fig. 8. The left column of the figure represents framework structures and some examples from known proteins are shown on the right in the corresponding rows. The framework structure in the upper row consists of an S-like β-sheet flanked by two right-handed bends. In accordance with the definition [6] in the right-handed bends the strands bend through 90 in the right-handed direction when passing from one β-sheet to the other. The overall fold of this framework structure can be represented as a right-handed superhelix if the S-like β-sheet is replaced by one imaginary strand. In the structures of the second row, the S-like β-sheet is flanked by the right-handed bend at the N-terminus and by an α-helix at the C-terminus. All together these elements form a right-handed superhelix if the S-like β-sheet is replaced by one imaginary strand. Addition to this superhelix of a C-terminal β-β-hairpin results in formation of the common structural motif observed in nuclease S, enterotoxin and verotoxin-1. This motif was recently described in detail and called the OB-fold [80]. In the structures of the third and bottom rows, S-like β-sheets are also involved in right-handed superhelices.

As seen, the S-like and Z-like β-sheets can be components of super-secondary structures of higher order which have a unique handedness themselves. These are right-handed

3β-corners (Fig. 7c), a complex variant of the abcd-unit having the right-handed bcd-superhelix (Fig. 5c) and several folds shown in Fig. 8 having different types of right-handed superhelices. A distinctive feature of these super-secondary structures is that some of them can involve only S-like β-sheets and the others Z-like β-sheets. This feature is of particular value in protein folding and modelling since it allows to determine the handedness of a triple-strand β-sheet if its structural "context" is known (for details, see [81]).

6 Discussion

As seen, globular proteins have a number of common features to be a consequence of chain connectivity. One of them is that there are commonly occurring folding units or super-secondary structures consisting of two or more α-helices and/or β-strands connected by loops. Super-secondary structures of each given type found in both homologous and non-homologous proteins have a very similar overall fold and packing of α-helices and/or β-strands. A distinctive feature of super-secondary structures is that many of them have a unique handedness. Figures 2, 4 and 6 (see also Fig. 4 in [66]) demonstrate that proteins involving a given type of super-secondary structure have similar overall folds and can be grouped into one structural family. On the other hand, these figures clearly show the differences between protein structures in each family.

One other important feature is that one of the two possible forms of some super-secondary structures can be selected at the level of structures of higher order. For example, β-β-hairpins can be both right-turned and left-turned in proteins but if a β-β-hairpin forms a strongly twisted and coiled structure or folds into a β-β-corner, it is practically always right-turned when viewed from the concave side. Similarly, triple-strand β-sheets having up-and-down topology can exist in two forms and the preference for one of these appears at the level of super- secondary structures of higher order that involve the β-sheets.

As shown above, α-α-corners, abcd-units in two-layer α/β-proteins and β-proteins, five- and second-segment supersecondary structures in α/β-proteins and 3β-corners are situated at edges of protein molecules and the remaining part of a molecule or domain is located on the definite side of each super-secondary structure. The packing of the other elements of secondary structure can be obtained by stepwise addition of α-helices and/or β-strands to a super-secondary structure taking into account the simple rules (see [28, 51, 66] and Figs. 2, 4 and 6). It is important to note that practically all the known protein structures of each family can be obtained in this way.

An analysis of the three-dimensional arrangement of α-helices and/or β-strands in known proteins and the features considered above suggest a hypothesis that super-secondary structures can fold independently of the remaining parts of molecules and can be nuclei of protein folding or ready "building blocks". The nuclei can grow, for example, in accordance with the schemes shown in Figs. 2, 4 and 6 (see also Fig. 4 in [66]). If this is not so and proteins fold in other ways, these schemes can be used for the prediction and modelling of protein structures.

References

[1] F.H.C. Crick *Acta Crystallogr.*, 6, 689–691 (1953).

[2] C. Chothia, M. Levitt and D.C. Richardson *Proc. Natl. Acad. Sci.*, **74**, 4130–4134 (1977).

[3] C. Chothia, M. Levitt and D.C. Richardson *J. Mol. Biol.*, **145**, 215–250 (1981).

[4] A.V. Efimov *Dokl. Akad. Nauk SSSR*, **235**, 699–702 (1977).

[5] A.V. Efimov *J. Mol. Biology*, **134**, 23–40 (1979).

[6] C. Chothia and J. Janin *Biochemistry*, **21**, 3955–3965 (1982).

[7] A.G. Murzin and A.V. Finkelstein *J. Mol. Biol.*, **204**, 749–770 (1988).

[8] A.V. Efimov *Protein Eng.*, **4**, 245–250 (1991).

[9] P. Argos, M.G. Rossmann and J.E. Johnson *Biochem. Biophys. Res. Commun.*, **75**, 83–86 (1977).

[10] S.R. Presnell and F.E. Cohen *Proc. Natl. Acad. Sci.*, **86**, 6592–6596 (1989).

[11] R. Huber, J.Römisch and E.-P. Paques *EMBO J.*, **9**, 3867–3874 (1990).

[12] J. Li, J. Carroll and D.J. Ellar *Nature*, **253**, 815-821 (1991).

[13] R. Raag, K. Appelf, N.-H. Xuong and L. Banaszak *J. Mol. Biol.*, **200**, 553-569 (1988).

[14] M. Juy, A.G. Amit, P.M. Alzari, R.J. Poljak, M. Claeyssens, P.Béguin, J.-P. Aubert *Nature*, **357**, 89-91 (1992).

[15] E.M.S. Harris, A.E. Aleshin, L.M. Firsov and R.B. Honzatko *Biochemistry*, **32**, 1618-1626 (1993).

[16] R.E. Stenkamp, L.C. Sieker, L.H. Jensen and J.E. McQueen Jr. *Biochemistry*, **17**, 2499-2504 (1978).

[17] S. Sheriff, W.A. Hendrickson and J.L. Smith *J. Mol. Biol.*, **197**, 273-296 (1987).

[18] B.C. Finzel, P.C. Weber, K.D. Kardman and F.R. Salemme *J. Mol. Biol.*, **186**, 627-643 (1985).

[19] F. Lederer, A. Glatigny, P.H. Bethge, H.D. Bellamy and F.S. Mathews *J. Mol. Biol.*, **148**, 427-448 (1981).

[20] F.S. Mathews, M. Levine and P. Argos *J. Mol. Biol.*, **64**, 449-464 (1972).

[21] S.J. Remington, L.F. Ten Eyck and B.W. Matthews *Biochem. Biophys. Res. Commun.*, **75**, 265-270 (1977).

[22] D.W. Rice, B. Dean, J.M.A. Smith, J.L. White, G.C. Ford and P.M. Harrison *FEBS Lett.*, **181**, 165-168 (1985).

[23] A. Wlodaver, A. Pavlovsky and A. Gustchina *FEBS Lett.*, **309**, 59-64 (1992).

[24] D.B. McKay *Science*, **257**, 412-413 (1992).

[25] J. Pandit, A. Bohm, J. Jancarik, R. Halenbeck, K. Koths and S.-H. Kim *Science*, **258**, 1358-1362 (1992).

[26] N.M. Soeter, P.A. Jekel, J.J. Beintema, A. Volbeda and W.G.J. Hol *Eur. J. Biochem.*, **169**, 323-332 (1987).

[27] M.R.N. Murthy, T.J. Reid III, A. Sicignano, N. Tanaka and M.G. Rossmann *J. Mol. Biol.*, **152**, 465-499 (1981).

[28] A.V. Efimov *FEBS Lett.*, **166**, 33-38 (1984).

[29] A.V. Efimov *Prog. Biophys. Mol. Biol.*, **60**, 201-239 (1993).

[30] L.J. Beamer and C.O. Pabo *J. Mol. Biol.*, **227**, 177-196 (1992).

[31] A. Mondragón, C. Wolberger and S.C. Harrison *J. Mol. Biol.*, **205**, 179-188 (1989).

[32] A. Mondragón, S. Subbiah, S.C. Almo, M. Drottar and S.C. Harrison *J. Mol. Biol.*, **205**, 189-200 (1989).

[33] W.F. Anderson, D.H. Ohlendorf, Y. Takeda and B.W. Matthews *Nature*, **290**, 754-758 (1981).

[34] R.W. Schevitz, Z. Otwinowski, A. Joachimiak, C.L. Lawson and P.B. Sigler *Nature*, **317**, 782-786 (1985).

[35] C.R. Kissinger, B. Liu, E. Martin-Blanco, T.B. Kornberg and C.O. Pabo *Cell*, **63**, 579-590 (1990).

[36] J. Drenth, J.N. Jansonius, R. Koekoek and B.G. Wolthers *Adv. Protein Chem.*, **25**, 79-115 (1971).

[37] O. Dideberg, P. Charlier, G. Dive, B. Joris, J.M. Frére and J.M. Ghuysen *Nature*, **299**, 469-470 (1982).

[38] P.J. Hart, A.F. Monzingo, M.P. Ready, S.R. Ernst and J.D. Robertus *J. Mol. Biol.*, **229**, 189-193 (1993).

[39] V. Ramakrishman, J.T. Finch, V. Graziano, P.L. Lee and R.M. Sweet *Nature*, **362**, 219-223 (1993).

[40] K.L. Clark, E.D. Halay, E. Lai and S.K. Burley *Nature*, **364**, 412-420 (1993)

[41] D.B. McKay, I.T. Weber and T.A. Steitz *J. Biol. Chem.*, **257**, 9518-9524 (1982).

[42] M.W. Parker, J.P.M. Postma, F. Pattus, A.D. Tucker and D. Tsernoglou *J. Mol. Biol.*, **224**, 639-657 (1992).

[43] S. Choe, M.J. Bennett, G. Fujii, P.M.G. Curmi, K.A. Kantardjieff, R.J. Collier and D. Eisenberg *Nature*, **357**, 216-222 (1992).

[44] M.A. Holmes and B.W. Matthews *J. Mol. Biol.*, **160**, 623-639 (1982).

[45] S. Remington, G. Wiegand and R. Huber *J. Mol. Biol.*, **158**, 111-152 (1982).

[46] C.I. Brädén *Q. Rev. Biophys.*, **13**, 317-338 (1980).

[47] J.S. Richardson *Adv. Protein Chem.*, **34**, 167-339 (1981).

[48] C. Chothia and A.V. Finkelstein *Annu. Rev. Biochem.*, **59**, 1007-1039 (1990)

[49] S.T. Rao and M.G. Rossmann *J. Mol. Biol.*, **76**, 241-256 (1973).

[50] M.J.E. Sternberg and J.M. Thornton *J. Mol. Biol.*, **105**, 367-382 (1976).

[51] A.V. Efimov *Mol. Biol. (Moscow)*, **16**, 799-806 (1982).

[52] M. Levitt and C. Chothia *Nature*, **261**, 552-557 (1976).

[53] V.I. Lim, A.L. Mazanov and A.V. Efimov *Mol. Biol. (Moscow)*, **12**, 206-213 (1978).

[54] S. Knight, I. Andersson and C.I. Brändén *J. Mol. Biol.*, **215**, 113-160 (1990).

[55] C. Oefner and D. Suck *J. Mol. Biol.*, **192**, 605-632 (1986).

[56] K. Nagai, C. Oubridge, T.H. Jessen, J. Li and P.R. Evans *Nature*, **348**, 515-520 (1990).

[57] R.S. Hegde, S.R. Grossman, L.A. Laimins and P.B. Sigler *Nature*, **359**, 505-512 (1992)

[58] C. Dumas, I. Lascu, S. Moréra, P. Glaser, R. Fourme, V. Wallet, M.-L. Lacombe, M. Véron and J. Janin *EMBO J.*, **11**, 3203-3208 (1992) .

[59] A. Guasch, M. Coll, F.X. Avilés and R. Huber *J. Mol. Biol.*, **224**, 141-157 (1992).

[60] W.C. Stallings, S.S. Abdel-Meguid, L.W. Lim, H.-S. Shieh, H.E. Dayringer, N.K. Leimgruber, R.A. Stegeman, K.S. Anderson, J.A. Sikorski, S.R. Padgette and G.M. Kishore *Proc. Natl. Acad. Sci.*, **88**, 5046-5050 (1991).

[61] J. Zheng, D.R. Knighton, L.F. Ten Eyck, R. Karlsson, N.-H. Xuong, S.S. Taylor and J.M. Sowadski *Biochemistry*, **32**, 2154-2161 (1993).

[62] K.H. Kim, Z. Pan, R.B. Honzatko, H.-M. Ke and W.N. Lipscomb *J. Mol. Biol.*, **196**, 853-875 (1987).

[63] O. Herzberg, P. Reddy, S. Sutrina, M.H. Saier Jr., J. Reizer and G. Kapadia *Natl. Acad. Sci.*, **89**, 2499-2503 (1992).

[64] F.A. Saul, L.M. Amzel and R.J. Poljak *J. Biol. Chem.*, **253**, 585-597 (1978).

[65] A.V. Efimov *FEBS Lett.*, **284**, 288-292 (1991).

[66] A.V. Efimov *FEBS Lett.*, **298**, 261-265 (1992).

[67] J. Deisenhofer, P. Epp, K. Miki, R. Huber and H. Michel *Nature*, **318**, 618-624 (1985).

[68] A. Musacchio, M. Noble, R. Pauptit, R. Wierenga and M. Saraste *Nature*, **359**, 851-855 (1992).

[69] R. Thieme, E.F. Pai, R.H. Schirmer and G.E. Schulz *J. Mol. Biol.*, **152**, 763-782 (1981).

[70] A. Arnone, C.J. Bier, F.A. Cotton, V.W. Day, E.F. Hazen Jr., D.C. Richardson, J.S. Richardson and A. Yonath *J. Biol. Chem.*, **246**, 2302-2316 (1971).

[71] T.K. Sixma, S.E. Pronk, K.H. Kalk, E.S. Wartna, B.A.M. van Zanten, B. Witholt and W.G.J. Hol *Nature*, **351**, 371-377 (1991).

[72] P.E. Stein, A. Boodhoo, G.J. Tyrrell, J.L. Brunton and R.J. Read *Nature*, **335**, 748-750 (1992).

[73] G.D. Brayer and A. McPherson *J. Mol. Biol.*, **169**, 565-596 (1983).

[74] L. Chen, J.P. Rose, E. Breslow, D. Yang, W.-R. Chang, W.F. Furey Jr., M. Sax and B.-C. Wang *Proc. Natl. Acad. Sci.*, **88**, 4240-4244 (1991).

[75] E.T. Baldwin, I.T. Weber, R.St. Charles, J.-C. Xuan, E. Appella, M. Yamada, K. Matsushima, B.F.P. Edwards, G.M. Clore, A.M. Gronenborn and A. Wlodawer *Proc. Natl. Acad. Sci.*, **88**, 502-506 (1991).

[76] R.St. Charles, D.A. Walz and B.F.P. Edwards *J. Biol. Biol.*, **264**, 2092-2099 (1989).

[77] P.M.D. Fitzgerald, B.M. McKeever, J.F. Van Middlesworth, J.P. Springer, J.C. Heimbach, C.-T. Leu, W.K. Herber, R.A.F. Dixon and P.L. Darke *J. Biol. Chem.*, **265**, 14209-14219 (1991).

[78] T.L. Blundell, J.A. Jenkins, B.T. Sewell, L.H. Pearl, J.B. Cooper, I.J. Tickle, B. Veerapandian and S.P. Wood *J. Mol. Biol.*, **211**, 919-941 (1990).

[79] K. Katayanagi, M. Miyagava, M. Matsushima, M. Ishikawa, S. Kanaya, H. Nakamura, M. Ikehara, T. Matsuzaki and K. Morikawa *J. Mol. Biol.*, **233**, 1029-1052 (1992).

[80] A.G. Murzin *EMBO J.*, **12**, 861-867 (1993).

[81] A.V. Efimov *FEBS Lett.*, **334**, 253-256 (1993).

Design of Model Fast-Folding Proteins

Eugene I. Shakhnovich

Department of Chemistry, Harvard University, 12 Oxford Street, Cambridge MA 02138, USA

Abstract

In order to address the problem of how (and if) proteins fold to their global minima we generate sequences which provide very low energy in a given structure. Statistical-mechanical arguments suggest that the design process provides sequences for which this structure is a global energy minimum. Then lattice Monte-Carlo simulation of the folding of these designed sequences are done for model proteins of 36, 48, 64, 80 and 100 monomers. In all cases polypeptides with designed sequences folded from random coil to global minimum conformations, effectively 'solving' the multiple minima problem. The rate of folding did not exhibit dramatic dependence on chain length but was strongly dependent on the energy of the native state. The same experiments with two-letter sequences (polar-nonpolar) were unable to find minimum energy conformations, they were always 'stuck' at non-native conformations with higher energy. These results are compared with recent experimental findings on kinetics and intermediates in protein folding.

1   Introduction:
Bottlenecks in the Protein Folding Problem

Why is the protein folding problem so complicated for scientists when it is easily solved by proteins? One possible reason is that proteins use involved folding algorithms (via special pathways etc) [1], and these algorithms have not been discovered yet. Another possibility is that proteins 'know' their force-field exactly while simulations necessarily use approximate force fields [2, 3] for which the native structure may be neither a global nor a pronounced local minimum. It is then hard to expect any folding algorithm to converge to a 'native' state which is not distinguished by energy from many other conformations.

The usual argument in favor of using any specific parameter set is that it distinguishes the native structure as having minimal energy among 100-1000 alternative structures [3]. However, it is not very convincing since proteins have at least $10^{20} - 10^{30}$ conformations available [1, 4], and a correct parameter set must distinguish the native structure as having the minimal energy among this number of conformations.

One approach which disentangles the problem of folding kinetics from the problem of parameters is based on lattice models with exhaustively enumerated conformations [5, 6, 7] where the conformation of the global energy minimum is known. Monte-Carlo folding simulations of a model 27-mer chain on a cubic lattice showed that there exist 'folding' and 'non-folding' sequences [5]. Detailed study of this model [6] revealed that folding sequences have their native structure as a pronounced energy minimum.

However the computational complexity of enumeration grows dramatically as chain length increases, which makes this approach hardly feasible for longer than 27-mer chains in 3 dimensions. On the other hand what is really necessary for folding simulations is to know the ground state conformation; in this sense enumeration may be not necessary if we have sequences for which the structure of global energy minimum is known.

This suggests the idea of using protein design to study folding. If one can choose a 3D target structure of a longer chain and then design a sequence which provides a very low energy to this structure then one can expect that for this sequence the target conformation is the global minimum of energy. Folding simulation with the same force-field as was used at the design stage will then reveal whether this ground state can be reached in a reasonable time.

2   Design of Stable Protein Sequences

An effective sequence design algorithm has been proposed and described in detail recently [8]. This algorithm is a Monte-Carlo optimization in sequence space of the energy of the native conformation. An auxiliary 'selective temperature' T_{sel} is introduced to overcome energy barriers in sequence space. T_{sel} sets the energy with which sequences generated are fitted to the target structure.

In the following discussion we will consider a simple model of a protein on a cubic lattice (in C_α representation). To go beyond short 27-mer [5, 6, 9] chains, we consider 36-mers (maximally compact conformation is a parallelepiped which fills $3 \times 3 \times 4$ fragment of cubic lattice), 48-mers ($3 \times 4 \times 4$), 64-mers ($4 \times 4 \times 4$), 80-mers ($4 \times 4 \times 5$) and 100-mers ($4 \times 5 \times 5$). (see Fig. 1 for an example).

For each of these models we start by drawing some arbitrary compact conformation which then serves as the target structure for sequence design.

Low energy sequences are designed using the sequence-space MC procedure described above. The energy function is taken in contact approximation [2]:

$$E_0(\{\sigma_i\}) = \frac{1}{2} \sum_{i,j}^{N} U(\sigma_i \sigma_j) \Delta(r_i^{tar} - r_j^{tar}) \tag{1}$$

where N is the total number of monomers and Δ defines the contact potential between them: $\Delta(r) = 1$ if $r_{low} < r < r_{high}$ and 0 otherwise. We consider our model proteins positioned on a cubic lattice with bond length 3.8A. Then any two monomers which are 3.8A apart (so that, say, $r_{low} = 3.7A$ and $r_{high} = 3.9A$) are considered to be in contact. This model is more compact than natural proteins (where $r_{low} = 6.5A$ and $r_{high} = 8.5A$ [2]) but is free of many technical details and is easier to analyze.

For the set of potentials $U(\sigma_i \sigma_j)$ we use parameters determined by Myazawa and Jernigan (denoted MJ) [2] from the statistical distribution of contacts in native proteins. Using the sequence design algorithm we generate for each target structure (at low $T_{sel} = 0.1$)a set of sequences with low energy (less than -0.5 kcal/m per monomer) in the target structure.

We also used a simplified set of potentials (denoted SMJ) which was obtained from the MJ set by replacing the strongest attractions (those with $U < -0.2$) by energies of -1 and all other interactions by $U(\sigma_i \sigma_j) = 0$. This set allows deeper optimization of the energy of the ground state conformation (see below).

The third set of parameters which we tried to use is a simple two-letter code analogous to that used in [10]. Monomers are considered to be either polar ($\sigma_i = 0$) or non-polar

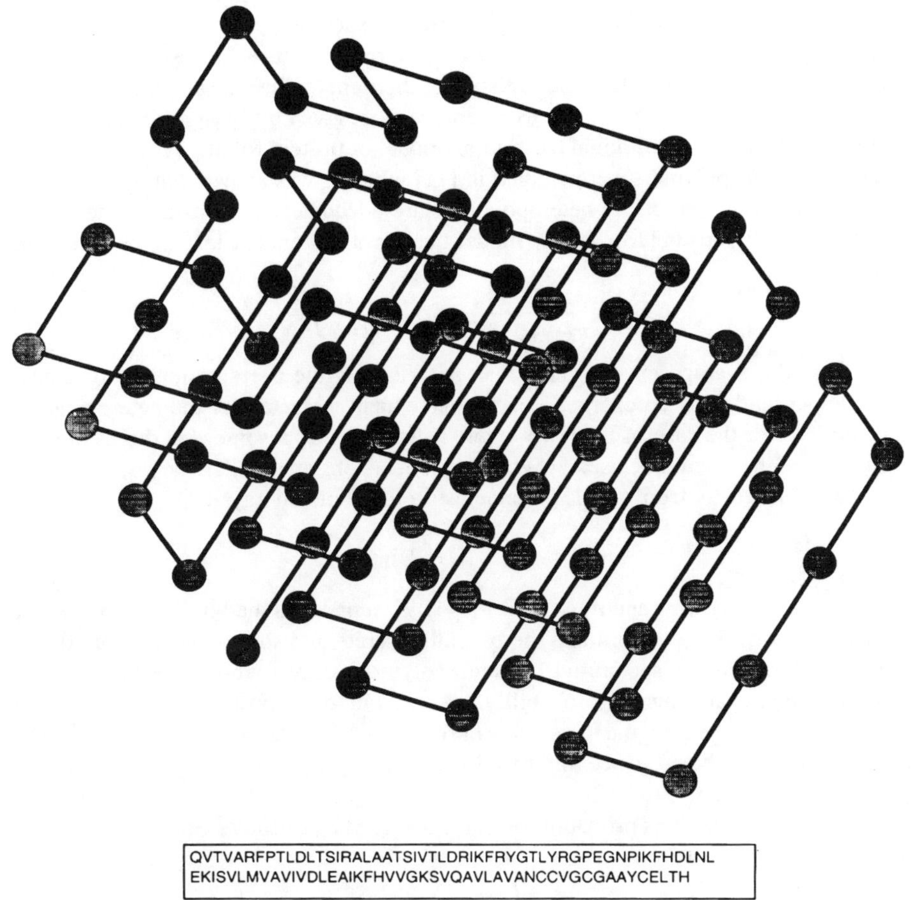

Figure 1: An example of a compact conformation of a 100-mer on a cubic lattice and the optimized sequence (with SMJ parameters). This conformation was often used as a native structure in our studies.

$(\sigma_i = 1)$. The interaction energy between monomers i and j is given by $U(\sigma_i, \sigma_j) = -\sigma_i \sigma_j$. This means that non-polar monomers attract each other while polar monomers have no interactions.

3 Target Structures are Pronounced Global Minima for Designed Sequences

With a high degree of certainty we can say that the target structures of sequences designed at low selective temperatures correspond to global minima.

We may divide the set of conformations (which number γ^N where γ is the number of available conformations per monomer) into two groups: conformations which have significant similarity with the target structure and the remaining vast majority of conformations which have marginal or no similarity with the target structure.

For conformations which are not similar to the target structure the designed sequence is effectively random and therefore the statistics of their energy levels are similar to those of a random heteropolymer. (A similar argument was first given by Bryngelson and Wolynes in their discussion of the 'minimal frustration' model of protein folding [11])

Random heteropolymers were studied in [12] where it was shown that the statistics of conformational energies of 3D heteropolymers are adequately described by the Random Energy Model of Derrida [13]. In this model the density of energy levels (the 'spectrum') has the form

$$n(E) = \frac{\gamma^N}{(2\pi N J^2)^{1/2}} \exp\left(-\frac{(E - E_0)^2}{2J^2 N}\right) \qquad (2)$$

where E_0 is the average energy and $J^2 = \rho B^2/2$ (ρ is the average number of contacts per monomer, and B is an energetic parameter which characterizes the heterogeneity of interactions). For the case of the "two-letter" heteropolymer a somewhat different model, the Discrete REM, should be used [15] but the conclusions are similar.

As follows directly from (2) [13] there exists a threshold energy

$$E_c = E_0 - JN\left(2\ln\gamma\right)^{1/2} \qquad (3)$$

such that $n(E < E_c) \ll 1$ and $n(E > E_c) \gg 1$. E_c represents the borderline between the continuous part of the spectrum and the bottom, discrete and sparse part (a more detailed description of the energy spectrum of a heteropolymer can be found in the caption to Fig. 1 in [16]). This means that the probability to find a random (unrelated to target structure) conformation with energy equal to or below the energy of the target structure E_N will be very small provided that the sequence optimization is deep, i.e., provided the energy difference $E_N - E_c$ is significant.

In order to estimate this probability we, following [17], linearize eq.(2) around E_c and get:

$$P|_{E \leq E_N} \sim \exp\left((E_N - E_c)\left(2\ln\gamma\right)^{1/2}/J\right) \qquad (4)$$

To determine J we simply generate a set of random sequences and fit them into the target structure (See Fig. 2).

Then $< E >= E_0$ and

$J^2 = (< E^2 > - < E >^2)/N$ where $<>$ denotes averaging over the set of random sequences. For γ we take the maximal (for a polymer with excluded volume on a cubic lattice) value of $\gamma = 5$.

Now we compare the estimated position of the boundary of the continuous spectrum E_c (3) with the energies of the designed sequences fit into the target ('native') conformation E_N. We take sequences which provide the lowest energy fit to the target conformation during the sequence design procedure at $T_{sel} = 0.1$. (Everywhere in this paper units kcal/m are used for energy and temperature). Below we give specific numbers for 100-mers (the target conformation is shown on Fig. 1) normalized per monomer: $\epsilon_N = E_N/N$, $\epsilon_c = E_c/N$. The results for chains of other lengths (we worked with 36,48,64,80 and 100-mers) are very similar.

1) Two-letter heteropolymers: $\epsilon_c = -0.88$, $\epsilon_N = -0.85$. The model is not specific e-nough to allow deep optimization of the native structure. It is not expected to fold (see below)

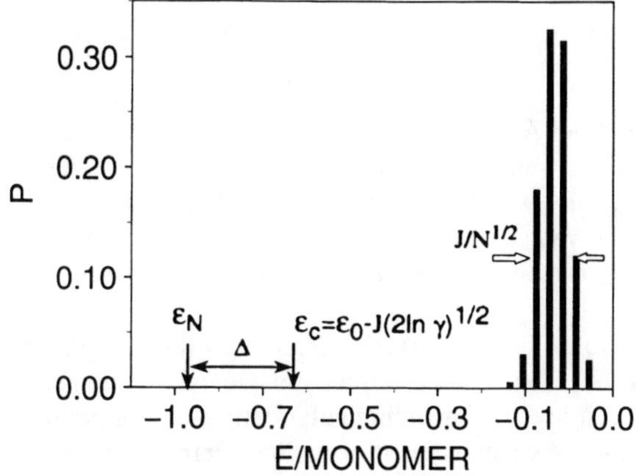

Figure 2: Estimate of the energy gap between the energy of the native state ϵ_N and the lower boundary of the continuous spectrum of conformations ϵ_c. We used an analytic expression based on the correspondence between heteropolymers and the REM [12] to estimate ϵ_c. The important parameter σ which is the standard variation of interaction energies was calculated as follows. We fitted 1000 random sequences (all having the same amino-acid composition) into the target structure (shown in Fig. 1 for 100-mer) and plotted a histogram of the energies of these fits (shown in this Figure). From this statistics we estimated the standard variance J (as shown).

2) MJ parameters: $\epsilon_c = -0.45$, $\epsilon_N = -0.57$. The estimated gap is statistically significant but relatively small.

3) SMJ parameters: $\epsilon_c = -0.62$, $\epsilon_N = -0.93$.

The probability that a random conformation will have energy equal or below E_N is $\exp(100(\epsilon_N - \epsilon_c) * 5.2) \sim e^{-160}$ for a 100-mer with SMJ parameters. (The specific numbers are from calculation with the given set of parameters in accord with eq.(4) with J determined as explained above.)

Therefore in the cases where 20 monomer types are used (the MJ and SMJ sets) the design procedure guarantees that the target structure is the global minimum of energy for designed sequences.

4   Folding of Designed Proteins

Now we have model polypeptide chains (with lengths up to N=100) for which the structure of the global energy minimum is known and may ask a direct question: will the chain find this minimum in a folding simulation? To address this we use a simple Monte-Carlo folding algorithm which was described in [5]. The energy function for simulation is the following:

$$E\{r\} = \frac{1}{2}B_0 \sum_{i,j} \Delta(r_i - r_j) + \frac{1}{2}A \sum_{i,j} U(\sigma_i \sigma_j)\Delta(r_i - r_j) \tag{5}$$

Here parameters B_0 and A shift and scale potentials, these values are adjusted to obtain the fastest folding in the same way as as it was done in [5, 6]. This transformation is non-specific and the structure which was the ground state before shifting and scaling remains such afterwards. The values of parameters used are: $B_0 = -1.5$, $A = 5$. The three sets described in the previous section (MJ, SMJ and 2-letter) were used for $U(\sigma_i \sigma_j)$. Simulations were run at $T \approx 3$.

Now we try to fold our test 'proteins' (36,48,64,80, and 100-mers). To create test 'protein structures' for each of the chain lengths we use an enumeration algorithm [9, 18] to generate 10000 compact self-avoiding cubic lattice conformations. Of course for each of these chains this set of generated conformations represents a tiny fraction of all possible conformations. From this set of 10000 conformations we pick randomly one conformation (one example is shown in Fig. 1). Now for this 'native conformation' and for each of three sets of parameters we designed low-energy sequences by running the design algorithm at low selective temperature, $T_{sel} = 0.1$.

Now we simulate folding of the designed sequences. Starting from random coil conformations we run lattice Monte-Carlo folding simulations with potential function (5) at different temperatures. When 20 types of monomers were used for design and folding (MJ and SMJ parameter sets) each chain in each run reached its global minimum conformation.

Attempts to fold model proteins with two types of residues (polar-nonpolar) were all unsuccessful, and resulted in a multitude of conformations with energies higher than the energy of the target structure and with no structural similarity with the target structure. We tried several different possible sets of (see Fig. 3) interaction energies U_{pp}, U_{nn}, U_{pn} but none of them made it possible to fold 2-letter model. Inspection of Fig. 3 teaches a few lessons. First of all we see that energywise simulations converge reasonably well, so that the difference in energy between the native conformation and the lowest conformation reached in simulations is relatively small. However there is no "structural convergence" at all, i.e. the conformations which are reasonably close to the native one by energy appear to be quite dissimilar with it structurally as comparison between Fig. 3a and Fig. 3b suggests. The latter implies that the reason why folding is unsuccessful in a 2-letter model is just because the energy gap is too small: when chain reaches low energy being still far from the native state structurally, the "driving force" to the native state (which is related to energy differences between the native state and the current conformation) disappears. That means that independent of the length of a run or a move set the chain will not reach the native conformation for the 2-letter code. Similar results for two-letter heteropolymers were obtained in recent simulations [19].

Sequences optimized with SMJ parameters provided a larger energy gap, correspondingly they fold ~ 10 times faster than sequences optimized with the MJ set. In the following we will the data obtained with the SMJ parameter set as the statistics are better in this case.

A typical folding trajectory (for a 100-mer) is shown in Fig. 4.

Analysis of energy changes with Monte-Carlo time shows that native (target) structures are likely to be global minima for designed sequences as no structure with energy lower than the energy of the native state has been encountered.

In all cases an intermediate which is a partly compact non-native state (with number of contacts corresponding to 50-70% of such in the fully compact state but having only 20-25

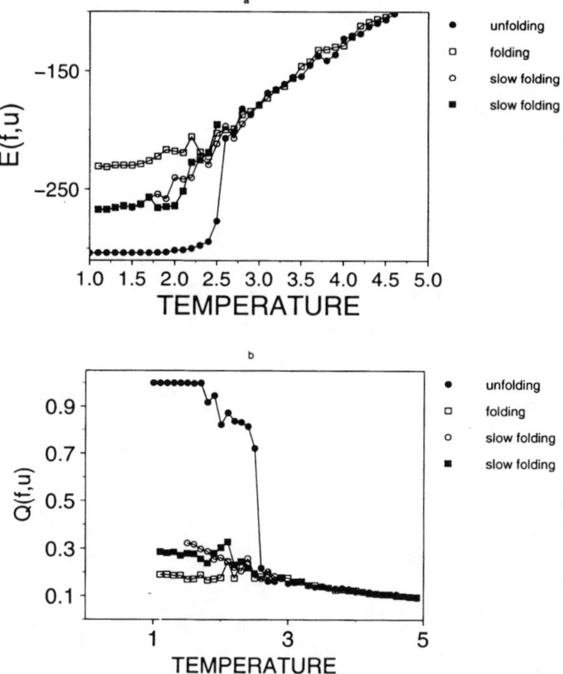

Figure 3: Simulation of folding and unfolding of a 2–letter 64–mer model protein. (a) Dependence of energy on temperature for unfolding and for folding at different rates of annealing. (b) Structural test. Annealing at any rate does not return the system to the target conformation.

% of native contacts) was formed first with subsequent formation of the native conformation on the background of this semi-compact state.

Typically it takes $10^6 - 10^8$ steps to reach the target structure starting from a random coil. For each length of the chain we ran 50 folding simulations to determine folding time distributions and the mean first passage time (MFPT) for the native state. The dependence of MFPT on chain length is shown on Fig. 5.

The folding time depends somewhat on the chain length but this dependence is not at all dramatic.

5 Discussion

The reported study is the first, to our knowledge, of an unbiased folding simulation for a protein model of realistic size which is complete — from random coil to unique native structure. The protein sequence is taken as an input for simulation, and a usual parameter set [2] is used. No forces which bias the chain towards its native secondary or tertiary structure are introduced in the energy function.

The main result of this work is the demonstration of folding of long chains (up to 100

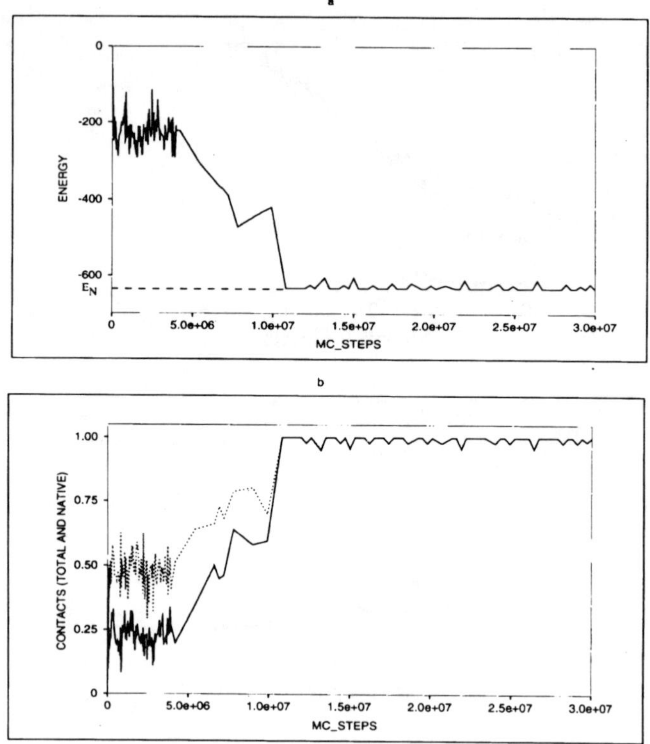

Figure 4: MC trajectory of folding simulations for 100-mer. (a) dependence of energy on MC-step. The energy of the native conformation is shown as E_N. (b) dependence of normalized number of all contacts (dotted line) and native contacts (solid line) on MC-step. The maximal number of contacts $N_{total} = 136$ for a compact 100-mer. For each conformation we normalize the number of all contacts, C and number of the native contacts, Q by N_{total}. C=1 corresponds to maximally compact conformation, Q=1 corresponds to the native conformation.

monomers) to their global minima in case when these minima are deep enough as provided by design procedure. This prediction of the theory can be tested experimentally by using site-directed mutagenesis to destabilize natural sequences and then measuring the rate of formation of native-like intermediate in the mutants.

The simulations reported in this study were successful because the sequences were designed to guarantee that the native structure is a pronounced global energy minimum with the force field used in the simulations. However, the design procedure required knowledge of the native (target) structure. This requirement is due to ignorance of the real protein force-fields for which natural sequences may have been optimized in the evolutionary process. This suggests that the bottleneck in folding of real proteins with their natural sequences may be our lack of knowledge of the correct potentials. When sufficient progress is made in this direction, relatively simple algorithms will be able to fold real-size proteins, at least to native-like molten globule conformations.

Our results also suggest that random sequences are not able to fold for kinetic reasons although thermodynamically their native structure may be reasonably stable at physiological

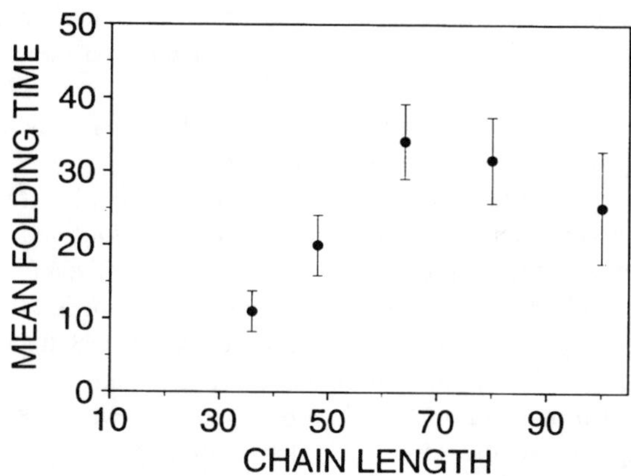

Figure 5: Dependence of mean first passage time (MFPT, mln of MC steps) of folding on chain length. For each chain length we made 50 folding simulations defining the MC step at which the native conformation was reached as the first passage time. Then from the distribution of passage times the mean and standard variance (error bars) were determined.

temperatures [16]. The implication is that a significant degree of evolutionary optimization is necessary to enable sequences to find their native structure. This result is in accord with elegant experiments on isolated F2 fragment of Tryptophane Synthase reported recently [20]. Within the native protein the F2 fragment is mainly stabilized by contacts with F1 fragment, these contacts are likely to have been optimized in this protein. Isolated F2 is supposed to form only internal contacts and is effectively a random sequence for that purpose. Analysis of the folding of F2 [20] suggests that this fragment forms a slightly compact state with non-native non-localized secondary structure and possible interconvertion between different conformations. This description is very close to our first compact intermediate, and our results suggest that a random sequence will not go beyond this stage. Some secondary structure is observed in isolated F2 and is likely to be a consequence of compactisation which forces the formation of intramolecular H-bonds to substitute lost main-chain–water H-bonds [21, 22].

The "hydrophobic zippers" hypothesis of folding was published recently [23] where the chains were assumed as composed of monomers of two types with contacts between hydrophobic groups being formed irreversibly. The folding of two proteins, crambin and BPTI was attempted within hydrophobic zippers approach and the results were reported that energy close to the native state energy was reached by structural similarity with the native state was marginal. The authors of [23] suggested that this is due to time limitations of simulations. Our analysis shows that the failure of "hydrophobic zippers" approach to fold model proteins is due to inconsistency of the approach rather than to the time limitation of simulations. Indeed, analysis of the data in Fig. 3 suggests that one should be extremely cautious judging the success of a simulation by convergence in energy. In fact that may mean nothing in the sense that there may be no structural convergence (which is only of interest for protein folding!). Our analysis suggests that the intrinsic deficiency of the approach taken in [23] is that it models protein sequences in "2-letter" code which is inappropriate to study folding.

The probability to find an optimized folding sequence by random choice is vanishingly small for long chains. It is already extremely small for the chain lengths 36-100 studied here.

However, directed design, which may serve as a rough model of evolutionary selection, allowed us to find optimized folding sequences in time which is $\sim N$. A very important feature is that the energy landscape in sequence space is 'smooth' [8], which makes sequence optimization effective. The fact that the vast majority of mutations decrease the stability of proteins [24] suggests that natural protein sequences were optimized to make the energy of the native state sufficiently low, though this optimization may not be absolute.

In order to make proper comparison with experiment we note that a simplified model of a protein was studied. The important feature which was not taken into account is the tight packing of side-chains in native proteins. It is believed, however, that side-chain packing occurs at the final stage of protein folding [25]. Correspondingly, our model aims to describe first stages of folding only until a native-like kinetic intermediate (the 'molten globule' [25]) is reached.

In all our simulations slightly compact non-native intermediate was formed first and very quickly. This is in accord with our earlier studies of shorter 27-mer chains [5, 26], with studies of folding in 2D lattice models [27] and, most importantly, with growing experimental evidence that such intermediates are formed at the first stage ($< 5ms$) of folding of different proteins [28]. It is important to mention that the first compact intermediate observed in experiments does not have any unique fold and/or stable localized secondary structure. This exactly corresponds to the intermediate which we observe in our simulations which is relatively compact but represents a multitude of interconverting conformations.

No intrinsic energretic preferences for secondary structure were introduced in our model. Correspondingly, no hierarchical mechanism of folding through formation of stable native-like elements of secondary structure in unfolded chain with their subsequent coalescence as suggested by framework and related models [29] was observed. That does not mean, however, that our results exclude framework-type scheme of folding. However, the successful resolution of Levinthal paradox by our model proteins suggests that framework mechanism may be not necessary and not only way to solve folding problem. In any case we are unaware of microscopic unbiased simulations where framework mechanism converged to the native state of any model protein. At present framework model is purely phenomenological and still requires a microscopic study which will validate it.

Our results suggest that the formation of the unique structure of the backbone may take place after the first compact intermediate is formed. Unique structure is likely to be formed from structureless globule via a nucleation-growth mechanism. The numerically exact character of our model will allow these nuclei to be trapped and followed and therefore give an exhaustive description of the folding process. Of course more sophisticated models (better lattice, realistic compactness, secondary structure) of proteins should be used for that purpose. This is the subject of our current research.

Acknowledgements

I am grateful to Alexander Gutin, Alexey Finkelstein, Martin Karplus, Peter Leopold, Oleg Ptitsyn and Andrej Sali for numerous interesting discussions. Joe Watson provided significant editorial help. Graphic program ASGL by Andrej Sali was used to generate some of the plots. This work was supported by the Packard Foundation.

References

[1] C. Levinthal, *J. Chim. Phys*, **65**, 44 (1968).

[2] S. Myazawa and R. Jernigan, *Macromolecules,*, 18, 534 (1985).

[3] M. Sippl, *J. Mol. Biol.*, **213**, 859, (1990); A. Godzik, A. Kolinski, and J. Skolnick, *J. Mol. Biol.*, **227**, 227 (1992).

[4] K. Dill, *Biochemistry*, **24**, 1501 (1985).

[5] E. I. Shakhnovich, G. M. Farztdinov, A. M. Gutin, and M. Karplus, *Phys. Rev. Lett.*, **67**, 1665 (1991).

[6] A. Sali, E. I. Shakhnovich, and M. Karplus *J. Mol. Biol., Submitted* (1993).

[7] R.Miller, C.Danko, M.J.Fasolka, A.C.Balazs, H.S.Chan, and K.A.Dill, *J. Chem. Phys.*, **96**, 768 (1992).

[8] E. Shakhnovich and A. Gutin, *Proc.Natl. Acad. Sci. USA*, **90**, 7195-7199 (1993); E.Shakhnovich and A.Gutin, *Protein Engineering* **6**, 793-800 (1993)

[9] E. I. Shakhnovich and A. M. Gutin, *J. Chem. Phys*, **93**, 5967 (1990).

[10] H. S. Chan and K. A. Dill, *J. Chem. Phys.*, **95**, 3775 (1991).

[11] J. D. Bryngelson and P. G. Wolynes, *Proc. Natl. Acad. Sci., USA*, **84**, 7524 (1987).

[12] E. I. Shakhnovich and A. M. Gutin, *Biophysical Chemistry*, **34**, 187 (1989); T. Garel and H. Orland, *Europhys. Let*, **6**, 307 (1988); E. I. Shakhnovich and A. M. Gutin, *Europhys.Let*, **8**, 327 (1989); E. I. Shakhnovich and A. M. Gutin, *J. Phys*, **A22**, 1647 (1989).

[13] B. Derrida, *Phys. Rev.*, **B24**, 2613 (1981).

[14] E. I. Shakhnovich and A. M. Gutin, *J. Theor. Biol.*, **149**, 537 (1991).

[15] A. M. Gutin and E. I. Shakhnovich, *J.Chem.Phys*, **98** 8174 (1993); C.Sfatos, A.M.Gutin, and E.I.Shakhnovich, *Phys. Rev. E, in press* (1993).

[16] E. I. Shakhnovich and A. M. Gutin, *Nature*, **346**, 773 (1990).

[17] G. Parisi, M. Mezard and M. Virasoro, *J. Physique (France) Lett*, **46**, L217 (1985).

[18] D. Covell and R. Jernigan, *Biochemistry*, **29**, 3287 (1990).

[19] E. M. O'Toole and A. Z. Panagiotoupoulos, *J. Chem. Phys.*, **97**, 8644 (1992); J.D.Honeycutt and D.Thirumalai, *Biopolymers*, **32**, 695 (1992).

[20] A. Chaffotte, Y. Guillou, M. Delepierre, H.–J. Hinz, and M. Goldberg, *Biochemistry*, **30** 8067 (1991).

[21] M. Levitt and A. Warshel *Nature*, **253**, 694 (1975).

[22] E. I. Shakhnovich and A. V. Finkelstein, *Biopolymers*, **28**, 1667 (1989).

[23] K. Dill, K. A. Fiebig and H. S. Chan, *Proc. Natl. Acad. Sci.* **90** 1942-1946 (1993).

[24] D. Shortle, W. Stites, and A. Meeker, *Biochemistry*, **29**, 8033 (1990); M. Bycroft, R. N. Sheppard, F. Lau, and A. R. Fersht, *Biochemistry*, **30**, 8697 (1990).

[25] O. B. Ptitsyn, *J. Protein Chem*, **6**, 273 (1985); A. Chaffotte, C. Cadieux, Y. Guillou, and M. Goldberg, *Biochemistry*, **31**, 4303 (1992); O. B. Ptitsyn et al, *FEBS Lett.*, 262, 20 (1990); O. B. Ptitsyn, *Protein Folding* (W.H.Freeman and Company, New York, 1992) chapter 6, pages 243–300.

[26] M. Karplus and E. Shakhnovich, *Protein Folding* (W. H. Freeman and Company, New York, 1992) chapter 4, pages 127–195.

[27] C. Camacho and D. Thirumalai, *Proc. Natl. Acad. Sci. USA*, **90**, 6369-6372 (1993).

[28] G. Elove, A. Chaffotte, H. Roder, and M. Goldberg, *Biochemistry*, **31**, 6876 (1992); P. Varley, A. Gronenborn, H. Christiensen, P. Wingfield, R. Pain, and M. Clore, *Science*, **260**, 1110 (1993).

[29] O. B. Ptitsyn, *Vestn. Acad. Nauk SSSR*, **5** 57 (1973); M. Karplus and D. Weaver, *Biopolymers*, **18**, 1421 (1979); P. Kim and R. Baldwin, *Ann. Rev. Biochem*, **59**, 631 (1982).

Modelling and Predicting Protein Structure using Distance Geometry

William R. Taylor and András Aszódi

Laboratory of Mathematical Biology, National Institute for Medical Research, The Ridgeway, Mill Hill, London NW7 1AA, U.K.

Abstract

The structure of a general method of molecular modelling and tertiary structure prediction is outlined and contrasted with more conventional approaches. Given specific constraints, the method will satisfy these as well as possible while maintaining the basic physico-chemical properties of the protein in question. However, in the absence of specific constraints the method will still apply general principles of globular protein structure and even when given a random sequence of residues will produce a recognisable protein structure. Although the method is still under development, we expect that both its speed and versatility will make it an attractive tool for many applications.

1 Introduction

The similarity of a protein sequence of unknown structure to one with a known structure allows a molecular model for the former to be constructed based on the latter. This approach of 'modelling by homology' extends from the most trivially similar sequences to those where the similarity between the sequences is almost undetectable. The non-trivial extreme of this range has two qualitatively distinct manifestations: a similarity, however tentative, might have been identified that places the protein of unknown structure within a specific family or super-family; on the other hand, the similarity might only encompass a generic motif (such as a fragment involved in calcium or DNA binding) which places an incomplete constraint on the protein fold. Within this catogory, the similarity might extend to a complete domain (which becomes equivalent to a specific similarity) to, at worst, might correspond only to isolated elements of secondary structure — the most general of all motifs in globular proteins.

It is desirable to have a methodology that can span this wide range of molecular modelling, rather than a series of different programs for each stage (or 'resolution') of the modelling. Not only must such a program be able to cope with these different degrees of resolution in different problems, but also must handle their occurrence within a single problem in which some parts may be well constrained and can be modelled almost at atomic resolution while other parts are unconstrained and must obey only the generic rules of protein folding. (See [10] and [16], for an examples).

In this review we outline the specification for such a general program based on the techniques of Distance Geometry. Details of the method will be described elsewhere in this volume (Aszódi and Taylor) and here we concentrate on the general background to the problem and the choice of method, the current scope of the method and its potential applications.

2   Choice of Method

Protein molecular modelling and prediction are problems of constraint satisfaction. Both contain very exact local constraints (bond lengths and angles and steric repulsion) that must be satisfied (otherwise the model quickly ceases to be recognizable as a protein). Progressing to a higher level of organisation, both contain elements of secondary structure that can impose slightly longer range constrains (although with less certainty where these have been predicted). Above this, a modelling problem is distinguished by having some global constraints that can, at least restrict, or more typically define a unique chain fold.

These various constraints are most commonly expressed as relationships between pairs of objects (typically atoms, residues or secondary structures) and if the relation is not already a simple distance it can often easily be expressed as such. This leads to Distance Geometry as the most general method to formulate the problem. However, the term covers two quite distinct techniques. One takes a set of distances which may be incompatible (or non-metric) in three dimensions and fits (or embeds) them into three dimensions such that the sum of squares of the errors from their ideal lengths is minimal. This is commonly referred to as the method of sub-space projection (or multi-dimensional scaling) but will be referred to here simply as the *projection* approach [7, 8]. Alternatively, a starting configuration of points can be specified in 3 dimensions which is then refined, through incremental steps, towards an optimal resolution of the distance constraints. This approach is a standard minimisation problem and can be tackled using a very wide variety of algorithms and will be referred to here as the *refinement* approach [2].

2.1   Projection vs. Refinement

The two approaches introduced above have differing advantages and disadvantages which has often resulted in their application to qualitatively different problems.

2.1.1   Starting configuration dependence

Perhaps the most fundamental difference in the approaches is that the Projection approach does not require a starting model from which to begin its refinement but (like the starship "Enterprise") simply materialises out of multidimensional hyperspace. This is a great advantage since with a 3D-space refinement approach the effect of the starting configuration must be evaluated and if there is no *a priori* guide to this, the best that can be done is to repeat the refinement with a reasonable number of random starting configurations.

If the distance estimates are very uncertain it can be desirable to use random starting models as a tool to explore the conformational space and test the stability of solutions. In the projective approach, however, this ability is not lost as the distance matrix can be perturbed by a random value or using the protocol outlined below (Section 2.2.2), started with a random set of distances [1].

2.1.2 Distance weighting

In the refinement approach, individual distances can be given different weights reflecting the degree of certainty with which they are expected to obtain their ideal value. Properly, each weight should be proportional to the reciprocal of the variance of the observed length, but in practice values are often chosen to reflect an intuitive feel for the relative importance of the property. Thus, for example, bond lengths can be given a relatively high weight (reflecting their small variance) to maintain their ideal lengths in the final model. By contrast, in the projective approach, individual distances cannot be weighted and although a mass can be assigned to each point, this affects all its associated distances. Clearly, where we desire a final model with undistorted bond lengths angles and steric exclusion, this behaviour is very undesirable.

2.1.3 Chirality

A further disadvantage of the projective approach is a lack of control over chirality. Any point set and its mirror image generate the same set of distances so in the inverse transformation the enantiomer that emerges from the projection cannot be predetermined. With such obviously chiral structures as proteins, it is important to have the correct hand. With a good (metric) set of distances this is not a problem since the mirror image can be recognised at atomic resolution by the hand of the asymmetric α-carbon, or at the residue level, by the hand of the α-helix. However, at the residue level if there are no helices, or there is an equal mix of right and left handed forms, then a problem is encountered. This problem, however, is not restricted to the projection approach as the refinement approach also initially lacks this chiral information. The difference lies in the mode of solution: the projection approach calculates the final conformation directly giving no scope for interference whereas the gradual refinement approach allows local chiral structures to be imposed which (it is hoped) propagate their asymmetry to favour the formation of the correct enantiomer. If no asymmetric structures can be imposed, however, then the approaches are equivalent.

2.1.4 Kinetic traps

To achieve a final fold the stepwise refinement approach must be able to get there along an energetically favourable path and it is possible that low energy folds exist that have no favourable approach path. These would typically involve structures that have some non-local cooperative interaction (such as knots). This problem can be overcome by random sampling (Monte Carlo minimisation) but this is a very inefficient method that is unlikely to randomly jump to conformations that involve multiple dependencies. (Efficency can be improved using a more 'intelligent' method such as a genetic algorithm — see Argos in this volume). By contrast, with its direct materialisation from hyperspace, the projective approach is independent of any kinetic (pathway) bias. This property can be viewed either as an advantage or disadvantage. It can be argued that it allows the projective approach to sample undesired conformations (such as knots), alternatively, it can be argued that the limited simulation time of the refinement approach (relative to real folding times) cannot allow sufficient exploration of the conformational space, resulting in a bias towards structures with sequentially local interactions.

2.2　Method Specification

2.2.1　Combined approach

For the purpose of *ab initio* structure prediction it is attractive to have a method that is independent of the starting configuration (since there is no preferred initial state) and also free of any kinetic dead-ends (since we want to avoid prolonged run times). For the predictive aspect of the modelling problem we therefore adopted the projective approach as our basic method. The problems associated with this approach (weighting and chirality) were then dealt with by a modification to the projection protocol (described below) and a subsequent phase of (real-space) refinement to correct bond lengths and steric violations and impose correct chirality. Viewed from a different point it might be seen that we are simply using the projective approach to provide a starting configuration for refinement — so avoiding (or minimising) the need to perform repeated refinements from random starting configurations.

This combined approach is similar to a protocol now commonly used in the refinement of NMR derived distance constraints, however, despite the similarities, the two problems are distinct. NMR can provide a sparse matrix of quite accurate distance estimates and one of the main problems is to interpolate these scattered values to generate a full metric matrix. By contrast, a prediction/modelling application will provide a full matrix of distances but some (if not all) of these will be based only on the generic affinity of two residues (e.g. two hydrophobic residues might be assigned a short distance). This will result in a matrix of highly non-metric data.

2.2.2　Dealing with highly non-metric data

The projection of highly non-metric data typically results in a jumbled mass of points that bears little resemblance to a protein structure. A major contribution to this unrecognizable state is the violation of bond length and steric volume constraints. Unfortunately, (as discussed above) these cannot simply be given higher weights. To overcome this problem, and the general problem of weighting individual distances, we projected our data not directly into three dimensions, but into a higher dimensional space. In this space the distances between points (represented as multi-component vectors) can be refined towards their ideal values. The resulting new positions were then used to generate another distance matrix that can subsequently be projected. Furthermore, in this matrix, any distance values that were not refined in real-space can be reset to their desired values (so maintaining a 'soft' bias towards the desired packing). Generally this was done with a degree of strictness reflecting the importance of the effect. This process was repeated with the dimensionality of the projection reduced in each subsequent cycle until three dimensional space was reached [1].

The introduction of intermediate cycles of projection allowed weighting to be introduced but it does not, unfortunately, allow chirality to be refined as the handedness we anticipate in three dimensions is ambiguous in higher dimensions — in the same way that left and right are ambiguous in three dimensions where the direction of view can be altered (see Aszódi and Taylor in this volume). This refinement was, therefore, carried out entirely in three dimensional space.

2.2.3　Real-space refinement

Since the data with which we were primarily concerned was expressed as target (or ideal) distances, a full energy refinement method was not implemented. Conventionally, this would involve calculating potential functions and their derivatives to determine the direction and magnitude of motion for each point. Instead, the shifts required to move each point towards its ideal separation for each pairwise interaction were accumulated in a resultant vector and the point moved along this by a small fraction. The process was then repeated using the new locations. This crude, but fast, method was considered sufficient since the interactions being satisfied were primarily local in nature — either adjacent bonded or non-bonded (steric) interactions.

For the longer range interactions involving extended motifs, a more sophisticated version of this basic algorithm was used. Using the bonds from the α-carbon of one residues to its sequential neighbours to define a local reference frame, a set of inter-atomic vectors can be defined from the current residue to all others in the motif. When this vector set is rotated and translated into an equivalent local frame in the model then a set of shift vectors is defined to move the model atoms to their corresponding positions in the ideal motif. As with the simple bump and bond refinement, atoms were shifted only by a fraction of these vectors and the process was iterated. To aid convergence, smaller shifts were made further away from the site of local superposition. This approach is both rapid and robust — indeed, when the motif of an immunoglobulin fold (about 100 residues) was erroneously applied to a mirror image projection, the refinement method was able to invert the molecule and refine it.

3 Sources of Predicted Distances

3.1 Generic preferences

3.1.1 Hydrophobic packing

The propensity for hydrophobic residues to pack together is probably the most basic interaction found in globular proteins. It can be modelled in a very simple way as a binary effect (which is either present or not) and when present in a pairwise interaction implies that the pair of residues should have a shorter ideal separation which should be shortest for hydrophobic/hydrophobic pairs and greatest for hydrophilic/hydrophilic pairs (See [1] and Aszódi and Taylor in this volume for examples).

A slightly more sophisticated model might take into account the degree of hydrophobicity of each residue (using one of the many physico-chemical or empirical scales). If multiple aligned sequences are available then the hydrophobicity can also be averaged over the sequence family or, in addition, modified by the degree of conservation [13].

Any of the above schemes, using different scales with different numbers of sequences of differing degrees of similarity, will give rise to a different range of values (or packing preferences). To convert these values to distances that are in the right range for the expected size of the protein requires a general scaling method. This can best be achieved by chosing a scaling method that transforms the packing preferences into a set of distance that have the same distribution as would be expected for a spherical compact protein of the same size as that being modelled. Most simply this can be done by refining the parameters of the transform function to match the moments of the calculated and theoretical distance functions. Previously, this was achieved matching only the first two moments [13], however, a more general method has been described which can match the complete distribution (Aszódi and Taylor, in preparation).

3.1.2 Empirical potentials

The hydrophobic scaling described in the previous subsection is a general method to match a single property onto an individual protein. If only short range interactions are considered then it is not necessary to perform any scaling as these will be independent of the size of the protein in which they occur. At its most simple this might be a single value specifying the desired packing distance for all hydrophobic residues. However, it can easily be elaborated to take account of, initially, the size of the residue and, subsequently, to consider the packing neighbourhood of the residue. Such preference (or empirical) potentials have been derived from the databank of known protein structures and used to assess the stability of folds — either native or modelled [13, 6].

3.2 Specific interactions

3.2.1 Conserved motifs

Recurring substructures in proteins (often referred to as *motifs*) constitute potential units of structure that might be recognised in a sequence and incorporated into a model. Motifs can range in uniqueness from an element of secondary structure to a whole protein domain. They might, respectively, have a theoretically known ideal form or be derived from a single occurrence. More typically, the situation will be intermediate, with no exact theoretical form but rather an average derived from several up to hundreds of examples. Embodiment of the average structure in a practical form is, in principle, not difficult and might be achieved by the superposition of the occurrences of the motif followed by the averaging of the coordinates of equivalent positions.

With remotely related motifs, however, sufficient differences might have accumulated to make superposition difficult. This problem has been overcome using locally defined frames of superposition — effectively defining a local structural environment for each residue. This environment takes the form of a set of interatomic vectors (in the local frame) to every other atom in the protein [17, 9]. With the multiple superposition of equivalent frames each vector becomes a bundle of vectors (one for each protein) and the coherence of this bundle indicates whether the interaction is conserved or variable [14]. This can be made exact by measuring the sum of the squares of the deviations of the vectors from their average and the resulting value can then be used directly as a weight in the refinement of a segment of the model towards the motif (as described above).

3.2.2 Stick folds

The motifs described above have been derived from either complete domains or parts of real proteins. However, the method can equally well be used with outline sketches of protein folds — indeed, this was the application to which it was originally applied [12, 13]. The outline folds used in that application take the form of simple 'stick' figures where each stick represents the axis of a secondary structure (see also [3, 4]). This simple representation has the advantage that it allows many folds to be combinatorially generated — which in current terminology, might be viewed as a secondary structure based lattice approach.

For the purposes of further assessment, each of the stick folds must be converted into a more realistic representation — to at least the level of residue (α-carbon) resolution. In doing this the secondary structures need to be refined (as motifs) towards their ideal forms while the connecting loop regions are grown between. The approach outlined above was used to incorporate a propensity for hydrophobic packing as a desired distance matrix. However, the matrix used for projection was a linear combination of the this matrix with a distance matrix derived from the ideal (stick) fold. The component from the stick model gives a bias to maintain the overall fold while local hydrophobic packing is encouraged by the desired distance matrix. As the refinement cycles progressed, the effect (weight) of both the generic hydrophobic packing propensity and the stick distances were reduced relative to the weight on motifs, resulting in a final model with ideal motifs and geometry (which was continually refined).

3.2.3 Pairwise sequence correlations

It is often thought that specific pairwise interactions can be predicted from the covariance of residues at different positions in a multiple sequence alignment. This belief has been based mainly on isolated examples — and while some of these involving functional residues in active or binding sites may be correct, it is difficult (if not impossible) to separate the effect from the general strong trend of conserved residues to lie close together anyway. A recent survey of a reasonably large number of proteins has found little support for any signal associated with correlated (compensating) amino acid substitutions when the effect of conservation is taken into account ([15]; compare also the findings of Sander *et al.*, in this volume).

3.3 Sources of Real Data

While the development of the methods described here have been motivated largely to incorporate predicted sources of data, they can equally be used to construct models to incorporate physical or biochemical based distance estimates.

3.3.1 NOE distance estimates

The nuclear Overhauser effect (NOE) is exploited in NMR spectroscopy to obtain evidence of sequentially long range interactions. As mentioned above, not only are these generally sparse but also specify interactions between hydrogen atoms. With some increase in uncertainty they can be converted to estimates of inter-residue distances and used directly in our modelling program. Initial studies indicate that with dihydrofolate reductase (which is a relatively large protein for NMR studies) better convergence can be obtained starting from a stick model of the correct fold and applying our method compared with the standard EXPLOR (simulated annealing) method [5].

3.3.2 Surface estimates

Estimates of surface exposure can be obtained from NMR spectroscopy by inclusion of a paramagnetic relaxation agent in the solution. These estimates can either be indirect (generic for residue type) or made specific through site-directed mutagenesis. However, under certain conditions, even the generic measure can yield specific exposure information (see Lesk in this volume).

Both natural and artificial post-translational residue modification provide an additional source of information on surface exposure. Polysaccharide attachment sites and protease cleavage sites provide examples of these sources and in their incorporation in the method described above, such sites can be treated as very hydrophilic residues.

4 Conclusions

A method has been outlined that can incorporate distance constraints in a flexible way to produce a molecular model. In the absence of specific constraints the method will apply general principles of globular protein structure and even when given a random sequence of residues will still produce a recognisable protein structure.

One of the most interesting of these areas of application for our method is to investigate the range of possible folds that can be generated from random protein sequences and to discover the effect of hydrophobic periodicity on the formation of secondary structure. Our method provides an ideal environment for this type of investigation as it is unhampered not only by the artificial constraints of a regular lattice but also does not have any strong kinetic pathway bias. This latter property might also help in an evaluation of the importance of the folding pathway in attaining the native fold.

While the above aims are concerned with the basic properties of protein folding, a more practical use of our modelling program is simply for the construction of molecular models 'by homology' with a sequence of known structure. While it is possible to adapt our method to the construction of atomic detail, we foresee its use more at the residue level where rough models (perhaps containing a substantial predictive component) can be rapidly constructed. Those that appear promising can then be more fully assessed and further refined by other methods.

References

[1] A. Aszódi and W. R. Taylor. Folding polypeptide α-carbon backbones by distance geometry methods. *Biopolymers*, 1994. in the press.

[2] A. T. Brünger and M. Nilges. Computational challenges for macromolecular structure determination by X-ray crystallography and solution NMR-spectroscopy. *Quaterly Rev. Biophs.*, **26**, 49–125 (1993).

[3] F. E. Cohen, M. J. E. Sternberg, and W. R. Taylor. Analysis and prediction of protein β-sheet structures by a combinatorial approach. *Nature*, **285**, 378–382 (1980).

[4] F. E. Cohen, M. J. E. Sternberg, and W. R. Taylor. Analysis of the tertiary structure of protein β-sheet sandwiches. *J. Mol. Biol.*, **148**, 253–272 (1981).

[5] J. deVlieg, R. M. Scheek, W. F. van Gunsteren, H. J. C. Berendsen, R. Kaptein, and J. Thomason. Combined procedure of distance geometry and restrained molecular dynamics techniques for protein structure determination from nuclear magnetic resonance data: Application to the DNA binding domain of Lac repressor from *Escherichia coli*. *Prot. Struct. Funct. Genet.*, **3**, 209–218 (1988).

[6] D. T. Jones, W. R. Taylor, and J. M. Thornton. A new approach to protein fold recognition. *Nature*, **358**, 86–89 (1992).

[7] I. D. Kuntz, G. M. Crippen, P. A. Kollman, and D. Kimelman. Calculation of protein tertiary structure. *J. Mol. Biol.*, **106**, 983–994 (1976).

[8] I. D. Kuntz, J. F. Thomason, and C. M Oshiro. Distance geometry. *Meth. Enzymology*, **177**, 159–204 (1989).

[9] C. A. Orengo and W. R. Taylor. A local alignment method for protein structure motifs. *J. Mol. Biol.*, **233**, 488–497 (1993).

[10] L. H. Pearl and W. R. Taylor. A structural model for the retroviral proteases. *Nature*, **329**, 351–354 (1987).

[11] M. J. Sippl. Calculation of conformational ensembles from potentials of mean force. an approach to the knowledge-based prediction of local structures in globular proteins. *J. Mol. Biol.*, **213**, 859–883 (1990).

[12] W. R. Taylor. Towards protein tertiary fold prediction using distance and motif constraints. *Prot. Engng.*, **4**, 853–870 (1991).

[13] W. R. Taylor. Protein fold refinement: building models from idealised folds using motif constraints and multiple sequence data. *Prot. Engng.*, **6**, 593–604 (1993).

[14] W. R. Taylor, T. P. Flores, and C. A. Orengo. Multiple protein structure alignment. *Prot. Sci.*, 1994. submitted.

[15] W. R. Taylor and K. Hatrick. Compensating changes in protein multiple sequence alignments. *Prot. Engng.*, **7** (1994). in the press.

[16] W. R. Taylor, D. T. Jones, and A. W. Segal. A structural model for the nucleotide binding domain of the cytochrome b_{-245} β-chain. *Protein Science*, **2**, 1675–1685 (1993).

[17] W. R. Taylor and C. A. Orengo. Protein structure alignment. *J. Molec. Biol.*, **208**, 1–22 (1989).

Modelling Secondary Structure Formation by Distance Geometry Techniques

András Aszódi and William R. Taylor

Laboratory of Mathematical Biology, National Institute for Medical Research, London, United Kingdom

Abstract

Model polypeptide chains were folded into 3-dimensional compact conformations using Distance Geometry techniques. Interresidue distances were predicted from the hydrophobicity of the monomers and were refined by repeated projections into lower-dimensional spaces. Main-chain hydrogen bond networks were constructed and propagated through the structure by adjusting local conformations to comply with ideal distance constraints around hydrogen bonds. The resulting folds were compact globules with distinct hydrophobic cores and contained secondary structure elements like real protein molecules. In addition to similarity in appearance, several global properties of the model chains were also very close to those of native folded polypeptides. The method can serve as a starting point for the development of a novel structure prediction algorithm.

1   Introduction

The native folded conformation of a soluble protein molecule is a compact globule that possesses a distinct *hydrophobic core* and a varying amount of *secondary structure*. In a previous paper [1] we modelled the *hydrophobic effect* using Distance Geometry techniques. A novel algorithm was devised which generated compact three-dimensional conformations of α-carbon backbones by gradually projecting interresidue distance matrices into lower-dimensional spaces. The distances were predicted from the binary hydrophobicity assigned to the residues and geometric refinements were performed to prevent steric clashes. Compact and globular conformations were obtained which possessed distinct hydrophobic cores but no secondary structure.

The algorithm has been extended recently by the inclusion of main-chain hydrogen bonding [2]. Hydrogen bond candidates were allowed to form in the structure where chain segments approached each other below a pre-set distance threshold. In this way the hydrophobic effect, which drove hydrophobic residues close to each other, indirectly influenced the formation of hydrogen bonds. The positive cooperativity characteristic of native hydrogen-bond networks was simulated by a clustering algorithm. The resulting model chain folds were then compared to real monomeric proteins and the extent of secondary

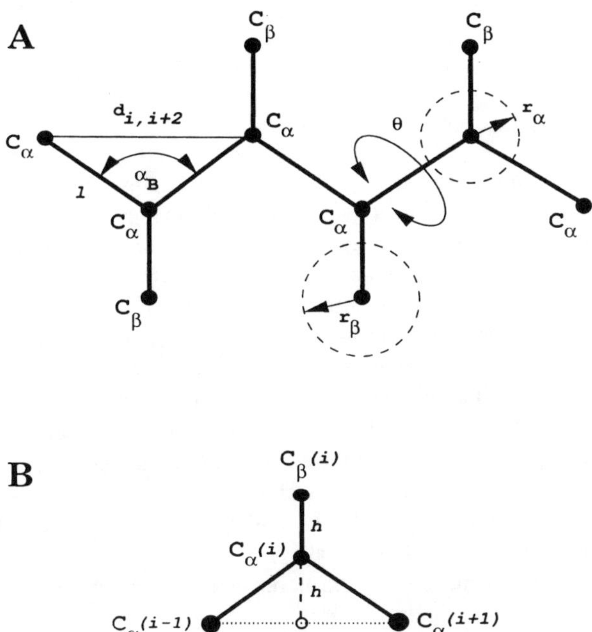

Figure 1: The geometry of the model polypeptide chain. **A**, The chain was composed of a backbone of C_α-atoms joined by virtual bonds, and of dummy C_β-atoms. The virtual bond length was $l = 3.8$ Å , the average virtual bond angle α_B was 1.82 rad corresponding to an average second-neighbour distance $d_{i,i+2}$ of 6.0 Å. No constraints were imposed upon the bond torsion angle θ. Volume exclusion was modelled by setting the van der Waals radii r_α and r_β of the C_α- and C_β-atoms to 2.0 Å and 2.5 Å , respectively. **B**, The position of the i-th dummy C_β-atom was calculated by reflecting the midpoint M between the $i - 1$-th and $i + 1$-th C_α-atoms through the i-th C_α atom position.

structure formation was assessed. Other aspects of protein chain conformation were also monitored using geometric criteria.

2   Methods

2.1   The Polypeptide Chain

Polypeptide chains were modelled by an α-carbon *backbone* and single C_β atoms representing *side-chains* (Fig. 1A). The side-chains were represented by pseudo-atoms (referred to as dummy C_β atoms), the coordinates of which were calculated from the local conformation of the backbone (Fig 1B). The fake C_β coordinates or their distances could be calculated from the C_α positions or distances alone.

The monomers were geometrically identical, the only difference between them was their *'hydrophobicity'*, a binary property, which could either be present ('hydrophobic' monomers) or absent ('hydrophilic' monomers) [1]. In the present study, model chains with random binary hydrophobicity sequences were generated. For each position the probability of being occupied by a 'hydrophobic' residue was 1/2, independent of the occupancy of the other positions.

2.2 Folding Strategy

The objective of our simulations was to fold the model chains into realistic conformations using Distance Geometry techniques [1]. The distance matrix of the model chain was predicted, the various constraint violations were eliminated, and then the method of subspace projection [4, 10] was used to embed the distance matrix into Euclidean space. Some of the non-metricities in the distance matrix were checked and removed by processing the metric matrix but in general it was not guaranteed that the metric matrix had only non-negative eigenvalues. If $M < N-1$ eigenvalues were positive, then the points were projected into an M-dimensional subspace. If all eigenvalues were positive, then the M largest eigenvectors whose sum was a preset fraction of the sum of all eigenvalues were used. Projection into less than 3-dimensional subspaces was not allowed. Further refinements were carried out in the embedding space, and, if the dimension was higher than three, a new distance matrix was constructed and the cycle repeated until a 3D embedding was achieved. In 3D a separate adjustment cycle was iterated to refine chirality.

The possible set of conformations of the model chains were determined by the following interactions between the monomers:-

1. **Steric constraints:** The minimal and maximal distances between any two C_α atoms were determined by the van der Waals radii and the maximal length of the totally extended connecting chain, respectively.

2. **Hydrophobic interactions:** it was assumed that in general hydrophobic residues tend to cluster together, therefore their expected distances would be smaller than for hydrophilic residues.

3. **Hydrogen bonding:** The geometrical consequences of interpeptide hydrogen bonding was modelled by simple distance constraints.

These interactions were applied to all residues in the model chain, regardless of their sequential positions.

The model chains also had global properties which could not be simply described as a sum of pairwise interactions. The global properties in the present model were:-

1. **Residue density:** the number of C_α atoms per volume has a characteristic value for real polypeptides. This usually changed during projection steps and needed repeated re-adjustment.

2. **Shielding:** although solvent molecules were not included in the model, the effect of a uniform polar environment was simulated by maximising the number of buried hydrophobic and exposed hydrophilic residues.

3. **Chirality:** since the model chains were built of symmetric monomers, the chirality of the chain conformation (especially for secondary structures) had to be adjusted separately.

2.3   Steric Constraints

The maximal allowed separation between any two C_α atoms was the length of the chain connecting them in a fully extended conformation. To model repulsion and volume exclusion, the atoms were not allowed to approach each other any closer than the sum of their respective van der Waals radii. For any $C_\alpha - C_\alpha$ pair an adjustment factor was computed so that by multiplying their distance by this factor would eliminate the steric clash, no matter whether it occurred between them or their respective C_β atoms. In distance space, the C_α - C_α matrix entries were multiplied by the factor matrix entries. In Euclidean space, the C_α atoms which lay too close or too far apart were moved symmetrically along the line connecting them so that their new distance should be changed by the corresponding entry in the factor matrix.

2.4   Residue Density

The average residue density of protein molecules is about $\rho = 6 \cdot 10^{-3} \mathring{A}^{-3}$ [6]. The density had to be re-adjusted to compensate for the shrinkages accompanying projection. It must be kept in mind that the residue density was defined for three-dimensional space only and its value was meaningless in higher dimensions. Consequently, for adjustments performed in distance space and immediately after projection in higher dimensions, indirect scaling methods were applied based on the observed distribution of distances within three-dimensional spheres of uniform density. In the 3D adjustment cycle, the residue densities were approximated by the inertial ellipsoid method [17].

2.5   Hydrophobic Interactions

The binary nature of the 'hydrophobicity' of the monomers defined three possible kinds of pairwise interactions: hydrophobic / hydrophobic, hydrophobic / hydrophilic and hydrophilic / hydrophilic. To each interaction a *desired distance* was assigned, and the entries in the distance matrix were updated in every cycle by these desired distances using the following linear combination:

$$d_{ij}^{(new)} = (1 - s_{ij})d_{ij}^{(old)} + s_{ij}d_{ij}^{(des)} \tag{1}$$

where the 'strictness' values $0 \le s_{ij} \le 1$ regulated the amount of desired modification. The hydrophobic / hydrophobic desired distances were set to a lower value at greater strictness than for the other two interactions to model the tendency of hydrophobic residues to cluster together inside the hydrophobic core. The actual choice of the distances was somewhat arbitrary as these served only as guidelines to the program rather than hard constraints to be strictly adhered to.

2.6   Solvent Accessibility

The interaction with solvent was modelled by burying exposed hydrophobic residues and bringing buried hydrophilic residues to the molecular surface. The accessibility of a residue (represented as a C_α atom) was determined by placing a cone on the C_α so that the apex of the cone was centered on the C_α atom position, the axis of the cone went through the centroid of the molecule and the cone angle was the minimal angle necessary to enclose all the other C_α points in the molecule (Fig. 2). Intuitively, an exposed C_α would have a cone

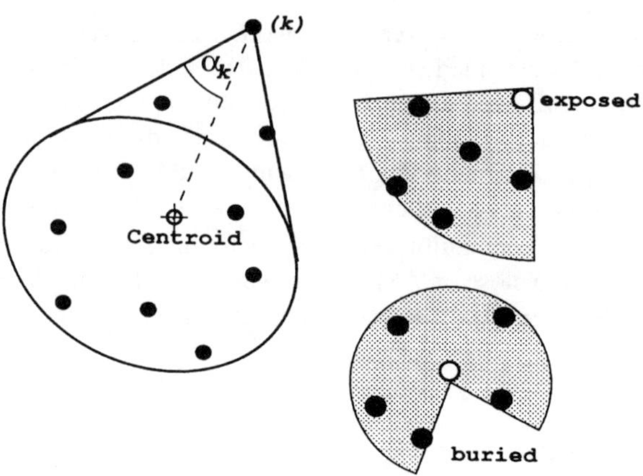

Figure 2: The conic accessibility calculation. A cone is positioned onto each residue in turn so that it should encompass all the other residues in the structure. The size of the cone angle α_k increases with decreasing accessibility.

with a relatively narrow cone angle (less than 180°) whereas buried positions would give 'wide open' inverted cones that have large angles $> 180°$, equivalent to narrow ordinary cones clear of any atoms. In distance space, pairs of residues which were far apart, both exposed and at least one of the partners was hydrophobic, were moved closer by a factor depending on exposure and distance. In Euclidean space, exposed hydrophobics were moved towards the centroid, while buried hydrophilics were moved outward by a suitable factor.

2.7   Hydrogen Bonds

The C_α backbone of our model chains did not represent the peptide groups involved in main-chain hydrogen bonding in real polypeptides. Instead, fake hydrogen bonds were allowed to form between the midpoints of the segments connecting two neighbouring C_α atoms and each midpoint could participate in two hydrogen bonds. The definition of the desired geometry of these fake H-bonds involved the specification of the positions of two pairs of neighbouring C_α atoms which together formed a *H-bond ladder* of length one (Fig. 3A).

2.7.1   Prediction of Hydrogen Bond Topology

H-bond formation exhibits positive cooperativity in two directions: firstly, if a H-bond is formed, giving rise to a unit-length ladder, the neighbouring bonds (extending the ladder) are more easily formed *along the chain*, producing longer parallel or antiparallel ladders. (Helices could be treated as special cases of parallel ladders where the two strands overlap in the sequence.) Secondly, an existing non-helical ladder also aligns the peptide units so that it becomes easier for other ladders to join both sides, thus forming sheets *across* chain segments (Fig. 4). For our model chains, first an initial set of *candidate ladders* were chosen.

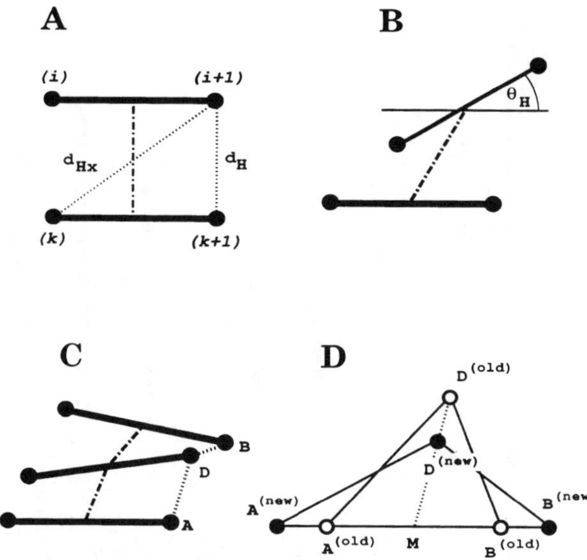

Figure 3: The geometry of the main-chain hydrogen bonds. The black dots symbolise C_α-atoms, the thick lines the virtual $C_\alpha - C_\alpha$ bonds. Hydrogen bonds are shown as dotted-dashed lines. **A**, The H-bond is perpendicular to the virtual bonds and the C_α atoms are set to d_H distance apart when on the same side of the H-bond, and d_{Hx} apart when on the opposite side of the H-bond. **B**, The asymmetric arrangement of the chain segments around a H-bond is characterised by the H-bond torsion angle θ_H. **C**, The position of the C_α atoms which are involved in the correction of the angle between two H-bonds connected through a common peptide unit. **D**, The adjustment of the H-bond — H-bond angle.

The 'rungs' of these ladders were all shorter than a preset threshold value, the *candidate length*. The construction of candidate ladders automatically included the simulation of positive cooperation of H-bond formation along the chain. The non-helical ladders were then joined by a single-linkage clustering algorithm, equivalent to that used in multiple sequence alignment [14], to select the best set of self-consistent ladder arrangement.

2.7.2   Hydrogen Bond Geometry Adjustment

The geometry of H-bonds were described in terms of $C_\alpha - C_\alpha$ distances between the four C_α atoms surrounding a H-bond (Fig. 3A). Two of the six distances corresponded to the $C_\alpha - C_\alpha$ while the remaining four distances were set to average values calculated by [18]. These distances within ladders were adjusted in distance and Euclidean spaces exactly in the same way as the steric collision distances. The angle subtended by the two H-bonds linked through a common 'peptide' was also adjusted so as to approach $180°$ (Fig. 3C,D).

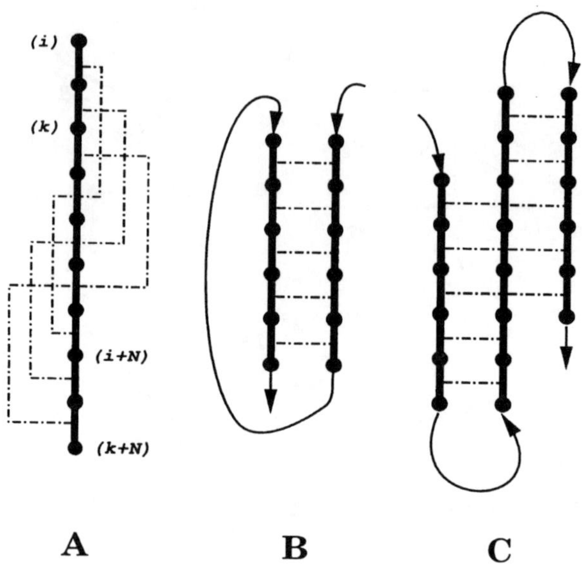

Figure 4: The H-bond ladder topologies. For symbols, see the legend for the previous figure. **A**, A helical topology. **B**, The topology of a parallel ladder. **C**, Two antiparallel ladders joined in a sheet through a common strand.

2.8 Chirality

If the distance matrix of an asymmetric point set is projected into Euclidean space, both enantiomers can be obtained, since multiplying any eigenvector by -1 transforms the solution into its mirror image, and handedness consistency cannot be guaranteed through a series of projections. A further complication arises in multidimensional spaces. Whereas N-dimensional chiral objects cannot be aligned with their mirror images in N dimensions, this alignment is perfectly possible in an $(N + 1)$–dimensional space — for example, 2D chiral triangles can be aligned in 3D without difficulty. The consequence is that the very notion of 3D chirality becomes meaningless in more than 3 dimensions as there will be no means by which the mirror images could be distinguished. Due to these obstacles, the chirality of our model chains was assessed and adjusted in 3D Euclidean space only.

The ultimate source of asymmetry in real proteins are the chiral C_α atoms, while in our model chains the monomers were symmetric. The asymmetric influences that led to the preference of one mirror image over the other were incorporated in the handedness adjustment of hydrogen bonds. The steric arrangement of the four C_α atoms flanking a main-chain hydrogen bond is chiral and the handedness of the configuration is consistent among right-handed helices and β-sheets. The asymmetry can be described by negative H-bond torsion angles $\theta_H < 0$ as defined in Fig. 3B. In our model the configuration of these 4 C_α atoms around the H-bonds were set so that their handedness be consistent, *i. e.* all torsion angles should have the same sign.

The inversion of handedness in helices required special treatment. A seven-residue long ideal right-handed α-helix was centered on each residue in helical strands which was locally better buried than its neighbours, then the model coordinates were moved towards the ideal coordinates [15]. This operation did not change the position of the locally buried helical residues, ensuring that the hydrophobic moment of the helix remained unchanged and the local geometry was preserved.

2.9   Scoring

The quality of the final three-dimensional conformations could be assessed according to several criteria and it would be very hard to create a single score that reflected accurately all aspects of the structure. In the current implementation the following scoring schemes were used:-

1. Constraint score: the relative violations of the van der Waals distance constraints were averaged in this score (its ideal value was 0).

2. Hydrogen bond score: similarly to the constraint score, the relative violations of distance constraints imposed upon the structure by the hydrogen bond geometry were averaged.

Both scores were simultaneously minimised during refinement. In addition to these, the fraction of hydrogen bonds with incorrect handedness was also monitored to assess the chirality of the model.

3   Results

Simulations of 90 model chains (all 100 residues long) with hydrogen bonds and 90 model chains without hydrogen bonds were carried out and various structural properties of the two sets were compared to each other and to a representative set of non-homologous (less than 25 % homology), well-resolved (better than 2.6 Å) monomeric protein structures. Examples of the structures obtained are shown in Fig. 5.

3.1   Geometric Properties

The model chains folded into compact globules which had almost the same average residue densities as real proteins. The density distribution of the simulated chains (irrespective of H-bonding) was similar to that of the reduced reference set containing real proteins of chain length between 80 and 120 amino acids (Fig. 6).

The molecular shape of the simulated chains was studied by fitting an ellipsoid to the molecule [17]. The ellipsoids were then represented as points on the plot of the normalised semiaxes. The majority of the protein molecules in the representative set were found to be in the prolate region, which was well reproduced by the H-bonded model chains. On the other hand, the non-H-bonded molecules populated the spherical region as well, showing less preference for the prolate ellipsoidal shape (Fig. 7).

3.2   Hydrophobic Core Formation

The assessment of hydrophobic core formation in the artificial chains was carried out by calculating the *hydrophobic radius of gyration* [12]. To facilitate comparison with the

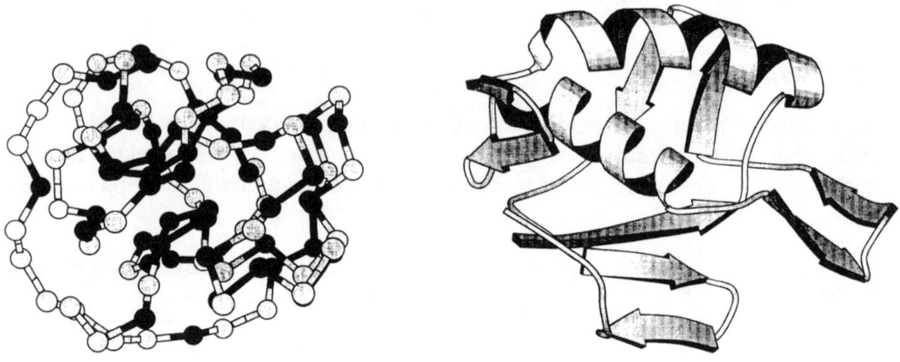

Figure 5: Typical model chain conformations. **A**, A structure generated without main-chain H-bonds. Only the C_α trace is shown; the shaded atoms correspond to 'hydrophobic' residues. Drawn by the `xmol` program (© Minnesota Supercomputer Center, 1993). **B**, With main-chain H-bonding enabled, three antiparallel β-hairpins and a mixed sheet were formed in this model chain, protected by two helices from one side. Drawn by the `MolScript` program [9].

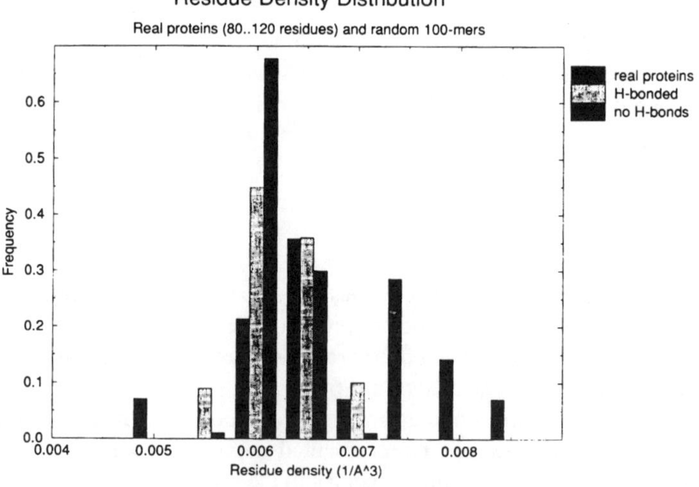

Figure 6: The residue density distribution of simulated chains with H-bonds (light columns) and without H-bonds(medium grey columns) compared to that of the subset of the reference proteins (black columns) with chain lengths between 80 and 120 residues. Hydrogen-bonded model chains gave an average residue density of $\overline{\rho_H} = (6.5 \pm 0.4) \cdot 10^{-3}$ Å^{-3}, the corresponding value for non-H-bonded chains was $\overline{\rho_{xH}} = (6.4 \pm 0.2) \cdot 10^{-3}$ Å^{-3}.

binary hydrophobicity of the monomers in the model, in real proteins the amino acids A,

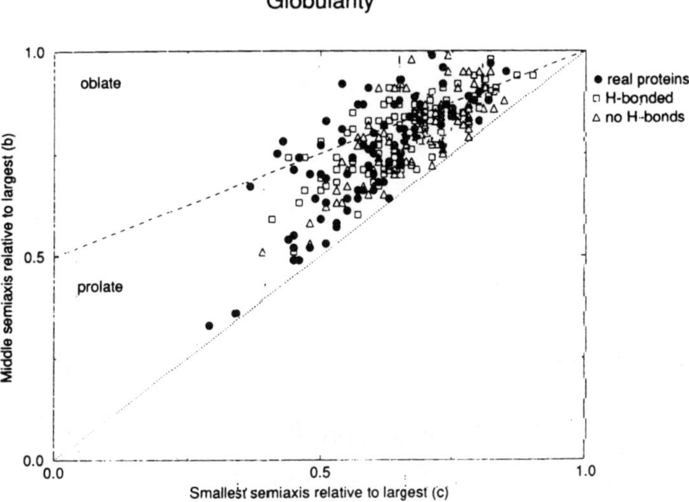

Figure 7: Globularity plot of polypeptide chains. A perfect sphere would occupy the upper right square of the diagram. Oblate ellipsoids fall over, prolate ellipsoids fall below the broken line indicating $b = (1+c)/2$. Model structures without H-bonds ($\triangle$) are generally closer to the spherical region, whereas H-bonded simulated chains ($\square$) prefer the prolate region, similarly to real proteins ($\bullet$).

C, F, G, I, L, M, P, V, W and Y were regarded as 'hydrophobic' and the rest 'hydrophilic'. The hydrophobic radius of gyration of real proteins was almost always slightly less than the overall radius of gyration (Fig. 8). The H-bonded model chains followed this general tendency well. In the case of non-H-bonded chains, more marked hydrophobic cores were formed as judged by the comparatively lower hydrophobic radii of gyration.

3.3   Secondary Structure Formation

Although the non-H-bonded model chains folded into compact globules with distinct hydrophobic cores, no secondary structure was formed in these simulations. The H-bonded chains, on the other hand, *always* contained H-bond networks which were visually very similar to those found in real proteins. No special sequence patterns which would be prerequisites of secondary structure formation were found; all random sequences in our simulations folded up into conformations containing helices and/or sheets.

For real proteins, the expected number of main-chain H-bonds (as defined by the DSSP program of [8]) varies approximately linearly with the chain size. The mean density of hydrogen bonds in our model chains was 41.2 bonds per 100 residues, less than the expected value of about 68 bonds. Helices were formed more easily than extended structures: 61 % of all H-bonds were found in α-helices, π-helices contained an additional 14 %, whereas antiparallel and parallel β-sheets accounted for 15 % and 10 % of H-bonds, respectively. The various secondary structure types occurred in almost all combinations, except for all-β proteins. The handedness of secondary structure elements were well reproduced in the simulations: most β-sheets exhibited a pronounced left-handed twist (viewed across the strands), and the overwhelming majority of helices were right-handed.

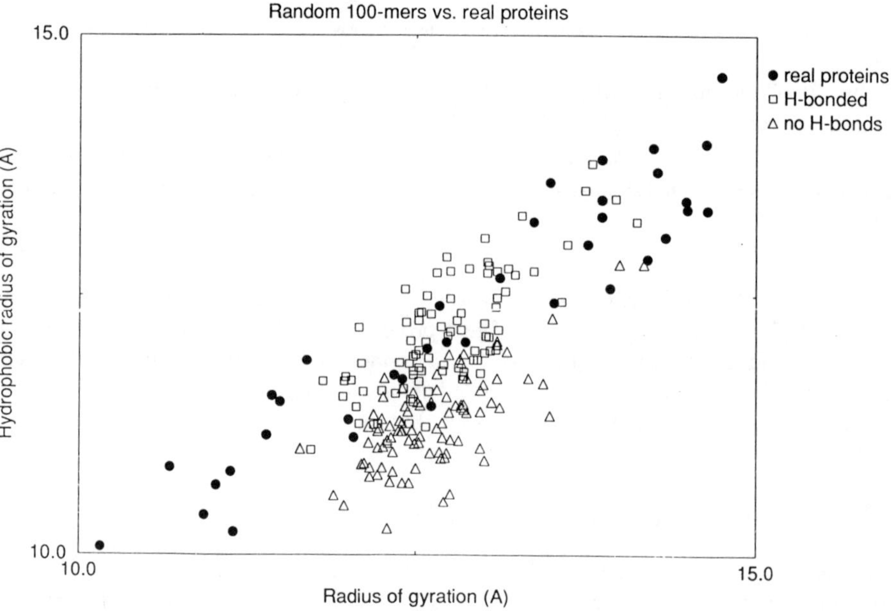

Figure 8: Hydrophobic core formation in polypeptide chains. The hydrophobic radius of gyration is plotted against the overall radius of gyration. Hydrogen-bonded model structures (□) follow the distribution of the proteins in the representative set (•). Model chains without H-bonds (△) possess more distinct hydrophobic cores than the native proteins as indicated by the comparatively lower values of hydrophobic radii of gyration.

4 Discussion

4.1 Structure Formation

The folding of model polypeptide chains by Distance Geometry methods incorporated the simulation of two major structure forming factors: the hydrophobic effect and main-chain hydrogen bonding. Realistic structures were formed, similar to monomeric protein molecules. The model chains reproduced the globularity and compactness of the native chains, possessed distinct hydrophobic cores and secondary structure. The hydrophobic effect and hydrogen bonding were responsible for the generation of these features to different extent.

4.1.1 Analysis of Molecular Shape

Compact globules were formed in all simulations, irrespective of the inclusion of H-bonds. The desired distances (especially between hydrophobic residue pairs) provided enough 'attraction' to form a compact structure. The average residue density of the model chains was close to the average density of the real chains indicating that the packing of amino acids was well reproduced even by the simple geometry of the model structures. Without main chain hydrogen bonding the model structures were more spherical than the real proteins, whereas the H-bonded models more often assumed prolate ellipsoidal shapes. This could be explained by considering that secondary structure elements, helices as well as extended strands, are elongated and often pack along a common axis, giving rise to prolate ellipsoidal molecules. Our H-bonded model folds reproduced this tendency of real protein molecules very well.

4.1.2 Hydrophobic Core Formation

With H-bonding disabled, the program generated structures with distinct hydrophobic cores (Fig. 5A). In the H-bonded models, however, the hydrophobic cores were less compact, similar to the situation encountered in native polypeptides. In the model chains, the formation of secondary structure was the result of a compromise between the strict geometric requirements of hydrogen bonding patterns and the less oriented drive towards perfect hydrophobic packing. In real proteins the sequences probably evolved so that both criteria can optimally be satisfied — clearly not the case with the random sequences in our simulation.

4.1.3 Secondary Structure

The inclusion of main chain H-bonds in the model was required to produce secondary structures suggesting that although the hydrophobic effect is necessary to determine the overall shape of the molecule, it cannot provide the precise geometric ordering which seems to be a prerequisite of secondary structure formation. The main chain H-bonds can, on the other hand, impose a local order on the chain geometry which is then propagated through the H-bond network in the molecule.

The relative abundance of helical structures in the simulated chains was probably due to the essentially local nature of helix formation. On the other hand, β-sheet formation requires the global ordering of several chain segments, often far away from each other in the sequence. In our simulations local H-bond formation was encouraged, in line with other local folding theories [5], giving rise to mainly helical structures.

The density of H-bonds was lower in our model chains than what could be expected from real proteins of similar size. There is an optimal amount of secondary structure which can be accommodated in native protein molecules but this optimum was not reached in our simple models. One reason for this could be that as we imposed no constraints on the torsion angles of the virtual $C_\alpha - C_\alpha$ bonds, this additional flexibility could make it more difficult for the algorithm to single out an optimal conformation with the highest possible number of H-bonds.

4.1.4   Hydrophobicity and Secondary Structure

Characteristic patterns of hydrophobicity are often associated with secondary structure elements. Numerous attempts have been made to predict the secondary structure of proteins based on the recognition of periodic hydrophobicity patterns in the sequence [3, 16], with variable success. In our simulations, the hydrophobic moments of amphipathic helices often correctly pointed towards the protein interior when the program picked up a stretch with the characteristic 3 . . . 4 residues/turn periodicity formed by chance in the random sequence. However, the correlation between sequence patterns and secondary structure was not strong enough when non-random sequences with α-helical hydrophobic patterns were supplied to the program.

4.1.5   Handedness

In real proteins, the chirality of the amino acid monomers is the source of the handedness of the molecule. Chirality is present at almost all levels of protein structure [11], and one enantiomer is usually preferred over its mirror image. At a local level, we could generate secondary structure elements with the correct hand from symmetric monomers by forcing the C_α atoms around H-bonds into the configuration observed in real proteins. This simplification was necessary in order to avoid the computational burden associated with a more detailed representation where the chiral arrangement of all main-chain atoms would have ensured the correct handedness of helices and β-sheet twists through steric clashes.

4.2   Outlook

The results discussed above suggest that the Distance Geometry techniques encoded into the present program constitute a powerful toolkit that enables the detailed study of the factors responsible for protein structure formation. The method should be able to calculate the main chain conformation of any protein if reliable predictions of the interresidue distances and H-bond topologies could be made.

4.2.1   Distance Prediction

The current scheme of distance matrix prediction applied in the program ignored many details. The most obvious refinement would be the modelling of more realistic chains in which monomers would have graded hydrophobicities, different sizes (to simulate packing more accurately) and the use of pair potentials which are derived from statistics of short-range interresidue distance distributions [13, 7]. It is also possible to include experimental distance constraints derived from NMR spectra with appropriate strictness values based on estimates of their reliability. However, the full prediction of the distance matrix cannot be carried out by statistical methods since the larger interresidue distances do not reflect any specific interactions and are essentially random.

4.2.2 Hydrogen Bond Topology Prediction

The H-bond ladder construction and clustering applied in the program modelled the positive cooperativity and some of the steric requirements inherent in secondary structure formation in native proteins well, and produced plausible H-bond networks. However, the sub-optimality of overall H-bond density and the bias towards helix formation suggest that the simple assumptions behind the algorithm should be refined before structure prediction could be attempted. A possible improvement could be to supply additional information from sequence-based secondary structure prediction algorithms. By incorporating further details we hope to construct a truly versatile tool for the exploration of the rules of secondary structure formation in native proteins.

References

[1] A. Aszódi and W. R. Taylor, Folding polypeptide alpha-carbon backbones by distance geometry methods *Biopolymers*, 1994(in press)

[2] A. Aszódi and W. R. Taylor, Secondary structure formation in model polypeptide chains *Protein Engng.* (submitted)

[3] F. E. Cohen, R. M. Abarbanel, I. D. Kuntz, and R. J. Fletterick, Secondary structure assignment for α/β proteins by a combinatorial approach *Biochemistry*, **22**, 4894–4904 (1983).

[4] G. M. Crippen and T. F. Havel, Distance Geometry and Molecular Conformation, Chemometrics Research Studies Press, Wiley, New York, 1988

[5] K. A. Dill, K. M. Fiebig, and H. S. Chan. Cooperativity in protein-folding kinetics *Proc. Natl. Acad. Sci. USA*, **90**, 1942–1946 (1993).

[6] L. M. Gregoret and F. E. Cohen, Protein folding: Effect of packing density on chain conformation *J. Mol. Biol.*, **219**, 109–122 (1991).

[7] D. T. Jones, W. R. Taylor, and J. M. Thornton, A new approach to protein fold recognition *Nature*, **358**, 86–89 (1992).

[8] W. Kabsch and C. Sander, Dictionary of protein secondary structure: Pattern recognition of hydrogen-bonded and geometrical features *Biopolymers*, **22**, 2577–2637 (1983).

[9] P. J. Kraulis, MolScript: A program to produce both detailed and schematic plots of protein structures *J. Appl. Cryst.*, **24**, 946–950 (1991).

[10] A. L. MacKay, The numerical geometry of biological structures *In:* M. J. Geisow and A. N. Barrett, editors, *Computing in biological science*, Chap. 10, pp. 349–392. Elsevier Biomedical, Amsterdam, 1983

[11] J. S. Richardson, The anatomy and taxonomy of protein structure *Adv. Prot. Chem.*, **34**, 167–339 (1981).

[12] B. Robson, E. Platt, R. V. Fishleigh, A. Marsden, and P. Millard, Expert system for protein engineering: Its application in the study of chloramphenicol acetyltransferase and avian pancreatic polypeptide *J. Mol. Graphics*, **5**, 8–17 (1987).

[13] M. J. Sippl, Calculation of conformational ensembles from potentials of mean force. An approach to the knowledge-based prediction of local structures in globular proteins *J. Mol. Biol.*, **213**, 859–883 (1990).

[14] W. R. Taylor, A flexible method to align large numbers of biological sequences *J. Molec. Evol.*, **28**, 161–169 (1988).

[15] W. R. Taylor, Protein fold refinement: Building models from idealised folds using motif constraints and multiple sequence data *Protein Engng.*, **6**, 593–604 (1993).

[16] W. R. Taylor and J. M. Thornton, Recognition of super-secondary structure in proteins *J. Mol. Biol.*, **173**, 487–514 (1984).

[17] W. R. Taylor, J. M. Thornton, and W. G. Turnell, An ellipsoidal approximation of protein shape *J. Mol. Graphics*, **1**, 30–38 (1983).

[18] H. Wako and H. A. Scheraga, Distance-constraint approach to protein folding. I. Statistical analysis of protein conformations in terms of distances between residues *J. Prot. Chem.*, **1**, 5–45 (1982).

A Library of Signature Pentapeptides for the Protein Data Bank

Ikuo Uchiyama, Atsushi Ogiwara, and Minoru Kanehisa

Institute for Chemical Research, Kyoto University, Uji, Kyoto 611, Japan

Abstract

We have developed an automatic method to extract signature patterns that characterize a group of related proteins. The method is applied here to a non-homologous set of proteins taken from the Protein Data Bank (PDB). For all the pentapeptides that appear in each sequence, we examine the occurrences of the same pentapeptides and related ones in homologous sequences using the BLAST search against the SWISS-PROT protein sequence database. The group of related proteins is then defined by the occurrences of conserved pentapeptides in homologous sequences. We measure the degrees of conservation and uniqueness of such signature pentapeptide patterns by the frequencies of occurrences within a homology group and in the entire database. The signature patterns are compared with secondary and other structural features and with PROSITE for functional identification. In addition to the well-known observations that conserved patterns tend to be located inside the globule and that they often correspond to functional sites, it is found that more conserved patterns are more unique in the entire collection of protein sequences. Furthermore, the pentapeptides conserved in more distantly related sequences are found to be closer in the three-dimensional structure.

1 Introduction

A functional site of a protein is generally well conserved in the amino acid sequences of related proteins. It is often composed of separate polypeptide chain segments which are apart in the sequence but are close in the three-dimensional structure. Thus, when conserved blocks are observed they may be utilized for structural and functional predictions. The multiple sequence alignment is a standard procedure to identify such conserved blocks, often called consensus patterns or sequence motifs, but this usually involves tedious manual tasks of collecting and analyzing groups of related sequences.

We have been developing automatic procedures to construct and maintain libraries of sequence motifs from the rapidly expanding body of protein sequence and structural data. In a previous paper [1], we reported a method to identify from the protein sequence database conserved amino acid patterns that were found exclusively within a group of functionally related proteins. Given the superfamily classification of the PIR database, we applied this procedure and constructed a dictionary of sequence motifs that related to specific superfamilies. However, because the PIR superfamily classification does not cover the entire protein sequence database currently available, and because the classification is not fully computerized, we have developed another method to perform both the grouping of related sequences and the search for conserved patterns at the same time (I. Uchiyama, A. Ogiwara, and M. Kanehisa, manuscript in preparation). Here we apply this method to the Protein Data Bank to construct a library of signature pentapeptides that bear structural and functional significance.

2   Materials and Methods

2.1   Database

We selected a nonhomologous set of 86 sequences from the January 1993 version of the Protein Data Bank (PDB). No pairs of sequences belong to the same superfamily according to our PDB-PIR cross-reference. The selected sequences also have descriptions of functional features in the corresponding entries of the SWISS-PROT database. The PDB identifier plus the chain identifier, if any, for each of the sequences is: 1YCC, 2CDV, 2CCYA, 256BA, 3B5C, 2CPP, 1FD2, 1HIP, 5RXN, 2TRXA, 1PAZ, 8ADH, 6LDH, 7ICD, 1GD1O, 3DFR, 3GRS, 2CYP, 7CATA, 2SODB, 1FNR, 2TSCA, 8ATCB, 8ATCA, 3CLA, 2AAT, 1PFKA, 3ADK, 1RHD, 1BP2, 2SNS, 2RNT, 7RSA, 2TAAA, 1LZ1, 5CPA, 2SGA, 4PTP, 1CSEE, 9PAP, 4PEP, 3TLN, 3BLM, 1PYP, 1ALD, 2CTS, 2CA2, 1WSYB, 1YPIA, 3PGM, 5PTI, 1TGSI, 2SSI, 7APIA, 2SECI, 4CPAI, 4SGBI, 1HOE, 5P21, 3INSA, 1TNFA, 1NXB, 1CRN, 1I1B, 2FB4H, 2MCG1, 2HLAA, 1MBD, 2MHR, 1UBQ, 1MSBA, 1RBP, 2PABA, 1UTG, 1HDDC, 2LTNA, 9WGAA, 2LBP, 8ABP, 2GBP, 3GAPA, 3WRP, 2CRO, 2PLV2, 3PHV, and 3HMGA.

2.2   BLAST algorithm

The search is based on the BLAST algorithm [2]. BLAST initially extracts all W-mers from a query sequence and enumerates all possible neighborhood words that have similarity scores of at least T, given an amino acid similarity matrix, such as the PAM and BLOSUM matrices. Using this word list, BLAST constructs a deterministic finite automaton and rapidly searches the database for these words. When a word in the list is found, it indicates that there is a potential similarity between the query sequence and the database sequence containing the word. BLAST then extends the alignment of the word to seek for a maximal-scoring segment pair (MSP). The similarity of two sequences represented by the combination of MSPs is considered to be significant if the similarity score is greater than a given threshold S, or the Poisson probability is less than a given threshold P. Our purpose here is to extract conserved pentapeptides among groups of homologous proteins. We choose the word size parameter $W = 5$. Starting from a query sequence we search for similar pentapeptides with scores equal to or above $T = 21$ that appear in homologous sequences with scores equal to or above $S = 40$. The similarity matrix used is BLOSUM62 [3].

2.3   Uniqueness and conservation

In our previous paper [1], the uniqueness U and the onservation C are defined by the following equations:

$$U = \frac{n(p,g)}{N(p)} \tag{1}$$

```
     ENTRY                      SCORE                        PENTAPEPTIDES

CYB5_HUMAN CYTOCHROME B5.        708   ***.*******..*********************************....**************
CYB5_HORSE CYTOCHROME B5.        650   *.....*****....04******************.1245****....**********22432
CYB5_RABIT CYTOCHROME B5.        648   .......****....04*******************************....**************
CYB5_RAT CYTOCHROME B5.          633   *......****..**......*21141*******************....**************
CYB5_PIG CYTOCHROME B5.          627   *.....*****..***100..*************************....**************
CYB5_BOVIN CYTOCHROME B5.        606   ......*****..***100..*****...20***************....**************
CYB5_CHICK CYTOCHROME B5.        557   ...............2310...1.2.2...42..10...124*****....**************
CYB5_ALOSE CYTOCHROME B5 (F      457   ....*******..******************************....******.....****
CYM5_RAT CYTOCHROME B5, OUT      314   .......................2......0......02*****.............***.
NIA_CUCMA NITRATE REDUCTASE      163   .......................2...............053.................
NIA1_ARATH NITRATE REDUCTAS      163   .......................2...............052.................
NIA_LYCES NITRATE REDUCTASE      159   .......................2...............051.................
NIA2_ARATH NITRATE REDUCTAS      159   .......................2................53.................
NIA1_TOBAC NITRATE REDUCTAS      157   .......................2...............053.................
NIA2_TOBAC NITRATE REDUCTAS      156   .......................2...............052.................
NIA_SPIOL NITRATE REDUCTASE      155   .......................1...............053.................
NIA_MAIZE NITRATE REDUCTASE      154   .......................1...............052.................
NIA_BETVE NITRATE REDUCTASE      154   .......................2...............052.................
NIA2_HORVU NITRATE REDUCTAS      153   .......................1...............052.................
NIA1_HORVU NITRATE REDUCTAS      153   .......................1...............052.................
NIA1_ORYSA NITRATE REDUCTAS      150   .......................1...............052.................
NIA7_HORVU NITRATE REDUCTAS      142   .......................1...............052..............1..
CYB2_YEAST CYTOCHROME B2 PR      135   .............................*0.......35.................
NIA_CHLKE NITRATE REDUCTASE      133   .......................................052.................
CYB2_HANAN CYTOCHROME B2 PR      130   ............................*........30.................
NIA_EMENI NITRATE REDUCTASE      125   .......................................012.................
NIA_NEUCR NITRATE REDUCTASE      121   .......................................012.................
CYBR_DROME PROTEIN TU-36B (       90   .......................................042.................
ACO1_YEAST ACYL-COA DESATUR       80   ......................................2**.................
SUOX_CHICK SULFITE OXIDASE        69   .......................................02.................
VG37_HSVI1 HYPOTHETICAL GEN       50   ..........................................................
YDP1_LACLC HYPOTHETICAL 30.       48   ..........................................................
PHBC_ALCEU POLY-BETA-HYDROX       48   ..........................................................
IF4B_HUMAN EUKARYOTIC INITI       48   ..........................................................
```

Figure 1: A sample output of the BLAST search for signature pentapeptides. The cytochrome b5 sequence is compared against the SWISS-PROT database and conserved pentapeptides are identified.

$$C = \frac{n(p, g)}{m(g)} \tag{2}$$

where n and N are the numbers of entries containing a given pattern p in a given group g and in the entire database, respectively, and m is the number of entries belonging to the group g. Here, since we allow each pattern to contain substitutions to a limited extent, we count the occurrences of a set of similar patterns instead of the occurrence of the single pattern. We define, somewhat arbitrarily, the uniqueness and conservation as follows:

$$U = \sum_p n(p, g) \Big/ \sum_p N(p) \tag{3}$$

$$C = \sum_p n(p, g) \Big/ m(g) \tag{4}$$

2.4 Maximal group

The result of BLAST search is a list of homologous sequences ordered according to the homology score S. Based on this list a group is defined by the occurrence of a specific pattern. Namely, given some thresholds for U and C values, a collection of homologous sequences having a specific pattern forms a group. In this work the threshold for the conservation was 75%, while no threshold was set for the uniqueness. Then, a maximal group is defined by the largest number of entries having a conserved pattern equal to or above the threshold.

Table 1: A part of the motif library.

Res. No.	Penta- peptide	Group Size	Log Pval	Cons. (%)	Uniq. (%)	Sec. Str.	In/ Out	Fea- ture	Pro- site
1	TEFKA	1	71	100	8	O	O	N	0
2	EFKAG	1	71	100	12	O	O	N	0
3	FKAGS	1	71	100	2	O	O	N	0
4	KAGSA	1	71	100	6	O	O	N	0
5	AGSAK	1	71	100	7	O	O	N	0
6	GSAKK	2	60	100	18	a	O	N	0
7	SAKKG	2	60	100	20	A	O	N	0
8	AKKGA	2	60	100	13	A	O	N	0
9	KKGAT	5	54	100	15	A	O	N	0
10	KGATL	5	54	100	25	A	O	N	0
11	GATLF	5	54	100	25	A	M	N	0
12	ATLFK	5	54	100	45	A	M	N	0
13	TLFKT	5	54	100	41	A	O	N	0
14	LFKTR	12	44	100	80	A	O	N	0
15	FKTRC	13	43	100	81	A	O	B	0
16	KTRCL	12	44	100	66	a	O	B	0
17	TRCLQ	12	44	100	80	t	O	B	0
18	RCLQC	81	23	90	84	T	O	B	0
19	CLQCH	105	0	88	60	T	O	B	1
20	LQCHT	86	7	83	92	t	O	B	1

3 Results

3.1 Sample output

Figure 1 shows a portion of a sample output of our procedure where the SWISS-PROT database release 25.0 was searched with the sequence of human cytochrome b_5 (SWISS-PROT code: CYB5_HUMAN) as a query. Each line represents a database entry sorted by the homology score S. The columns following the score represent the occurrences of all the pentapeptides of the query sequence from the N-terminus to the C-terminus. An asterisk indicates that the database sequence has a pattern exactly matched with the pattern in the query, a dot indicates that the sequence does not have the pattern, and a digit indicates that the sequence has a similar pattern where the score is represented in terms of the difference from the threshold T_0. For example, in this particular case we chose $T_0 = 24$, so that '5' means that the sequence has a similar pattern with the score of 29. Note that the dots in the query sequence itself shown in the top line imply that the self-similarity score of certain pentapeptides is below the threshold.

In the example shown in Figure 1, cyotochrome b_5, nitrate reductase, and sulfite oxidase form one group which is characterized by the conserved pentapeptide of EHPGG or HPGGE, or more precisely [EKSDN]-H-P-G-G or H-P-G-G-[EASPQV]. This is consistent with the consensus pattern in PROSITE, F-[LIV]-x(2)-H-P-G-G where H of position 5 is heme iron binding site [4]. The two positions denoted by 'x' in PROSITE are so variable that our procedure failed to detect the preceding hydrophobic residues. The group defined here is the same as that of PROSITE 10.1 except that the additional entries NIA_CHLKE

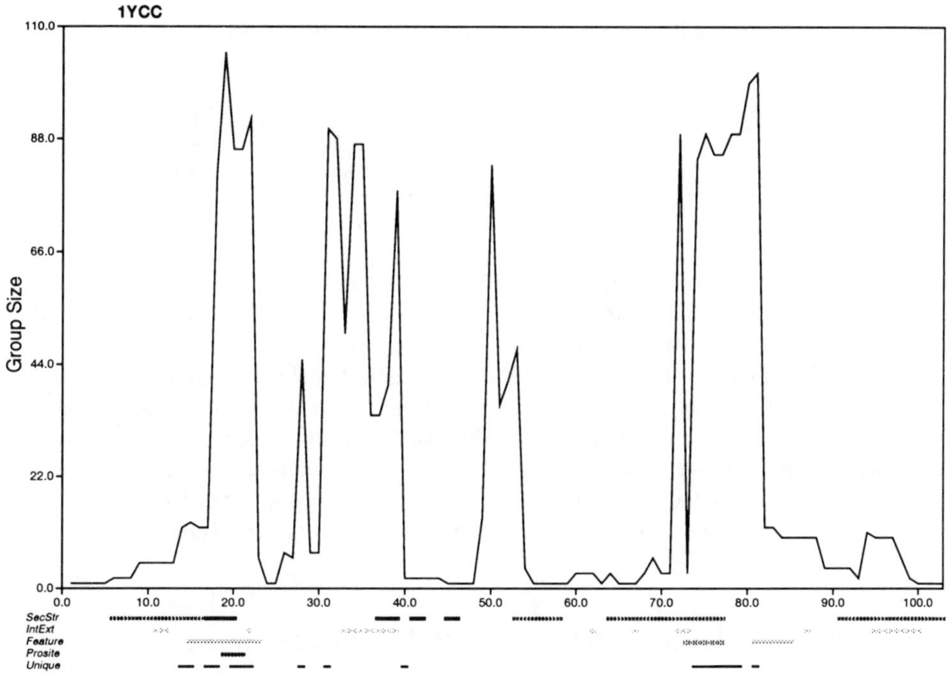

Figure 2: The group size formed is plotted against the residue number of each pentapeptide in cytochrome c, which is compared with structural and functional features using the data shown in Table 1. The group size roughly corresponds to the degree of conservation in distant sequences.

and ACO1_YEAST are included in our group. The former is missed by PROSITE 10.1 because this sequence is newly added to SWISS-PROT since release 25.0. The latter sequence is stearoyl-CoA desaturase of yeast, which has additional C-terminal region when compared with other stearoyl-CoA desaturase sequences. This region has weak similarity with cytochrome b_5 domain including HPGG signature sequence. Thus, we think this region potentially came from cytochrome b_5, although it was not mentioned in the original paper [5]. This sequence is missed by PROSITE because the first position of the pattern in PROSITE is substituted from F to Y.

3.2 Dictionary of conserved pentapeptides

We performed the BLAST search of conserved pentapeptides for each of the 86 PDB sequences against the SWISS-PROT database. The result can be summarized in a table such as shown in Table 1. For all the pentapeptide patterns from the N-terminus to the C-terminus of each query sequence, the table shows how many sequences are found in the homology group, the minus logarithm of the Poisson probability P-value, the degrees of conservation and uniqueness by Eqs. (3) and (4), the secondary structure assignment by DSSP, the inside/outside classification by the accessible surface area, functional sites as described in the SWISS-PROT feature table and in PROSITE [4].

Figure 2 is a graphical representation of the result in Table 1. The peaks in the group

Table 2: Summary of conserved pentapeptides identified.

Cutoff P-value	1	1e-05	1e-10	1e-15	1e-20	1e-25	1e-30	total
Entries	42	72	79	81	82	82	83	86
Pentapeptides	159	429	627	892	1105	1289	1586	16981

size correspond to conserved regions, which are compared with structural and functional features. As shown in this figure, it is often the case that conserved regions correspond to functional sites, and conserved patterns are unique as well.

Table 2 shows the number of conserved patterns identified with different cutoff P-value. The group size increases as the cutoff P-value is increased or as more distantly related sequences are incorporated in a group. Roughly half of the 86 sequences contain signature pentapeptides that are well conserved among large groups consisting of distantly related proteins, as shown for the cutoff P-value of 1. When smaller groups consisting of more closely related sequences are considered, the majority of sequences analyzed contain certain signature pentapeptides.

3.3 Correspondence to structural and functional features

The correspondence of conserved patterns with structural and functional features was analyzed according to the associated P-values. Note that when the P-value is very small, the group consists of very closely related sequences. Therefore, the pentapeptide identified may not actually be conserved patterns in a regular sense; they are conserved only within mostly identical sequences.

Figure 3(a) shows the relative frequency of occurrences of pentapeptides in helical, beta, and coil regions plotted against the minus logarithm of the P-value. The plots are more or less flat, indicating that there was no tendency for secondary structures to appear in whether conserved or non-conserved regions. In contrast, the inside/outside preference shown in Figure 3(b) clearly indicates that conserved patterns tend to appear inside of the protein globule.

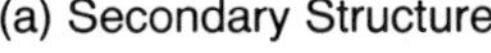

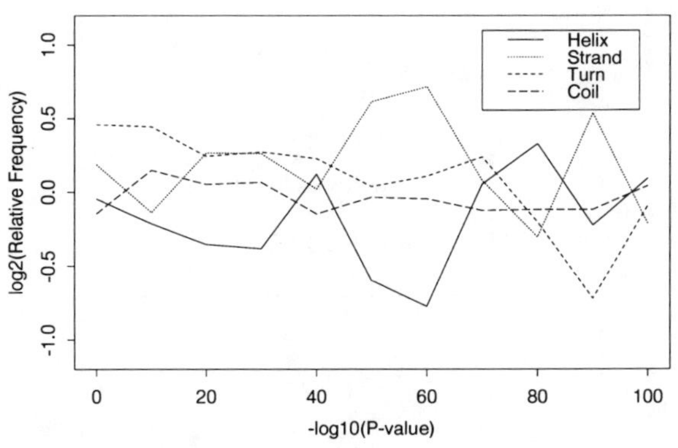

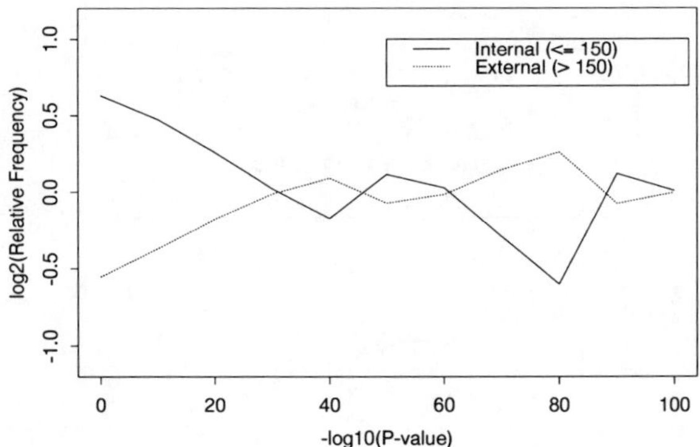

Figure 3: The relative frequency of occurrences of signature pentapeptides plotted against the minus logarithm of the P-value. Pentapeptides are classified by (a) the secondary structure and (b) the accessible surface area.

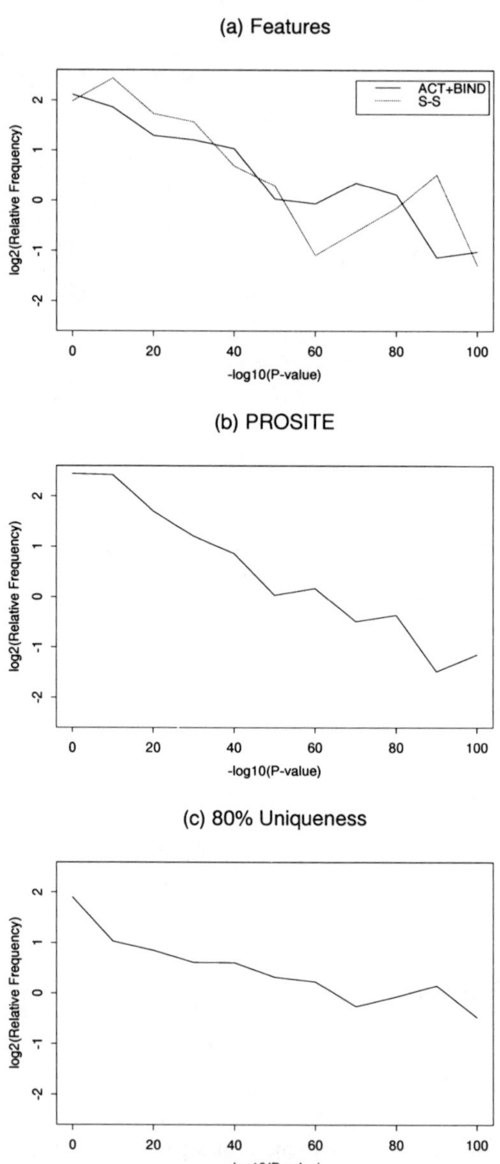

Figure 4: The relative frequency of occurrences of signature pentapeptides plotted against the minus logarithm of the P-value. (a) The pentapeptides with annotations of active sites, binding sites, and disulfide bonds in the SWISS-PROT feature table. (b) The pentapeptides corresponding to PROSITE patterns. (c) The pentapeptides that are more than 80% unique in the SWISS-PROT database.

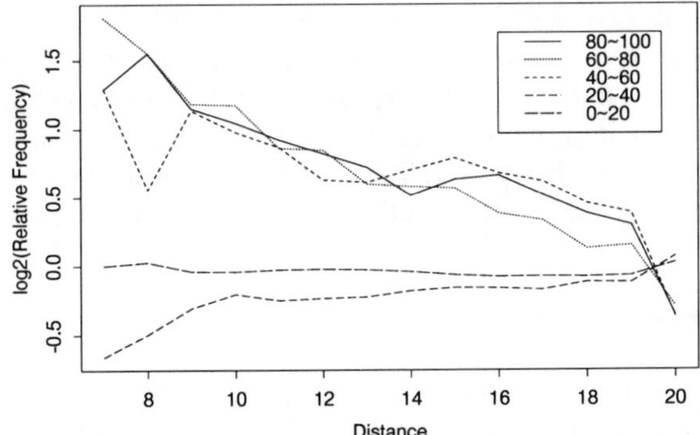

Figure 5: The relative frequency of occurrences of pairs of signature pentapeptides plotted against the minus logarithm of the P-value. The pairs are classified by the group size which is indicated by the percentage of the largest group.

Figure 4 shows the correlation with functional sites. Again, there is a clear tendency for conserved patterns to appear in functional sites annotated in the SWISS-PROT feature table (Fig. 4(a)) or in PROSITE (Fig. 4(b)). Of course, these observations are well known, but one additional observation here is the relationship between conservation and uniqueness. Figure 4(c) shows the relative frequency of occurrences of pentapeptides with 80% or more uniqueness according to Eq. (3), which indicates that more conserved patterns are more unique in the entire collection of protein sequences.

4   Discussion

The conserved sequence patterns are usually identified by the multiple alignment of a group of sequences, where a group must be predefined, which often involves intensive manual tasks. In contrast, the method presented here computationally perform both the grouping and the pattern identification at the same time. When the group formed is large enough to contain distant sequences, the conserved pattern identified may have structural and functional significance. For example, the same pentapeptides in general are known to take different three-dimensional conformations [6], but we expect that more conserved pentapeptides also have more conserved conformations.

Figure 5 shows an example of analyzing the group size dependence of a structural feature. The relative frequency of observing pentapeptide pairs is plotted against the distance between the pair, which is calculated by the mean square distance of all the Cα pairs. The pentapeptides are classified by the group size, up to 20% of the largest one, 20-40% of the largest one, and so on. This indicates that the pairs of more conserved pentapeptides, i.e., conserved pentapeptides in larger groups of distantly related proteins, are closer in the three-dimensional structure. Thus, by examining the groups formed, it may be possible to predict the closeness in the three-dimensional structure.

The advantage of automating the process of identifying sequence motifs is clear; newly determined sequences are rapidly increasing and frequent updates of motif libraries are now required. However, there are limitations to the present method. The sequence motifs identified by the current method covered roughly half of those reported in PROSITE (unpublished data). The requirement of conserved pentapeptides is a strong restriction and many motifs in PROSITE with x's were missed. It is perhaps necessary to search combinations of shorter patterns with various spacings, but then the efficient BLAST algorithm will no longer be applicable. We expect that with more refinement the current method will become more useful not only for identifying sequence motifs, but also for structural predictions.

Acknowledgments

This work was supported by the grant-in-aid for scientific research on the priority area 'Genome Informatics' from the Ministry of Education, Science and Culture, Japan. The computation time was provided by the Supercomputer Laboratory, the Institute for Chemical Research, Kyoto University.

References

[1] A. Ogiwara, I. Uchiyama, Y. Seto, and M. Kanehisa, Construction of a dictionary of sequence motifs that characterize groups of related proteins, *Protein Engineering*, **5**, 479-488 (1992).

[2] S.F. Altschul, W. Gish, W. Miller, E.W. Myers, and D.J. Lipman Basic local alignment search tool, *J. Mol. Biol.*, **215**, 403-410 (1990).

[3] S. Henikoff and J.G. Henikoff Amino acid substitution matrices from protein blocks, *Proc. Natl. Acad. Sci. USA*, **89**, 10915-10919 (1992).

[4] A. Bairoch PROSITE: a dictionary of sites and patterns in proteins, *Nucleic. Acids. Res.*, **20**, 2013-2018 (1992).

[5] J.E. Stukey, V.M. McDonough, and C.E. Martin The OLE1 gene of saccharomyces cerevisiae encodes the delta 9 fatty acid desaturase and can be functionally replaced by the rat stearoyl-CoA desaturase gene, *J. Biol. Chem.*, **265**, 20144-20149 (1990).

[6] W. Kabsch and C. Sander On the use of sequence homologies to predict protein structure: identical pentapeptides can have completely different conformations, *Proc. Natl. Acad. Sci. USA*, **81**, 1075-1078 (1984).

Glycosylation and Protein Conformation

Jan E. Hansen[*†], Ole Lund[*†‡], Khristoffer Rapacki[†], Henrik Clausen[§], Erik Mosekilde[‡],
Jens O. Nielsen[*] and John-Erik S. Hansen[*].

[*] Laboratory for Infectious Diseases, Hvidovre Hospital, University of Copenhagen,
DK–2650 Hvidovre, Denmark
[†] Center for Biological Sequence Analysis, The Technical University of Denmark,
DK–2800 Lyngby, Denmark
[‡] Department of Physics, The Technical University of Denmark, DK–2800 Lyngby,
Denmark
[§] Department of Oral Diagnostics, Royal Dental College, DK–2100 Copenhagen, Denmark

Abstract

Peptide linked carbohydrates affect such physical properties of glycoproteins
as solubility, stability and conformation. Biologically, glycosylation modulates
immunogenecity, tissue distribution and lifetime of glycoproteins. Increasing
evidence indicates that cell surface carbohydrates play a major role in recognition, development, oncogenesis and viral infection. It is therefore of interest to
investigate the relation between O or N-linked oligosaccharides and glycoprotein conformation. The paper briefly reviews the current knowledge in this area
and looks statistically on the local amino acid distribution and conformation at
glycosylation sites in glycoproteins as inferred from the data bases.

1   Introduction

Even though the importance of glycosylation is not yet fully acknowledged most secreted
proteins are glycosylated. The biological roles of these carbohydrates are quite diverse [1]:

For Human Herpes Virus (HSV) glycoproteins co-translational addition of carbohydrates is essential for formation of correct disulfide bridges, correct folding and proper
function [2]. The antigenic properties of HSV glycoproteins are modulated by carbohydrates since the antigenic site II of herpes gC-1 glycoprotein is dependent of an N-linked
oligosaccharide (β(1-4)-linked galactose). This carbohydrate is essential for maintaining
the proper antigenic conformation of the epitope even though it does not itself constitute a
part of the physical epitope [2].

The surface glycoprotein gp120 of Human Immunodeficiency Syndrome Virus (HIV) is
heavily glycosylated. The carbohydrates constitute more than 50% of the molecular weight
of gp120. Non-glycosylated gp120 cannot bind to its cell surface receptor CD4. The 19-
24 N-linked oligosaccharides on gp120 reduce the immunogenicity of peptide epitopes by
shielding them either directly or indirectly by conformational effects [3, 4]. A carbohydrate
neo-antigen Tn, which is also found in some cancers, is created by aberrant O-linked
glycosylation of gp120 [3].

The hepatic asialoreceptor recognizes glycoproteins which have lost their terminal carbohydrates (sialic acids) independently of their primary structure [5]. The glycoproteins

Figure 1: (a) & (b). See Color Plates 6 and 7, page xviii and xix.

Figure 2: (a) & (b). See Color Plate 8, page xx.

are endocytosed by hepatocytes and degraded after binding. The hepatic asialoreceptor function as a "wear and tear" recognizer of "old" glycoproteins, thereby regulating their lifetime in circulation.

During development of cancer, cell surface carbohydrate structure is altered, hence tumor associated carbohydrate-markers has been used in diagnosis of human cancers. Recently tumor cell surface carbohydrates have been associated with the metastatic potential (ability to spread) of tumor cells and even with the prognosis of cancer patients [6].

The carbohydrate content of neural cell adhesion molecules NCAM has recently been shown to be involved in a variety of neural functions including synaptic and axonally related processes, especially during the development of the central nervous system [7].

The above mentioned functional aspects of the carbohydrate part of glycoproteins have prompted reinforced interest in investigating the conformational aspects of glycoproteins as protein conformation is generally closely related to protein function. Nature has probably endowed eukaryotes with this complex glycosylation apparatus because it is a cheap way of modifying protein function without changing the genome. Thereby greater diversification - one of the major driving forces behind biological evolution- of protein function is created.

The difficulties in making crystals [8] of highly glycosylated proteins, possibly due to large flexibility [9] of some carbohydrate chains, hamper efforts to investigate the role of glycosylation on conformation. Examples of X-ray crystallographic data for glycoproteins including their carbohydrate part are few in the Brookhaven Data Base.

New techniques as FAB-MS, other mass spectrometric techniques [10] and NMR seem useful [11]. Theoretical protein dynamics using standard software with structural data from NMR or other techniques also look promising [12].

2   Biosynthesis of glycosylation

Initial N-linked glycosylation is a co-translational event [5, 13] occurring as the yet unfolded glycoprotein emerges from the ribosome in the rough endoplasmic reticulum (rER). A still uncharacterized oligosaccharyl transferase transfers en bloc a $Glc_3Man_9GlcNAc_2$ precursor from a lipid dolichol carrier to Asn on the peptide chain. The Asn has to be in the consensus triplet sequon Asn-X-Thr/Ser, where X must not be Pro. Neither may Pro follow the triplet sequon [14]. However, not all consensus triplets are glycosylated, possibly due to steric hindrance caused by rapid three-dimensional folding of the peptide chain [13]. It is believed that the oligosaccharyl transferase and the folding process compete in the process of glycosylation. Processing begins in rER with removal of the three terminal glucose residues and at least one mannose residue. Further processing occurs in golgi through a multistep pathway involving both glycosidases and glycosyltransferases which elongate the carbohydrate chain so that a carbohydrate tree with one or more antennae is formed. This delicate process of carbohydrate branching depends on the glycosylation apparatus of the cell expressing the glycoprotein and is concequently cell and species specific. An example of a bi-antennary carbohydrate is seen in Fig 1. Recently the carbohydrate specific chaperone calnexin was found [15]. It binds glycoproteins from which two of the three terminal glucose residues have been removed by glycosidases and the proteins remain bound to calnexin for a period of time dependent on their rates of conformational maturation. Misfolded proteins also associate with calnexin: unable to finish their maturation they remain bound and are eventually degraded within the endoplasmic reticulum. Once the glycoproteins are released a glycosyltransferase with specificity for misfolded proteins can add a glucose more. When this is removed by glycosidase II the protein will again bind to calnexin. This reglycosylation-deglycosylation cycle will continue as long as the protein is misfolded. Once the glycoprotein is correctly folded or assembled into oligomeric form, it is no longer reglycosylated, it is therefore released by the membrane bound calnexin and is free to leave the rER. Thus calnexin forms a part of the endoplasmic reticulum "quality control" procedure, which ensures that the glycoproteins leaving the organelle for transit to golgi are correctly folded and assembled [15].

The onset of O-glycosylation is less well defined. It is a post-translational process which is initiated in the cis-golgi complex after the glycoprotein is N-glycosylated, partly folded, and after oligomerization and disulfide-bridging has taken place [16]. The glycosyltransferase(s) forming the α-1 link between usually a GalNAc sugar and the sidechain of Serine or threonine is still uncharacterized. After glycosylation the glycoprotein is probably locally re-folded [16]. After the first GalNac is linked to Ser/Thr in the peptide, further sugars are added and the carbohydrate chain is elongated and branched by other glycosyltransferases in golgi in a way similar to N-linked glycosylation. No strict amino acid sequence requirements around O-glycosylated serines and threonines sites are known [17, 18]. This makes prediction of O-glycosylation of glycoproteins from their primary sequence difficult [19]. In Fig. 2 the relative abundances of amino acids on positions up- and downstream of O-glycosylated serines and threonines deduced from 57 mammalian glycoproteins are shown. The high abundance of Ser and Thr around the O-linked glycosylation site (position 0) reflects that Ser and Thr are often clustered in glycoproteins. The high abundance of Pro around O-linked glycosylation sites is wellknown. Proline is a good turn maker. Notice also that amino acids with small sidechains are favoured, with the exception of glutamic acid. This "Logo" was made using standard software made by Schneider and Stephens[20]

3 Conformational effects of N-linked glycosylation

The most direct demonstration of a change in secondary structure by N-linked glycosylation was performed by Otvos et al. [21]. Two 12 and 13-residue synthetic peptides with known glycosylation sites from rabies virus glycoprotein were used as model systems in non-glycosylated and glycosylated states. The conformations of the peptides before and after glycosylation were compared using circular dichroism (CD) and Fourier-transform infrared spectroscopi (FT-IR). The most significant conformational effects of N-glycosylation was breakage of the pronounced 3_{10}, or alpha-helicity, of the non-glycosylated peptides and formation of a type I(III) β-turn in the glycosylated peptides [21]. The spectra of the two N-glycosylated peptides were close to identical even though their amino acid sequence were quite different [21]. Phosphorylation of the 13-residue peptide resulted in a distorted conformation. The 13 residue peptide could thus exist in three entirely different conformations depending on the nature of the co/post-translational modification [21]. This underscores the necessity of including co/post-translational modification in methods for predicting the secondary structure of glycoproteins. However, the conformation of N-linked glycosylation sites is not invariably a turn. By X–ray analysis of hemmaglutinin Asn-linked glycosylation sites Wilson et al. [22] found that only 2 sites were within turns, 2 sites were within α-helices and 2 sites were within β-sheets [22]. We have examined the dihedral angles for

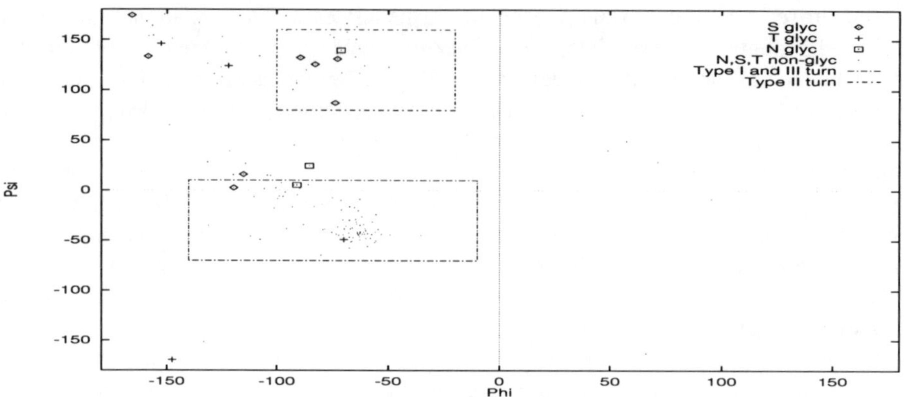

Figure 3: Ramachandran-plots for glycosylated and non-glycosylated asparagine (N), serine (S) and threonine (T) sites. Also shown are the standard regions for dihedral angles in type I, II and III turns [23].

glycosylated Asn and non-glycosylated Asn (3 sites) in endodeoxyribonuclease in complex with actin and 1,4 aspergillus glycanase. As seen in Fig. 3 glycosylated Asn and non-glycosylated Asn could not be distinguished. One of the sites has ψ around $130°$, and φ around $-75°$. According to the classification of turns by Garnier et al. [23] this could be in a type II β-turn. The two other sites ,with ψ less than $30°$ and φ around $-80°$, could be in a type I(III) β-turn.

Edge et al. [24] have studied the conformational effects of N-linked glycosylation using NMR on a 22-residue peptide from IgM tailpiece containing a single glycosylation site at Asn-9, a 124-residue glycoprotein ribonuclease B containing a single glycosylation site at Asn-34 and ribonuclease A - a deglycosylated form of ribonuclease B. They concluded that glycosylation has a modest but significant effect on protein dynamics and stability. For the IgM peptide an increase in conformational stability was found in the peptide backbone and in the amino-acid side chains in close proximity of the glycosylation site. The effect of N-glycosylation on ribonuclease conformation was modest and no major differences in secondary or tertiary structure were found between the glycosylated and non-glycosylated enzyme. However, increased global stability was observed as the hydrogen-deuterium solvent exchange rate was reduced. This could be caused either by reduced dynamic fluctuations in the glycosylated variant or by increased stability of the glycosylated protein in partially folded states [24]. It is tempting to speculate that N-linked glycosylation interferes with the nucleation process of the initial formation of secondary structure, as the N-linked precurser has a high molecular weight and could stabilize the folded peptide chain locally during folding.

Another example of conformational effects of N-linked conformation is the quaternary structure of the Fc part of IgG. Crystallographic data [25] has revealed that the complex N-linked carbohydrate attached to Asn-297 on each heavy chain stabilizes the lateral movement of the arms extending from the hinge region because the identical carbohydrates act as a spacer between the arms of IgG and interdigitate so that one of the α-1,3 antennae interacts with the trimannose of the opposing carbohydrate (see Fig. 1a). Furthermore, the other α-1,6 antenna interacts with the Phe-243, Pro-244,245 and Thr-260 giving rise to a long range conformational effect of N-linked glycosylation (see Fig. 1b). The stability in the hinge region provided by this carbohydrate-protein interaction is essential for binding to

Figure 4: See Color Plate 9, page xxi.

the Fc receptor (FcR) of monocytes and macrophages since the deglycosylated IgG loses its capability to bind to FcR. [5].

The influence of quarternary structure on glycosylation was studied by Dahms et al. [26] who found that combining a common β chain with to different α chains of Mac-1 and LFA-1 resulted in a different oligosaccharide structure of the common β chain [26].

Consequently, not only does glycosylation affect conformation but conformation also affects glycosylation.

4 Conformational effects of O-linked glycosylation

The role of O-linked oligosaccharides on protein conformation is best illustrated by mucins which are mucus proteins with tandem repeat sequences rich in serine, threonine, proline and alanine. This makes them heavily O-glycosylated. Ovine submaxillary Mucin (OSM) have been shown to contain numerous mono- and disaccharide chains. Electron microscopi, light scattering and 13^C-NMR of OSM have shown that it has a highly extended, filamentous flexible structure [27] which collapses to dimensions typical of denatured globular proteins on removal of the carbohydrate chains [28]. Mucins have peptide chain dimensions about 3 fold more expanded than found for deglycosylated mucins or denatured proteins [29]. Membrane bound mucins form extended rods protruding from the cell surface, and the O-glycosylated areas protect them from proteolysis [30]. An extended conformation of eight O-glycosylated peptides from the N-terminus of Interleukin-2 were found by Paulsen et al. using NMR [31, 32].

These studies demonstrate the profound effects of O-glycosylation on protein conformation [16, 1]. The extended structure of an O-glycosylated peptide stretch becomes apparent when looking at the Aspergillus Awamori 1,4 Glycanase in Fig. 4. The extendedness is a result of intramolecular H-bonding between the carbohydrates and the backbone [33].

By applying Chou and Fasman analysis Aubert et al. [34] predicted that the secondary structure of the peptide chain around carbohydrate-peptide linkage was a β-turn in all 9 cases of O-linked glycosylation and in 19 out of 28 cases of N-linked glycosylation [34]. This finding has recently been confirmed by using circular dichroism of synthetic glycopeptides by Hollosi et al. [33]. They found that O-glycosylated dipeptides containing a D-serine are likely to adopt a type II β-turn, while those containing a Pro-Ser or Val-Ser sequence feature a type I(III) β-turn. Glycosylation again seemed to stabilize the backbone conformation [33]. We have examined the dihedral angles for glycosylated serines and threonines versus non-glycosylated in 1,4 aspergillus glycanase as seen in Fig. 3. Again discretion of glycosylated and non-glycosylated was not possible. Most of the sites have ψ above 130° which according to to the turn classification by Garnier et al. [23] could be a type II β-turn.

5 Summary and Conclusions

N-linked glycosylation at asparagine can break the 3_{10} or α-helices and help formation of β-turns or turn-like structures. However, not all N-linked glycosylation sites are β-turns. N-linked glycosylation increases stability of the glycoprotein and is the ligand for a conformational proof-reading chaperone. Clustered O-linked glycosylation at serine and threonine extend and stabilize the glycoprotein in a rodlike structure and are mostly found within β-turns. Prediction of secondary structure of glycoproteins may not be possible if these co/post translational modifications are not included in the prediction method.

The importance of the biomedical applications possible with rational design of recombinant glycoproteins based on such a prediction method, can probably not be overestimated.

Acknowledgements

The authors wish to thank H. Bohr and S. Brunak for many helpfull comments on this work. The work was supported by The Danish 1991 Pharmacy Foundation, The Danish Insurance Foundation, The NOVO Nordisk Foundation, The John and Birthe Meyer Foundation, The Roikjer Foundation and The Danish National Research Foundation.

References

[1] G.W. Hart, Glycosylation, *Curr. Opin. Cell Biol.*, **4**, 1017–1023 (1992).

[2] S. Olofsson, Carbohydrates in herpes virus infections, *APMIS (suppl. 27)*, **100**, 84–96 (1992).

[3] J.-E.S. Hansen, Carbohydrates of Human Immunodeficiency Virus, *APMIS (suppl. 27)*, **100**, 96–109 (1992).

[4] I. Laczkó, M. Hollósi, L. Ürge, K.E. Ugen, D.B. Weiner, H.H. Mantsch, J. Thurin and L. Ötvös, Synthesis and conformational studies of N-glycosylated analogues of the HIV-1 principal neutralizing determinant, *Biochemistry*, **31**, 4282–4288 (1992).

[5] G. Opdenakker, R.M. Rudd, C.P. Pomting and R.A. Dwek, Concepts and principles of glycobiology, *FASEB J.*, **7**, 1330–1337 (1993).

[6] T. Muramatsu, Carbohydrate signals in metastasis and prognosis of human carcinomas, *Glycobiology*, **3**, 294–296 (1993).

[7] D.R. Wing, T.M. Rademacher, B. Schmidtz, M. Schachner and R.A. Dwek, Comparative glycosylation in neural adhesion molecules, *Biochem. Soc. Trans.*, **20**, 386–390 (1992).

[8] S.W. Homans, Conformation and dynamics of oligosaccharides in solution, *Glycobiology*, **3**, 551–555 (1993).

[9] D.V. Renouf and E.F. Hounsell, Conformational studies of the backbone (poly-N-acetyllactosamine) and the core region sequences of O-linked carbohydrate chains, *Int. J. Biol. Macromol.*, **15**, 37–42 (1993).

[10] F. Hillenkamp and M. Karas, Matrix-assisted laser desorption/ionization mass spectrometry of biopolymers, *Analyt. Chem.*, **63**, 1193–1202 (1991).

[11] A. Dell and K.H. Khoo, Covalent Structure determination of glycopolymers, *Curr. Opin. Struct. Biol.*, **3**, 687–693 (1993).

[12] E.F. Hounsell, M.J. Davies and D.V. Renouf, Studies of oligosaccharide and glycoprotein conformation, *Biochem. Soc. Trans.*, **20**, 259–263 (1992).

[13] H. Schachter, Glycoproteins: Their Structure, biosynthesis, and possible clinical implications, *Clin. Biochem.*, **17**, 3–14 (1984).

[14] A. Bause, Structural requirements of N-glycosylation of proteins, *Biochem. J.*, **209**, 331–336 (1983).

[15] C. Hammond and A. Helenius, A chaperone with a sweet tooth, *Curr. Biol.*, **3**, 884–886 (1993).

[16] G.J. Strous and J. Dekker, Mucin-type Glycoproteins, *Crit. Rev. Biochem. Molec. Biol.*, **27**, 57–92 (1992).

[17] Å.P. Elhammer, R.A. Poorman, E. Brown, L.L. Maggiora, J.G. Hoogerheide and F.D. Kzdy, The specificity of UDP-GalNAc:polypeptide N-Acetylgalactosaminystransferase as inferred from a database of in Vivo substrates and from the in vitro glycosylation of proteins and peptides, *J. Biol. Chem.*, **268**, 10029–10038 (1993).

[18] B.C. O'Connell, F.K. Hagen and L.A. Tabak, The influence of flanking sequence of the O-glycosylation of threonine in vitro, *J. Biol. Chem.*, **267**, 25010–25018 (1992).

[19] I.B.H. Wilson, Y. Gavel and G.V. Heijne, Amino acid distributions around O-linked glycosylation sites, *Biochem J.*, **275**, 529–534 (1991).

[20] T. D. Schneider and R. M. Stephens, Sequence logos: A new way to display consensus sequences, *Nucl. Acids Res.*, **18**, 6097-6100 (1990).

[21] L. Otvos, J. Thurin, E. Kollat, L. Urge, H.H. Mantsch and M. Hollosi, Glycosylation of synthetic peptides breaks helices, *Int. J. Peptide Protein Res.*, **38**, 476–482 (1991).

[22] I.A. Wilson, R.C. Ladner, J.J. Skehel and D.C. Wiley, *Biochem. Soc. Trans.*, **11**, 145–147 (1993).

[23] J. Garnier and B. Robson, The GOR method for predicting secondary structure in proteins, in: *Prediction of protein structure and the principles of protein conformation*, ed. G.D. Fasman, Plenum Press, New York, 417–465 (1989).

[24] C.J. Edge, H.C. Joao, R.J. Woods and M.R. Wormald, The conformational effects of N-linked glycosylation, *Biochem. Soc. Trans.*, **21**, 452–455 (1993).

[25] J. Deisenhofer, Crystallografic refinement and atomic models of human Fc fragment and its complex with fragment B of protein A from Staphylococcus Aureus at 2.9 and 2.8 Åresolution, *Biochemistry*, **20**, 2361–2370 (1981).

[26] N.M. Dahms and G.W. Hart, Influence of quartenary structure on glycosylation, *J. Biol. Chem.*, **261**, 13186–13196 (1986).

[27] J.R. Gum, Mucin genes and proteins they incode: Structure, diversity and regulation, *Am. J. Respir. Cell. Mol. Biol.*, **7**, 557–564 (1992).

[28] M.C. Rose, Mucins: Structure, function, and role in pulmonary diseases, *Am. J. Physiol.*, **263**, 413–429 (1992).

[29] K.J. Butenhof and T.A. Gerken, Structure and dynamics of Mucin-like glycopeptides. Examination of peptide chain expansion and peptide-carbohydrate interactions by stochastic dynamics simulations, *Biochemistry*, **32**, 2650–2663 (1993).

[30] K.L. Carraway and S.R. Hull, Cell Surface Mucin-like glycoproteins and mucinlike domains, *Glycobiology*, **1**, 131–138 (1991).

[31] H. Paulsen, A. Pollex-Krüger and V. Sinnwell, Konformationsanalytische untersuchung von N-terminalen O-glycopeptidsequenzen des interleukin-2, *Carbohydrate Research*, **214**, 199–226 (1991).

[32] H. Paulsen, R. Busch, V. Sinnwelland and A. Pollex-Krüger. Konformationsanalytische untersuchung von D-Xylosehaltigen O-glycopeptidsequenzen, *Carbohydrate Research*, **214**, 227–234 (1991).

[33] M. Hollosi, A. Perczel and G.D. Fasman, Coopoperativity of carbohydrate moity orientation and - turn stability is determined by intramolecular hydrogen bonds in protected glycopeptide models, *Biopolymers*, **29**, 1549–1564 (1990).

[34] J.P. Aubert,G. Biserte and M.H. Loucheaux-Lefebvre, Carbohydrate-peptide linkage in glycopreteins, *Arch. Biochem. Biophys.*, **175**, 410–418 (1976).

Database and Neural Network–based Predictions of
Protein Structure

1D Secondary Structure Prediction through Evolutionary Profiles

Burkhard Rost and Chris Sander

EMBL Heidelberg, Meyerhofstr. 1, D-69117 Heidelberg, Germany

Abstract

For only about one third of the new proteins, 3D structure can be predicted. For the remaining two thirds, a compromise has to be made. An extreme simplification is the projection of 3D structure onto a string of 1D secondary structure assignments. Here, we report how neural networks can be configured such that strand is predicted significantly better, and that the prediction looks like native proteins in terms of the length of predicted segments. Using evolutionary information contained in multiple sequence alignments as input to neural networks, secondary structure can be predicted at significantly increased accuracy. Pre–processing the alignment information by using a position–specific conservation weight and the number of insertions and deletions in each alignment position is found to be advantageous. Addition of the global amino acid content yields a further improvement, mainly in predicting structural class. The final network system has a sustained overall accuracy of more than 72% evaluated on 250 sequence-unique chains. Of particular practical importance is the definition of a position–specific reliability index. For 40% of all residues the method has a sustained three–state accuracy of 88%, as high as the overall average for homology modelling.

1 Introduction

What is the goal of protein structure prediction? Two different objectives can be roughly distinguished. First, understanding the basic principles of protein folding. Second, making predictions that can be used in various fields of research in molecular biology. These objectives tend to partition the field of theoretical biology into two major schools that use different 'tool-boxes' to attack 'the protein folding problem', i.e. the question of how the three–dimensional (3D[1]) structure of a protein is determined in detail from the sequence of amino acids. That the sequence determines the 3D structure is well established[1, 2, 3], and appears to hold even in view of chaperones which play a rôle in assisting protein folding [4, 5]. The tools of the first school are physical and biochemical first principles. On this ground it is attempted to predict 3D structure from the sequence. The difficulty lies in the

---

[1] Abbreviations used: 3D three–dimensional; Swissprot: data bank of protein sequences; PDB: protein data bank of known structures; DSSP: dictionary of secondary structures of proteins; HSSP: data base of homology derived structure of proteins; SOS: protein sequence of unknown three–dimensional structure; PHD: profile network system from Heidelberg (three levels of networks for the prediction of secondary structure).

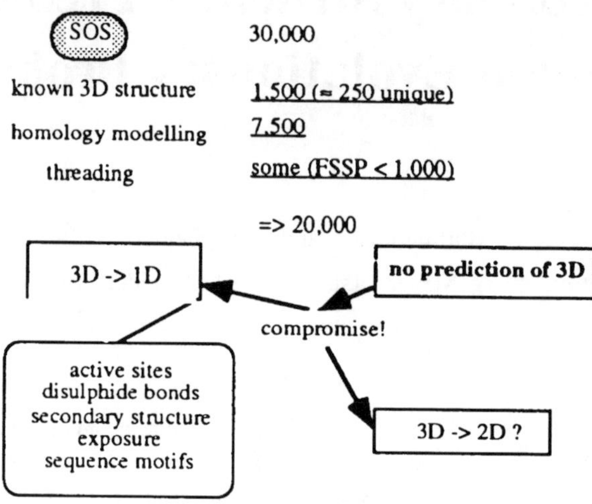

Figure 1: **The scope of protein structure prediction:** Numbers given are conservative estimates according to Swissprot (release 25) [51], PDB (December, 1993) [47, 52]. The number for homology modelling stems from the HSSP release of July, 1993 [53], that for threading from a data base of homologue structures, not exhibiting significant sequence homology [54].

complexity of the folding problem. Today, analyses based on first principles are at best restricted to very short protein sequences. The tool of the second school is application of statistics to a data bank of experimentally known protein structures. The difficulties mainly lie in that such attempts are restricted to features already common to the known examples, and that, so far, only simplified features of 3D structure can be dealt with. A third school might be made out as those who attempt combining the pros and cons of the two pillars, e.g. [6].

What is the scope of protein structure prediction in practice? Given a sequence of unknown 3D structure (SOS), how can theory help to predict features of its 3D structure? There is approximately a chance of 1:3 that the 3D structure of SOS can be predicted with sufficient accuracy (Fig. 1). However, for some two thirds of all known sequences prediction in 3D is not possible. These proceedings contributes various approaches to predict a 2D distance map from the sequence, but even such approaches do not yet operate in general. Thus, in most cases, we still have to accept an extreme compromise by projecting 3D structure onto 1D, i.e. by restricting the prediction to very simple features, like positions of active sites, disulphide bonds, secondary structure, surface exposure, sequence motifs or function of the protein.

Here, we shall present some aspects of a data base based prediction of secondary structure (assigned by DSSP [7], further projected onto 3 states: helix, strand, and rest — dubbed loop). The method uses a three–level system of neural networks, described in detail elsewhere [8, 9, 10, 11]. The main idea is that instead of single amino acid sequences, profiles of evolutionary conserved sequences are input to the network. The following questions will be addressed. How can neural networks be adapted to the special case of classifying sequence patterns into secondary structure? How can the information used as

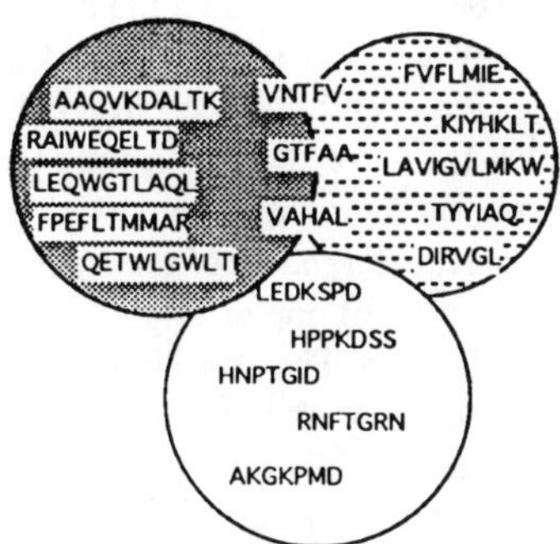

Figure 2: **Secondary structure prediction as a pattern recognition problem:** Certain oligopeptides have high preference to be in a particular secondary structure. Circles: upper left (dark shading): helix, upper right (light shading): strand, centre (no shading): loop. The 3 pentapeptides between the helix and strand circles are observed in both structures [55].

input be increased? Does pre-processing of the evolutionary profiles pay off in terms of prediction accuracy? Is the method accurate in comparison to the ultimate prediction, and in comparison to other prediction methods? What are the perspectives to predict the secondary structure of SOS?

2 Configuring a system of neural networks to predict secondary structure

2.1 Neural networks as tools for pattern classification

How can secondary structure be predicted from the database? Some stretches of sequence exhibit preferences for specific secondary structure states (Fig. 2). Consequently, the objective of a secondary structure prediction method is to classify sequences into e.g. three classes: helix, strand, and loop. A rather intuitive approach to pattern classification is drawing lines in between groups. A simple mathematical tool to perform such a task is the projection of vector $v1$ onto another $v2$, i.e. the computation in the product $v1 * v2$. Suppose, a stretch of w adjacent residues of a protein sequence is presented by a vector with $w * 20$ (20 different amino acids) components. Such a vector determines one point in a space of $w * 20$ dimensions. In analogy to the circles in Fig. 2, the task is to find a matrix that projects all points into three classes.

A simple tool that performs this task is a neural network [12, 13] (more precisely a multi–layered feed–forward network). The input is the sequence vector, i.e. each component of the vector determines the value of one input unit of the network. The output the secondary structure state (of the central residue in the stretch). In the simple case of only one layer, the output *out* generated by the input *in* is given by: $out = f(J * in)$, with J as the matrix for the junctions between each unit in the input and each unit in the output layer. The function f is a sigmoid trigger function (e.g. hyperbolic tangents), that assigns 0 to large negative values, 1 to large positive ones, and some value in between 0 and 1 else. From the data base we know which residue is found in which conformation. Given the junctions J, the output of the network is uniquely determined. Thus, the error of the network E can simply be computed as the square of the difference between network and observed output (summed over the three states). A simple way to reduce the error is by changing the junctions for each example presented such that the error decreases for each example presented, i.e. by gradient descent (as well known as back-propagation [14]): $\Delta J(t+1) \propto -\varepsilon \partial E(t)/\partial J$, in other words by computing the partial derivative of the error with respect to the junctions (t is the iterative time, i.e. the presentation of one example). The factor ε determines width of the stepwise convergence to a small error. The concept can be extended by using a layer of units between output and input, called hidden units. All results presented here were derived from networks with 15 hidden units. Training was done with conjugate gradient descent with the learning strength $\varepsilon = 0.05$, and the momentum term $(\Delta J(t+1)\alpha - \varepsilon \partial E(t)/\partial J + \alpha \Delta J(t-1))\alpha = 0.2$. The training was terminated once 75% of the training examples were classified correctly.

2.2 Better prediction of strand residues by balanced training

Evaluated on a data set of 130 unique protein chains (126 globular + 4 membrane) a network achieves a typical overall accuracy in 3 states of some 61–62%. In detail, the prediction is best for loop residues and far worst for strand (Fig. 3). The prediction accuracy for each of the 3 classes approximately mirrors the observed occurrence of these classes in the data set (Fig. 3). Consequently, the idea is to improve the prediction for strand residues by simply increasing the frequency in training examples for strand. (Note: such a procedure has been used previously for non-network predictions [15].) That is, instead of presenting in 1000 iteration time steps 220 examples for strand, 310 for helix and 470 for loop (dubbed 'unbalanced training'), now at each time step one example for each class is used for training (dubbed 'balanced training'). The first result is that all three classes are predicted almost equally well (Fig. 3). The second result is that the overall accuracy is reduced. (The latter stems from the evidence that the unbalanced prediction is better predicts loop residues, which tend to dominate the overall accuracy.)

2.3 Better prediction of segment length by 2nd level of structure–to–structure network

The network described so far, learns the classification of mutually independent patterns. The result is that the average length of a helix predicted is about 4 residues, compared to an average of 10 observed. The reason is that the network (as introduced here) cannot learn to correlate the secondary structure of adjacent residues, i.e. that e.g. helices span over at least three adjacent residues. A simple way to correct this shortage is the introduction of a 2nd level network (Fig. 4). A 1st level sequence-to-structure network predicts secondary structure from sequence. The output of the 1st level net is input to a 2nd level structure-to-structure network that predicts secondary structure from stretches of predictions. The 2nd level network has almost no effect in terms of overall accuracy. (This was probably the reason why the concept was forgotten after having been used already in one of the first applications of neural networks to secondary structure prediction [16].) However, the structure–to–structure network fulfils the purpose for which we introduced it: the average predicted helix extends over some 7 residues, i.e. the prediction looks considerably more protein like than that of the 1st level network (Fig. 3).

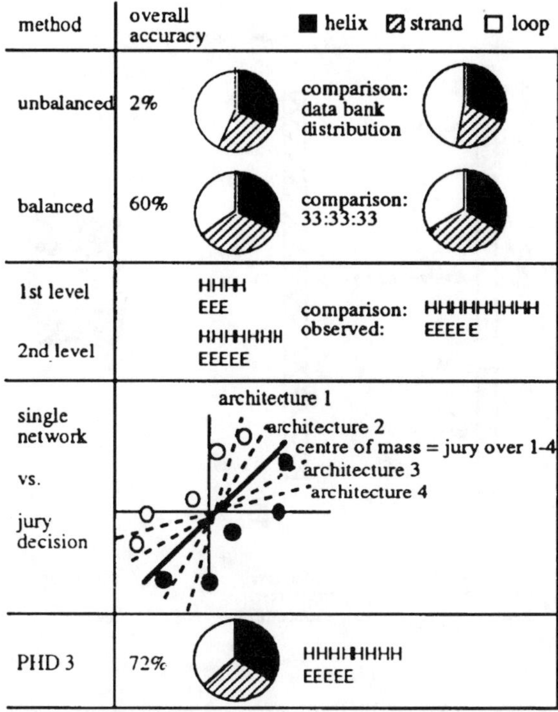

Figure 3: **Improvements on the network side:** Neural networks can easily be adapted to specific tasks. The first example is balanced training by which the performance for strand is improved (shown in pie charts, for comparison the relative distribution of helix, strand, and loop in the data set). The second example is the improvement in predicting the average length of secondary structure segments (for comparison the observed averages). Networks divide patterns by introducing lines. Differently trained networks make different errors (here the task is to distinguish between filled and open circles). An arithmetic average over different network outputs (jury decision) is comparable to computing the centre of mass for all lines. The error of the jury decision is smaller than that for any particular network. For comparison the results for the final network system (PHD3) as presented in chapter 3 are given.

2.4   Better overall accuracy by compiling a jury decision over different networks

A network classifies patterns separating them by lines. One particular training results in a particular classification, with a particular (partially random) error.

A different training (e.g. unbalanced vs. balanced training) results in a different realisation of the classification task. Which network to choose in practice? A simple way out of the dilemma is to compile an arithmetic average over different network realisations (dubbed 'jury decision'). The procedure can be expected to yield a better result than each particular network if the errors of each of the networks are to a certain degree uncorrelated (Fig.

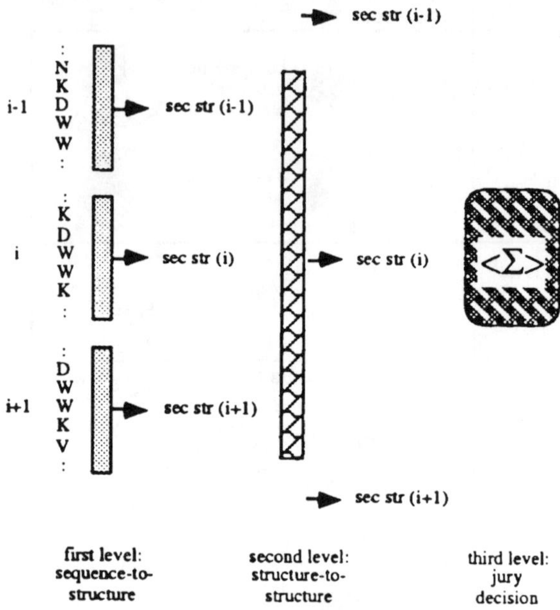

Figure 4: **Three level system for prediction of secondary structure:** First level, sequence–to–structure network: a window of 13 adjacent residues is shifted through all proteins. For each window the task of the network is to predict the secondary structure state of the central residue (here: D, W, W). Second level, structure–to–structure network: a window of 17 adjacent residues is shifted through all proteins. Again the task is to predict the secondary structure for the central residue. But the input is the output values, i.e. the predictions, of the 1st level net (as shown the 2nd level predicts the secondary structure for W at position i). Third level, jury decision: the output from differently trained networks (not sketched) for the same sequence position are summed. The secondary structure prediction for residue W at sequence position i is assigned to the unit with the maximal sum.

3). More precisely, the gain to be expected by the jury decision is inversely proportional to the correlation of the errors of particular networks. We use a jury decision on some 10 differently trained networks as the 3rd level of the network system (Fig. 4). The result is that the overall accuracy of the 3rd level is some 1–2 percentage points superior to any network on the 2nd level. Additionally, the 3rd level enables the combination of differently focused networks (e.g. unbalanced training: focus lies on predicting loop, balanced training: focus on predicting strand).

2.5 A tip to further improvements: core of helices better predicted than caps

Initiated by a discussion during the conference in Denmark, we investigated the following question: How does the prediction accuracy depend on the position of a residue in a helix? It

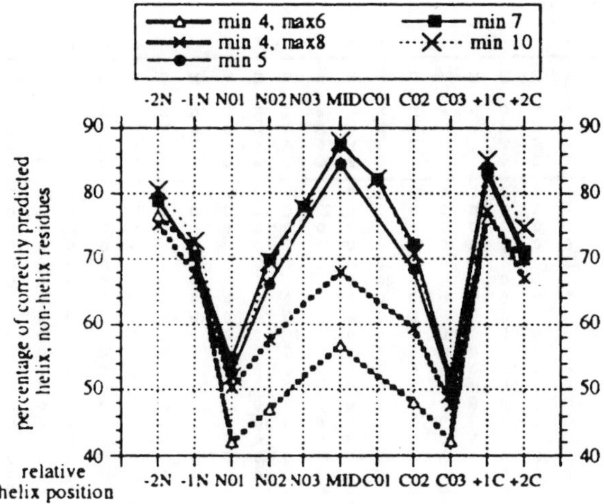

Figure 5: **Two-state accuracy in dependence of position relative to helices:** The two-state accuracy (helix, non-helix) is plotted for various positions relative to a helix: -2N, two residues before the begin of the helix; -1N, one residue before; N01, first residue in helix; N02, second residue; N03, third; MID, core of the helix = all other residues in the helix; C01, three residues before end of helix; C02, two before helix end; C03, last residue in helix; +1C, one residue after helix end; +2C, two residues after helix end. The curves differ according to the minimal and maximal lengths of helices analysed. Apparently, the network did not learn the N, and C-cap preferences inside a helix. This is less clear for shorter helices. One possible explanation is that for these the preferences must be more exposed, as only few residue types can bend a helix so sharply that a short helix is stable.

is known that there is a very strong preference for N-terminal caps of helices (first residues in a helix), and a still strong preference for C-terminal caps (last residues) [17]. The observation is slightly blurred by the difficulty to automatically determine begin and end of helices [18, 19, 20, 21, 22, 23]. But, the tendency is strong enough to allow the conclusion that cap residues exhibit a unique pattern. It should be expected that the network system learns such preferences.

However, the result indicates the opposite. The prediction accuracy is clearly worse for N– and C–caps than for the core of helices (Fig. 5). The tendency is that this is less so for shorter helices than for longer ones (Fig. 5). This observation indicates that the prediction can probably increased by combining in a jury decision either a network trained explicitly on caps with the one described here, or a prediction method explicitly focusing on cap preferences [17] (see as well contribution of Brunak & Engelbrecht in this issue) with the network system.

secondary structure	DSSP	E						E	E	E	E	E	E			E	E	E	E	E	E	H	H	H
	SH3	N	S	T	N	K	D	W	W	K	V	E	V	N	D	R	Q	G	F	V	P	A	A	Y
alignment	s1	N	K	S	N	P	D	W	W	E	G	E	L	N	G	Q	R	G	V	F	P	A	S	Y
	s2	E	E	H	.	G	E	W	W	K	A	K	s	s	K	R	E	G	F	I	P	S	N	Y
	s3	R	S	T	.	G	D	W	W	L	A	r	v	T	G	R	E	G	Y	V	P	S	N	F
	s4	F	S	.	.	.	.	F	F	G	V	•	v	D	D	L	Q	V	F	V	P	P	A	Y
profile	V	0	0	0	0	0	0	0	0	0	40	0	60	0	0	0	0	20	20	60	0	0	0	0
	L	0	0	0	0	0	0	0	0	20	0	0	20	0	0	20	0	0	0	0	0	0	0	0
	I	0	0	0	0	0	0	0	0	0	0	0	0	0	0	0	0	0	0	20	0	0	0	0
	M	0	0	0	0	0	0	0	0	0	0	0	0	0	0	0	0	0	0	0	0	0	0	0
	F	20	0	0	0	0	0	20	20	0	0	0	0	0	0	0	0	0	60	20	0	0	0	20
	W	0	0	0	0	0	0	80	80	0	0	0	0	0	0	0	0	0	0	0	0	0	0	0
	Y	0	0	0	0	0	0	0	0	0	0	0	0	0	0	0	0	0	20	0	0	0	0	80
	G	0	0	0	0	50	0	0	0	20	20	0	0	0	40	0	0	80	0	0	0	0	0	0
	A	0	0	0	0	0	0	0	0	0	40	0	0	0	0	0	0	0	0	0	0	40	40	0
	P	0	0	0	0	25	0	0	0	0	0	0	0	0	0	0	0	0	0	0	100	20	0	0
	S	0	60	25	0	0	0	0	0	0	0	0	20	20	0	0	0	0	0	0	0	40	20	0
	T	0	0	50	0	0	0	0	0	0	0	0	0	20	0	0	0	0	0	0	0	0	0	0
	C	0	0	0	0	0	0	0	0	0	0	0	0	0	0	0	0	0	0	0	0	0	0	0
	H	0	0	25	0	0	0	0	0	0	0	0	0	0	0	0	0	0	0	0	0	0	0	0
	R	20	0	0	0	0	0	0	0	0	0	20	0	0	0	60	20	0	0	0	0	0	0	0
	K	0	20	0	0	25	0	0	0	40	0	20	0	0	20	0	0	0	0	0	0	0	0	0
	Q	0	0	0	0	0	0	0	0	0	0	0	0	0	0	20	40	0	0	0	0	0	0	0
	E	20	20	0	0	0	25	0	0	20	0	60	0	0	0	0	40	0	0	0	0	0	0	0
additional information	N	40	0	0	100	0	0	0	0	0	0	0	0	40	0	0	0	0	0	0	0	0	40	0
	D	0	0	0	0	0	75	0	0	0	0	0	0	20	40	0	0	0	0	0	0	0	0	0
	N_{ins}	0	0	0	0	0	0	0	0	0	0	2	3	1	0	0	0	0	0	0	0	0	0	0
	N_{del}	0	0	1	3	1	1	0	0	0	0	0	0	0	0	0	0	0	0	0	0	0	0	0
	CW	1.0	0.8	0.7	0.8	0.6	1.1	1.5	1.5	0.8	0.9	1.0	0.7	0.7	0.9	0.9	0.7	1.5	1.0	1.2	1.5	0.9	0.7	1.5

Figure 6: **Extracting evolutionary information:** HSSP alignment of four sequences (s1–s4) to a region of SH3 [29]. Lower case letters in the alignment indicate insertions. The secondary structure is assigned with DSSP [7], the co-ordinates were kindly provided by Andrea Musacchio [56]. The HSSP profile is purely a computation of residue frequencies. The number of insertions and deletions are simple counts. The conservation weight is computed according to: $CW_i = \sum_{r,s}^{N} w_{rs} sim_{rs}^{i} / \sum_{r,s}^{N} w_{rs}$, with $w_{rs} = (1 - 0.01 * \%identity_{rs})$, where N is the number of sequences in the alignment, $identity_{rs}$ the percentage of sequence identity (over the entire length) of sequence r and s in the alignment, and sim_{rs}^{i} a value from the similarity matrix between sequence r and s at position i (e.g. Dayhoff [57])

3 Increasing input information by using profiles of evolutionary conservation

3.1 Input information insufficient to benefit from neural network

In principle, neural networks can achieve to classify highly correlated patterns. However, without hidden units the networks decompose only first order correlation. The network system as described so far, yields about the same results, no matter of whether or not hidden units are used [16, 24, 25]. In other words, the input information is not sufficient to use the capacity of the networks. But how can the input information be increased? One way that has been investigated without success was to increase the number of adjacent residue used for one example (here we chose: $w = 13$ for the 1st level, and $w = 17$ for the 2nd level). The problem with this appears to be the too small size of the data base, i.e. the longer the stretch, the less examples will be found in the data base for one particular stretch.

3.2 Specific information contained in pattern of evolutionary conservation

How can information be compiled that is specific about a certain structure? Studying multiple alignments reveals that structure is more conserved than sequence [26, 27, 28]. Different protein sequences can adopt the same 3D structure. By the pressure of selection (survival of appropriate function) evolution has explored the sequence variation that is possible without changing the 3D structure of a protein (and by that the function): a pair of native proteins has similar 3D structure if 30% residues are pairwise identical [27, 29]. Not any two residues can be exchanged. Instead, the substitution pattern is highly specific for a certain structure. The idea to use such information for prediction is not new [30, 26, 31, 32, 33]. Recently, the same idea has been used to predict secondary structure for single cases (for summary [10, 34]).

3.3 Compiling profiles of evolutionary conservation from multiple alignments

How can the information contained in the multiple alignment be extracted? The program used for generating the multiple alignment is MaxHom/HSSP [29]. The algorithm builds up the alignment in essentially two steps. In sweep 1, the sequences are aligned consecutively to the guide sequence by standard dynamic programming [35]. After each sequence has been added to the alignment an alignment profile is compiled and used to align the next sequence. In sweep 2, after all sequences have been added, the profile is recompiled, and the dynamic programming algorithm is repeated, this time using the constant profile, as derived after completion of sweep 1. The profile gives a simple count of the frequency of occurrence for each amino acid (Fig. 6). Additional information are the numbers of insertions and deletions in the alignment, and the compilation of a specific conservation weight (defined in caption of Fig. 6).

4 The more specific the input information, the better the prediction accuracy

4.1 Coarse-grained vs. fine-grained profile: from 68% to 70%

Do the details in the alignment matter? First, we projected the profile frequencies onto a grid of 4 with the intervals: 0–2, 3–33, 34–66, 67–100. A 2nd level network reaches an overall three-state accuracy of some 65%, compared to some 60% for a network similarly trained but using single sequences instead of multiple alignments as input (Fig. 7). The 3 level system reaches some 68%. Second, we used all details contained in the multiple alignment. The result is an increase of accuracy to almost 70%. This finding is slightly surprising. It bears the question: how does the improvement depend on the particular alignment, i.e. on the number of sequences aligned and on the variety? The answer is not clear cut. There is a tendency that the more various sequences in the alignment, the larger the improvement [11].

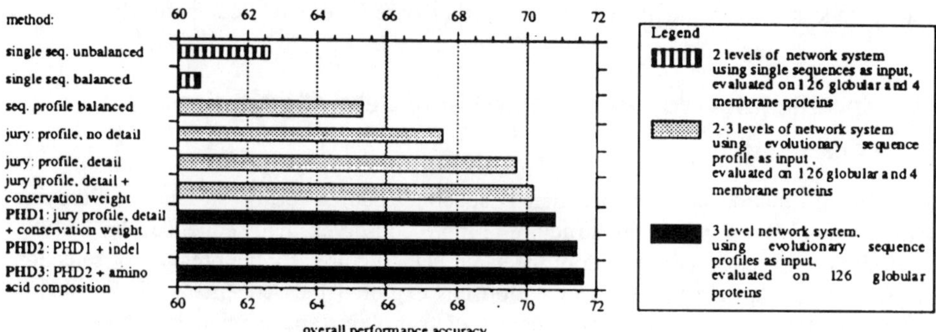

Figure 7: **Stepwise improvement of overall accuracy:** The stepwise improvement of secondary structure prediction based on multiple alignments in terms of overall accuracy.

However, this effect is partially blurred by a considerable variation of prediction accuracy with the protein chain (one standard deviation in order of 10%).

4.2 Adding conservation weight and indels: >71%

The position-specific conservation weight (Fig. 6) constitutes a knowledge-based focus on important sites. In principle, the network should be able to learn such a weight. The question at hand is: does it pay-off to assist the network by explicitly adding information to the input? It does. Using the conservation weight as additional input increases performance accuracy to above 70% (Fig. 7, the network system using detailed profiles and conservation weight will be dubbed 'PHD1').

Insertions and deletions are likely to occur in loop regions. Consequently, adding the number of insertions and deletions (dubbed 'indels') at each alignment position to the input should improve the performance for loop. Indeed, it does. The overall three-state accuracy becomes 71.4% compared to 70.8% without using indels (note: based on a set of 126 globular proteins). Furthermore, the tendency for over-prediction of helix and strand is reduced and the accuracy in predicting loop regions is increased.

4.3 Adding amino acid composition: better prediction of structural class

All information incorporated so far has been local in sequence, i.e. the input is compiled from a window of 13–17 consecutive residues. Of course, global information is introduced implicitly by the profiles, as a particular residue substitution pattern can depend on interactions between residues far apart in sequence. How can global information been explicitly used? A straightforward, technically simple way is to add units that code for the frequencies of amino acid occurrence in the protein outside of the stretch of 13–17 residues under investigation. The improvement in terms of overall accuracy is, at most, marginal. The network predictions do not differ much in terms of overall accuracy, average length predicted (and further measures, not given here) between a three-level network system using profile details, conservation weight, indels (dubbed 'PHD2') and a comparable system additionally using amino acid composition (dubbed 'PHD3'). However, the predictions differ in detail.

Proteins can be classified into four structural classes based on the relative content in secondary structure [36]: all–α, all–β, α/β, and rest. Such a classification is not clear cut [10]. Consequently, different schemes have been used in literature. The results given here use the scheme of [37]: all-α = α content $\geq$ 45%, β content < 5%; all-β = α content < 5%, β content $\geq$ 45%; and α/β = α content $\geq$ 30%, β content $\geq$ 20%.
The task is to predict the structural class from the amino acid sequence. (Similar work has been done before, unfortunately the results were either not at all cross-validated or did depend on data sets which did not exclude homologue proteins [38, 39, 40, 41, 42, 43, 37, 44].) PHD1 and PHD2, i.e. the networks not using global information successfully classify 70% of the 126 proteins into one of the four classes. PHD3, i.e. the network additionally using the global amino acid composition achieves 75%. The conclusion is that including global information does not pay–off in terms of local measures (per–residue accuracy asf.), but it does pay–off in terms of a global measure like the relative content of secondary structure.

5   Network system stands competition with alternatives

5.1   Prediction between 35% (random) and 88% (homology modelling)

What is the goal of predicting secondary structure? It appears to be simple: predict secondary structure such, that it is essentially compatible with the true 3D structure. But what is essentially compatible? One practical line to tackle the concept of compatibility is given by homology modelling. Homology modelling allows for predicting accurately the complete 3D structure of a protein, if applicable. This implies that it offers a perfect prediction of secondary structure [45, 46]. In an analysis of more than 100 homologue sequences, we found that the expected three-state accuracy for homology modelling is some 88%. What is the worst prediction one can get? One answer is given by random alignments: 35%. All prediction methods operate in between these two values: 35% — 88%. Consequently, it is meaningful to evaluate prediction accuracy additionally with respect to these two lines. In the following we shall use the term 'normalised overall three–state accuracy'. What we mean is a scaling of the prediction accuracy such, that a random prediction scores at 0, and homology prediction scores at 100%.

For a few selected methods the normalised accuracy is given along with the number of protein chains used for the evaluation (Fig. 8). Can these numbers be compared? Most of them were evaluated on different data sets. For some the data sets were too small. Only some of the methods did exclude homologue sequences in evaluation. Furthermore, as there is a considerable variation in prediction accuracy with protein chain, some methods might, by chance, have chosen chains that are easier to be predicted. Thus it is crucial to estimate prediction methods on the same data set.

5.2   Comparison of performance for different data sets

The levels of accuracy given so far based on a cross–validation test with 126 globular, and either with our without 4 additional membrane chains. How do the results depend on the data set? The set of 130 proteins was the maximal set of unique proteins in PDB [47] in 1992 [48]. Meanwhile, the number of unique proteins has risen to more than 250. This provides an excellent opportunity to investigate how PHD performs on a set of 124 new proteins

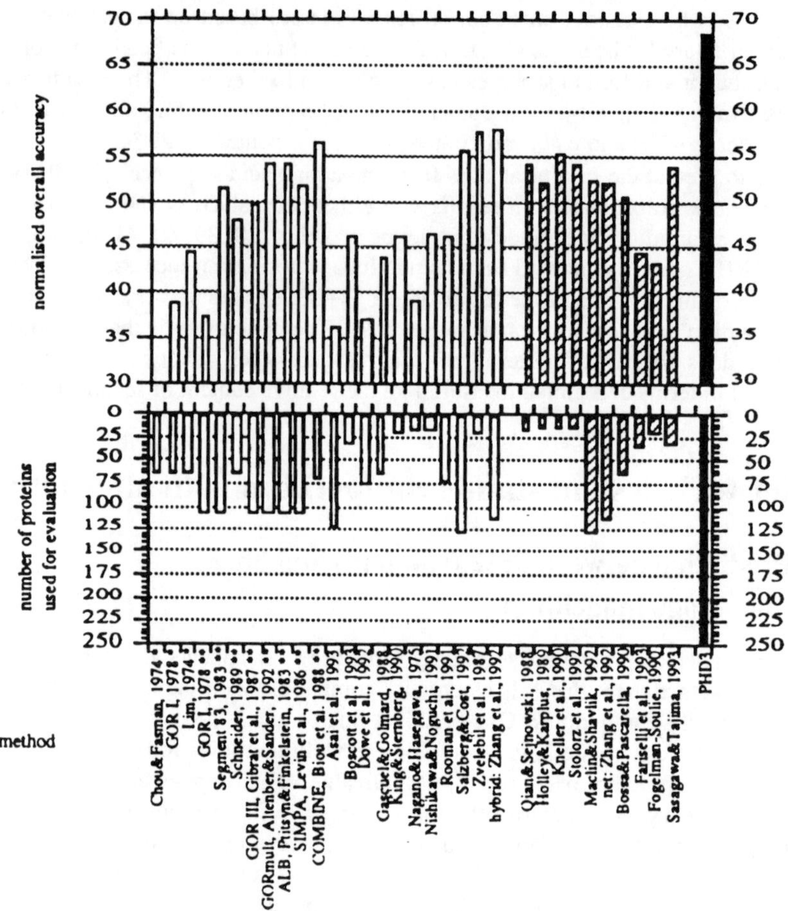

Figure 8: **Normalised overall accuracy for various prediction methods:** The results labelled by
'*' were taken from [58], those labelled by '**' from [59] (note: these are based on the same set,
but it is not excluded that some proteins had been used for setting up the methods, only the result
for ALB a method based on physical principles is reliable). The other results are taken from the
publications, referenced as in the literature list [49, 60, 61, 62, 63, 64, 65, 66, 33, 67, 15, 24, 17, 68,
69, 70, 71, 72, 73, 59, 74, 75, 76, 77, 78, 79, 80, 81, 82, 83]. The values are normalised such that
0% = random prediction, and 100% = homology modelling.

(dubbed 'Set 2') not homologue to the previously used set of 130. Furthermore, Set 2 can
be used to compare the network with methods like GORIII and COMBINE, as these base
on proteins not homologue to Set 2 (Fig. 9). To render a comparison with other methods,
(like the Chou-Fasman method [49, 50] that still is one of the most often used program in
practice) we performed a cross–validation experiment on the data sets used for evaluating
these method (Fig. 9). The conclusion is that PHD is some 10 percentage points superior in
terms of normalised overall accuracy to any other method, including those that use multiple
alignments on the ground of statistics. Furthermore, the additional tests indicate that the
network performance is likely to be >60% for more than 90% of the proteins.

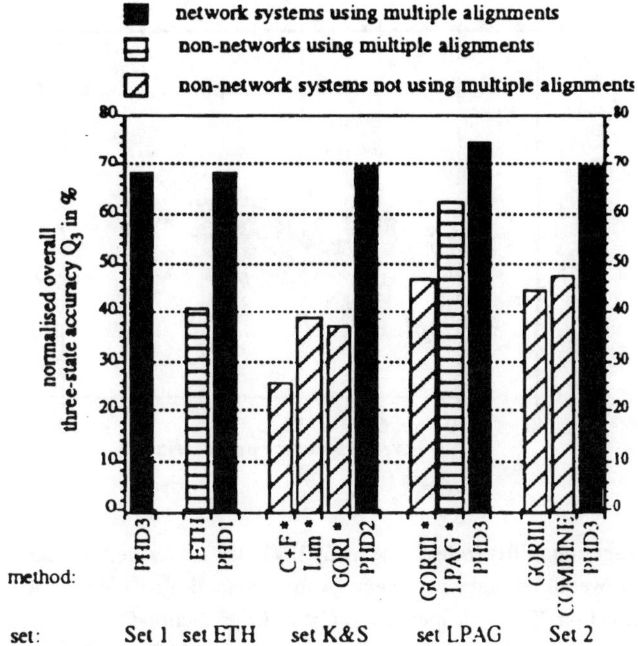

Figure 9: **Various prediction methods evaluated on the same data sets:** Normalised overall accuracy (0% = random prediction, 100% = homology modelling). The different sets are: 'Set 1', 126 unique globular protein chains (explicitly given in [11]); 'set ETH', 5 proteins for which expert predictions from Benner, Gerloff et al. were published [84, 85, 86, 87, 88]; 'set K&S', 62 proteins used for an analysis of secondary structure prediction methods a decade ago [58]; 'set LPAG', 82 protein fragments (from less than 20 structure families) used for evaluation of the GOR method with multiple alignments [81]; 'Set 2', 124 unique globular protein chains with no homology to Set 1 (explicitly given in [10]). The results labelled by '*' were taken from the literature.

6 Predictions useful for research in molecular biology

6.1 Variation of prediction accuracy with protein chain

How good is the prediction of PHD on SOS? After all the tests described, this question is still not an easy one. The expected overall accuracy is above 72% evaluated on all 250 unique globular protein chains. But, the variation with the protein is considerable, i.e. one standard deviation is some 9%. (And an analysis of the prediction accuracy of homology modelling suggests, that this value is not a disadvantage of the prediction method. On the contrary, the variation reflects the variation between different proteins. Thus, if a prediction method is reported to have a small variation this might indicate that the test set was too small.) The expected accuracy is therefore $72 \pm 9\%$. (But can be worse for exceptional cases.)

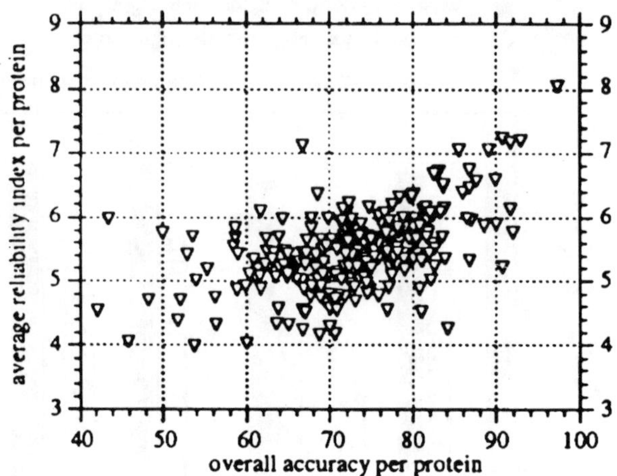

Figure 10: **Average reliability index:** The reliability index is averaged for each protein chain and plotted versus the overall accuracy per chain. A more complicated z–score–like measure did not yield a better distinction. The basic message is that if the prediction for SOS has a reliability index > 5, then the likelihood that the prediction is less accurate than 60% is approximately halved.

6.2 Definition of prediction reliability

Is it predictable, how accurate the prediction is for SOS? At first sight this question might look absurd. But, the answer is yes. Why? The PHD output consists of three values for helix, strand, and loop. Usually, the prediction is assigned to the output unit with maximal value. However, the difference between the output unit with highest value, and that of the unit with the next highest value, provides an estimate for the reliability of the prediction. Our experience is that this value correlates well with the prediction accuracy. About 40% of all residues are predicted at an accuracy of 88%, i.e. comparable to what would be possible if homology modelling could be done [10]. A similar index can be defined by statistical methods such as COMBINE. Compared on the same set of 124 proteins, COMBINE (without alignments) predicts 10% of the residues at a level of 80% accuracy, PHD3 more than 70% of the residues [10]. The average reliability per protein provides an additional help to estimate the quality of the prediction for SOS: the majority of proteins predicted at < 60% accuracy has an average reliability index < 5 (Fig. 9).

6.3 Availability of PHD prediction

The PHD predictions are available for fully automatic use. Send the word *help* (either as subject, or as only word in text) by electronic mail to the internet address *PredictProtein@EMBL–Heidelberg.de* for detailed instructions. Should the answer take more than 2 days, please repeat the procedure, or feel free to send a message to the internet address *Predict–Help@EMBL–Heidelberg.de*. (Note: *PredictProtein* is an automatic server with no ability to read letters, thus, send any personal message to *Predict–Help*.) The prediction will either be based on an alignment done by the server, or on an alignment provided by you in appropriate format. Is there an interest in obtaining secondary structure predictions? By early

December, 1993, more than 8,000 requests have been processed by the PHD server. The number of requests per month is increasing. Will the predictions be useful for 'scientific progress'? We shall see, meanwhile much can be done for a fruitful dialogue between theory and experiment.

Acknowledgements

First of all, we thank Søren Brunak and Henrik Bohr for the marvellous organisation of a vivid conference in a relaxed, inspiring and personal atmosphere. Thanks to two colleagues at the EMBL: Reinhard Schneider and Gert Vriend, for valuable ideas, discussion and assistance. Furthermore, we wish to thank Jean Garnier (INRA, Paris) for having supplied the programs GORIII and COMBINE. Last, not least, we wish to express our gratitude to all those who make co–ordinates of experimentally determined protein 3D structure available.

References

[1] C. B. Anfinsen, E. Haber, M. Sela and F. H. White Jr., The kinetics of formation of native ribonuclease during oxidation of the reduced polypeptide chain, *Proc. Natl. Acad. Sc. U.S.A.*, **47**, 1309-1314 (1961).

[2] C. J. Epstein , R. F. Goldberger, and C. B. Anfinsen, The genetic control of tertiary protein structure: studies with model systems, *Cold Spring Harbour Symp. Quant. Biol.*, **28**, 439-449 (1963).

[3] C. B. Anfinsen, Principles that govern the folding of protein chains, *Science*, **181**, 223-230 (1973).

[4] T. E. Creighton, Up the kinetic pathway, *Nature*, **356**, 194-195 (1992).

[5] J. J. Ewbank and T. E. Creighton, Protein folding by stages, *Curr. Biol.*, **2**, 347-349 (1992).

[6] M. J. Sippl, Calculation of Conformational Ensembles from Potentials of Mean Force. An Approach to the Knowledge-based Prediction of Local Structures of Globular Proteins, *J. Mol. Biol.*, **213**, 859-883 (1990.

[7] W. Kabsch and C. Sander, Dictionary of Protein secondary structure: Pattern recognition of hydrogen bonded and geometrical features, *Biopolymers*, **22**, 2577-2637 (1983).

[8] B. Rost and C. Sander, Exercising Multi-layered Networks on Protein Secondary Structure. In: O. Benhar, S. Brunak, P. DelGiudice and M. Grandolfo (eds.). Neural Networks: From Biology to High Energy Physics.Elba, Italy: International Journal of Neural Systems, 1992, 209-220.

[9] B. Rost and C. Sander, Improved prediction of protein secondary structure by use of sequence profiles and neural networks, *Proc. Natl. Acad. Sc. U.S.A.*, **90**, 7558-7562 (1993).

[10] B. Rost and C. Sander, Combining evolutionary information and neural networks to predict protein secondary structure. Proteins, *Journal of Chemical Biology*, submitted, September 24, 1993.

[11] B. Rost and C. Sander, Prediction of protein secondary structure at better than 70% accuracy, *J. Mol. Biol.*, **232**, 584-599 (1993).

[12] B. Rost, Neural networks and evolution — advanced prediction of protein secondary structure. Dep. of Physics and Astronomy, University of Heidelberg, F.R.G., 1993.

[13] B. Rost and G. Vriend, Neural Networks in Chemistry, *CDA News*, **8**, 24-27 (1993).

[14] D. E. Rumelhart, G. E. Hinton and R. J. Williams, Learning representations by back-propagating error, *Nature*, **323**, 533-536 (1986).

[15] O. Gascuel and J. L. Golmard, A simple method for predicting the secondary structure of globular proteins: implications and accuracy, *CABIOS*, **4**, 357-365 (1988).

[16] N. Qian and T. J. Sejnowski, Predicting the Secondary Structure of Globular Proteins Using Neural Network Models, *J. Mol. Biol.*, **202**, 865-884 (1988).

[17] R. Schneider, Sekundärstrukturvorhersage von Proteinen unter Berücksichtigung von Tertiärstrukturaspekten. Department of Biology, Univ. Heidelberg, FRG, 1989.

[18] W. R. Taylor and C. A. Orengo, A holistic approach to protein structure alignment, *Prot. Engin.*, **2**, 505-19 (1989).

[19] C. M. Wilmot and J. M. Thornton, β–Turns and their distortions: a proposed new nomenclature, *Prot. Engin.*, **3**, 479-493 (1990).

[20] S. Brunak, Non–linearities in training sets identified by inspecting the order in which neural networks learn. In: O. Benhar, C. Bosio, P. Del Giudice and E. Tabet (eds.). Neural Networks From Biology to High Energy Physics. Elba, Italy: 1991, 277-88.

[21] A. Perczel, K. Park and G. D. Fasman, Deconvolution of the Circular Dichroism Spectra of Proteins: The Circular Dichroism Spectra of the Antiparallel β-Sheet in Proteins, *Proteins*, **13**, 57-69 (1992).

[22] S. Woodcock, J.-P. Mornon and B. Henrissat, Detection of secondary structure elements in proteins by hydrophobic cluster analysis, *Prot. Engin.*, **5**, 629-635 (1992).

[23] N. Colloc'h, C. Etchebest, E. Thoreau, B. Henrissat and J.-P. Mornon, Comparison of three algorithms for the assignment of secondary structure in proteins: the advantages of a consensus assignment, *Prot. Engin.*, **6**, 377-382 (1993).

[24] H. L. Holley and M. Karplus, Protein secondary structure prediction with a neural network, *Proc. Natl. Acad. Sc. U.S.A.*, **86**, 152-156 (1989).

[25] S. B. Petersen, H. Bohr, J. Bohr, S. Brunak, R. M. J. Coterill, H. Fredholm and B. Lautrup, Training neural networks to analyse biological sequences, *TIBTECH*, **8**, 304-308 (1990).

[26] R. E. Dickerson, R. Timkovich and R. J. Almassy, The Cytochrome Fold and the Evolution of Bacterial Energy Metabolism, *J. Mol. Biol.*, **100**, 473-491 (1976).

[27] C. Chothia and A. M. Lesk, The relation between the divergence of sequence and structure in proteins. *EMBO J.*, **5**, 823-826 (1986).

[28] A. Pastore and A. M. Lesk, Comparison of the Structures of Globins and Phycocyanins: Evidence for Evolutionary Relationship, *Proteins*, **8**, 133-55 (1990).

[29] R. Schneider and C. Sander, Database of Homology-Derived Structures and the Structurally Meaning of Sequence Alignment, *Proteins*, **9**, 56-68 (1991).

[30] E. Zuckerkandl and L. Pauling, Evolutionary Divergence and Convergence in Proteins. In: V. Bryson and H. J. Vogel (eds.). Evolving Genes And Proteins.New York and London: Academic Press, 1965, 97-166.

[31] F. R. Maxfield and H. A. Scheraga, Improvements in the Prediction of Protein Topography by Reduction of Statistical *Errors. Biochem.*, **18**, 697-704 (1979).

[32] K. Nishikawa, Assessment of secondary structure prediction of proteins: Comparison of computerized Chou-Fasman method with others, *Biochim. Biophys. Ac.*, **748**, 285-299 (1983).

[33] M. J. Zvelebil, G. J. Barton, W. R. Taylor and M. J. E. Sternberg, Prediction of protein secondary structure and active sites using alignment of homologous sequences, *J. Mol. Biol.*, **195**, 957-961 (1987).

[34] B. Rost, C. Sander and R. Schneider, Progress in protein structure prediction?, *TIBS*, **18**, 120-123 (1993).

[35] T. F. Smith and M. S. Waterman, Comparison of biosequences, *Adv. Appl. Math.*, **2**j, 482-489 (1981).

[36] M. Levitt and C. Chothia, Structural patterns in globular proteins, *Nature*, **261**, 552-558 (1976).

[37] C.-T. Zhang and K.-C. Chou, An optimization approach to predicting protein structural class from amino acid composition, *Prot. Sci.*, **1**, 401-408 (1992).

[38] R. P. Sheridan, J. S. Dixon, R. Venkataghavan, I. D. Kuntz and K. P. Scott, Amino acid composition and hydrophobicity patterns of protein domains correlate with their structures, *Biopolymers*, **24**, 1995-2023 (1985).

[39] P. Klein, Prediction of protein structural class by discriminant analysis, *Biochim. Biophys. Acta*, **874**, 205-215 (1986).

[40] P. Klein and C. DeLisi, Prediction of protein structural class from the amino acid sequence, *Biopolymers*, **25**, 1659-1672 (1986).

[41] P. Klein, J. A. Jacquez and C. DeLisi, Prediction of protein function by discriminant analysis, *Mathematical Biosciences*, **81**, 177-189 (1986).

[42] G. Deleage and B. Roux, An algorithm for protein secondary structure prediction based on class prediction, *Prot. Engin.*, **1**, 289-294 (1987).

[43] G. Deleage and B. Roux, Use of class prediction to improve protein secondary structure prediction. In: F. G. D. (eds.). Prediction of protein structure and the principles of protein conformation. New York: Plenum Press, 1989, 587-597.

[44] B. A. Metfessel, P. N. Saurugger, D. P. Connelly and S. S. Rich, Cross-validation of protein structural class prediction using statistical clustering and neural networks, *Prot. Sci.*, **2**, 1171-1182 (1993).

[45] R. B. Russell and G. J. Barton, The limits of protein secondary structure prediction accuracy from multiple sequence alignment, *J. Mol. Biol.* submitted May 1993.

[46] B. Rost, R. Schneider and C. Sander, Redefining the goals of protein secondary structure prediction, *J. Mol. Biol.*, **235** in the press (1994).

[47] F. C. Bernstein, T. F. Koetzle, G. J. B. Williams, E. F. Meyer, M. D. Brice, J. R. Rodgers, O. Kennard, T. Shimanouchi and M. Tasumi, The Protein Data Bank: a computer based archival file for macromolecular structures, *J. Mol. Biol.*, **112**, 535-542 (1977).

[48] U. Hobohm, M. Scharf, R. Schneider and C. Sander, Selection of representative protein data sets, *Prot. Sci.*, **1**, 409-17 (1992).

[49] P. Y. Chou and U. D. Fasman, Prediction of protein conformation, *Biochem.*, **13**, 211-215 (1974).

[50] P. Y. Chou and G. D. Fasman, Prediction of the secondary structure of proteins from their amino acid sequence, *Adv. Enzymol.*, **47**, 45-148 (1978).

[51] A. Bairoch and B. Boeckmann, The SWISS-PROT protein sequence data bank, *Nucl. Acids Res.*, **20**, 2019-2022 (1992).

[52] E. E. Abola, F. C. Bernstein and T. F. Koetzle, The Protein Data Bank. In: E. Lesk A. M. (eds.). Computational molecular biology. Sources and methods for sequence analysis. Oxford: Oxford University Press, 1988, 69-81.

[53] R. Schneider and C. Sander, The HSSP data base of protein structure-sequence alignment, *Nucl. Acids Res.*, **21**, 3105-3109 (1993).

[54] L. Holm, C. Ouzounis, C. Sander, G. Tuparev and G. Vriend, A database of protein structure families with common folding motifs, *Prot. Sci.* **1**, 1691-1698 (1993).

[55] W. Kabsch and C. Sander, On the use of sequence homologies to predict protein structure: Identical pentapeptides can have completely different conformations, *Proc. Natl. Acad. Sc. U.S.A.*, **81**, 1075-1078 (1984).

[56] A. Musacchio, M. Noble, R. Pauptit, R. Wierenga and M. Saraste, Crystal structure of a Src-homology 3 (SH3) domain, *Nature*, **359**, 851-855 (1992).

[57] M. O. Dayhoff, Atlas of Protein Sequence and Structure. Washington, D.C., U.S.A.: National Biomedical Research Foundation, 1978.

[58] W. Kabsch and C. Sander, How good are predictions of protein secondary structure?, *FEBS Lett.*, **155**, 179-182 (1983).

[59] B. Altenberg and C. Sander, Current Quality of Secondary Structure Prediction, EMBL, 1992.

[60] V. I. Lim, Structural Principles of the Globular Organization of Protein Chains. A Stereochemical Theory of Globular Protein Secondary Structure, *J. Mol. Biol.*, **88**, 857-872 (1974).

[61] K. Nagano and K. Hasegawa, Logical Analysis of the Mechanism of Protein Folding, *J. Mol. Biol.*, **94**, 257-281 (1975).

[62] J. Garnier, D. J. Osguthorpe and B. Robson, Analysis of the Accuracy and Implications of Simple Methods for Predicting the Secondary Structure of Globular Proteins, *J. Mol. Biol.*, **120**, 97-120 (1978).

[63] W. Kabsch and C. Sander, Segment83. unpublished 1983.

[64] O. B. Ptitsyn and A. V. Finkelstein, Theory of protein secondary structure and algorithm of its prediction, *Biopolymers*, **22**, 15-25 (1983).

[65] J. M. Levin, B. Robson and J. Garnier, An algorithm for secondary structure determination in proteins based on sequence similarity, *FEBS Lett.*, **205**, 303-308 (1986).

[66] J.-F. Gibrat, J. Garnier and B. Robson, Further Developments of Protein Secondary Structure Prediction Using Information Theory. New Parameters and Consideration of Residue Pairs, *J. Mol. Biol.*, **198**, 425-443 (1987).

[67] V. Biou, J. F. Gibrat, J. M. Levin, B. Robson and J. Garnier, Secondary structure prediction: combination of three different methods, *Prot. Engin.*, **2**, 185-91 (1988).

[68] F. Bossa and S. Pascarella, PRONET: a microcomputer program for predicting the secondary structure of proteins with a neural network, *CABIOS*, **5**, 319-320 (1990).

[69] F. Fogelman-Soulié and C. Mejía, Incorporating knowledge in multi–layer networks: the example of proteins secondary structure prediction. In: F. Fogelman-Soulié and J. Hérault (eds.). Neuro Computing, Algorithms, Architectures and Applications. Berlin, Heidelberg: Springer, 1990, 185-194.

[70] R. D. King and M. J. Sternberg, Machine Learning Approach for the Prediction of Protein Secondary Structure, *J. Mol. Biol.*, **216**, 441-457 (1990).

[71] D. G. Kneller, F. E. Cohen and R. Langridge, Improvements in Protein Secondary Structure Prediction by an Enhanced Neural Network, *J. Mol. Biol.*, **214**, 171-182 (1990).

[72] K. Nishikawa and T. Noguchi, Predicting Protein Secondary Structure Based on Amino Acid Sequence, *Meth. Enz.*, **202**, 31-44 (1991).

[73] M. J. Rooman, J. P. Kocher and S. J. Wodak, Prediction of Protein Backbone Conformation Based on Seven Structure Assignments: Influence of Local Interactions, *J. Mol. Biol.*, **221**, 961-979 (1991).

[74] D. L. Dowe, J. Oliver, T. I. Dix, L. Allison and C. S. Wallace, A Decision Graph Explanation of Protein Secondary Structure Prediction. Dep. Computer Science, Monash Univ., Clayton 3168, Australia, 1992.

[75] S. Salzberg and S. Cost, Predicting Protein Secondary Structure with a Nearest-neighbor Algorithm, *J. Mol. Biol.*, **227**, 371-374 (1992).

[76] P. Stolorz, A. Lapedes and Y. Xia, Predicting Protein Secondary Structure Using Neural Net and Statistical Methods, *J. Mol. Biol.*, **225**, 363-377 (1992).

[77] X. Zhang, J. P. Mesirov and D. L. Waltz, Hybrid System for Protein Secondary Structure Prediction, *J. Mol. Biol.*, **225**, 1049-63 (1992).

[78] K. Asai, S. Hayamizu and K. Handa, Prediction of protein secondary structure by the hidden Markov model, *CABIOS*, **9**, 141-146 (1993).

[79] P. E. Boscott, G. J. Barton and W. G. Richards, Secondary structure prediction for modelling by homology, *Prot. Engin.*, **6**, 261-266 (1993).

[80] P. Fariselli, M. Compiani and R. Casadio, Predicting secondary structures of membrane proteins with neural networks, *Eur. Biophys. J.*, **22**, 41-51 (1993).

[81] J. M. Levin, S. Pascarella, P. Argos and J. Garnier, Quantification of Secondary Structure Prediction Improvement Using Multiple Alignments, *Prot. Engin.*, in press (1993).

[82] R. Maclin and J. W. Shavlik, Using Knowledge-Based Neural Networks to Improve Algorithms: Refining the Chou-Fasman Algorithm for Protein Folding, *Machine Learning*, **11**, 195-215 (1993).

[83] F. Sasagawa and K. Tajima, Prediction of protein secondary structures by a neural network, *CABIOS*, **9**, 147-152 (1993).

[84] S. A. Benner and D. Gerloff, Patterns of Divergence in Homologous Proteins as Indicators of Secondary and Tertiary Structure of the Catalytic Domain of Protein Kinases, *Adv. Enz. Reg.*, **31**, 121-181 (1990).

[85] S. A. Benner, Predicting de novo the folded structure of proteins, *Curr. Opin. Str. Biol.*, **2**, 402-412 (1992).

[86] S. A. Benner, M. A. Cohen and D. Gerloff, Correct structure prediction?, *Nature*, **359**, 781 (1992).

[87] S. A. Benner, M. A. Cohen and D. Gerloff, Predicted Secondary Structure for the Src Homology 3 Domain, *J. Mol. Biol.*, **229**, 295-305 (1993).

[88] D. L. Gerloff, T. F. Jenny, L. J. Knecht, G. H. Gonnet and S. A. Benner, The nitrogenase MoFe protein, *FEBS Lett.*, **318**, 118-124 (1993).

Fold–class Prediction by Neural Network

Martin Reczko, Henrik Bohr, Shankar Subramaniam*, Sudhakar Pamidighantam and
Artemis Hatzigeorgiou

German Cancer Research Center, Im Neuenheimer Feld 280, 69120 Heidelberg and
*University of Illinois at Urbana-Champaign, Urbana IL 61801 USA

Abstract

Predicting what fold-class a given protein belongs to on the basis of the protein's
sequence of amino acids is done with the help of neural network techniques.
In the case of fold-classes defined on rules based on the presence of similar
substructures or a certain percentage (30% - 60%) of sequence identity, the
network can determine what fold-class (out of a total of 42 classes) a novel
protein belongs to and with a correctness score of around 73 percent. Such
information may then be used to bias a proceeding homology modelling at
higher precision of the 3-dimensional protein backbone structure.

1   Introduction

It has recently been proposed[3, 1, 2] that all the known 3-dimensional protein structures
can be grouped into a smaller number of characteristic structural classes each consisting of
domains from more or less homologous proteins with a similar topological configuration of
their backbone. These structural domains or so-called folds of the proteins were introduced
in order to clarify the notion of structural similarity. Such fold classes could contain entire
proteins or well-defined sub-domains of proteins. A typical fold class contained the timbarrel
proteins of which e.g. 2TAA (Taka Amylase) was one class member. Although only around
100 fold classes up to now could be identified[1] among the existing crystallographic data
bank it was speculated that around 10 times more classes could be identified in future and
would represent an exhaustive set[3].

Most of the folds within one class would have more than 50 % sequence identity to
each other although folds with less sequence similarity also could belong to the same
class. It is important that each fold within one class would have a structure with a large
topological similarity and a similar packing pattern. This notion of fold classes is important
for homology modelling of new protein structures. By homology modelling an unknown 3-
dimensional protein structure is inferred from other known 3-dimensional protein structures
whose sequences are similar to the sequence of the protein in question. Indeed it is
well-known[7, 18] that one can predict or model protein structures to high accuracy by
combining knowledge of similar known protein structures. One of these methodologies that
successfully can utilize the information of the structure of homologous proteins are neural
networks methods. These networks can be trained exclusively on homologous proteins as a
basis for predicting a new protein structure from the corresponding sequence.

However, for proteins with very little homology to other proteins there exists no method
that can predict those protein's 3-dimensional structure to high accuracy just from their

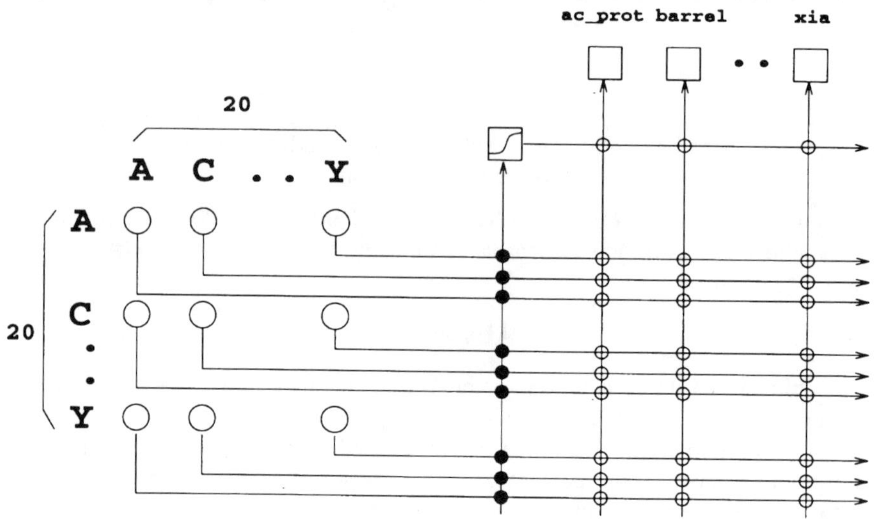

Figure 1: The Cascade Correlation architecture for predicting fold classes from dipeptide frequencies.

sequence data. On the other hand proteins with little sequence homology to others can be similar in structure to a whole class of other structures or domains. Structure similarity is thought to be far more important for predicting new protein structures since it contain the interaction patterns that partly control protein folding.

In the light of these facts it is therefore clear that a crucial step in predicting novel 3-dimensional structures of a protein from its sequence is first to determine its structural homology basis for further modelling, or in other words, what homologous fold class it belongs to and hence having a basis for an accurate prediction of its structure. Such a scheme is of course only making sense provided the protein in question has any relationship to some of the existing fold classes known to-day.

The scheme we propose for protein structure prediction consists of first using feed-forward neural networks for predicting what fold class a protein belongs to just from its sequence data. Once that is done the predicted fold class with its characteristic domains is used as a training set for a large recurrent neural network that on the basis of this set is predicting the given proteins distance matrix from its sequence. Such a distance matrix prediction will, due to the structurally homologous training set, be accurate enough for constructing the 3-dimensional backbone structure for the protein through an abstract energy minimization procedure[17] and subsequently positioning of the side-chains in order to obtain a full protein structure to high accuracy.

In the second chapter we shall discuss the neural network methodology for predicting the fold class a given protein belongs to and in the following chapter we present some results from actual neural network studies on fold-class prediction. Finally we discuss how the whole scheme of protein structure prediction can be assembled.

2 Methodology

The basic elements of a neural network, the neurons, are processing units that produce output from a characteristic non-linear function of a weighted sum of input data. A neural network is a group of such processing units, the individual members of which can communicate with each other through mutual interconnections between the neurons. The network will gradually acquire a global information processing capacity of classifying data when being exposed (trained) to many pairs of corresponding input and output data such that new output can be generated from new input. If a set of input is denoted by $\{x_j\}$ and the corresponding output is denoted by $\{y_i\}$ the processing of each neuron i in the net can be described as

$$y_i = f(\sum_j W_{ij} x_j + \eta_i) \tag{1}$$

where W_{ij} are the weights of the connections leading to the neuron i, η_i and f is the characteristic non-linear function for the neuron. As this equation already tells such type of network can be considered as a non-linear map between the input ond output data.

The most straightforward type of neural networks employed for this study were feed forward networks of the multi-layered perceptron type or more complicated recurrent neural networks equivalent to the networks used with real time recurrent learning (RTRL) [12]. The former networks have a unique direction of the data stream such that input are passed through the consecutive layers towards a specific layer of neurons that produce the output while the latter networks have a set of extra feed-back connections. We shall hereafter denote these layers of neurons as, mentioned in the consecutive order, the input layer, the hidden layers and the output layer. The reason for chosing these networks among many other types is due to their renown ability to generalize molecular biology data[6, 7, 8, 9] and their rather simple structure both with respect to processing of data and training. The back-propagation error algorithm[11] is the most commonly used training procedure and the one we shall employ. The training procedure is performed until a cost function C has reached a local minimum e.g. by a gradient descent. The cost function C is normally written as;

$$C = \frac{1}{2} \sum_{\alpha,i} (t_i^\alpha - z_i^\alpha)^2 \tag{2}$$

which is simply the squared sum of errors, t_i being the correct target value and z_i the actual value of the output neurons.

In earlier papers we have been studying various aspects of the use of perceptron layered nets to predict secondary structure or contacts in proteins on the bases of their sequence of amino acids. The network task has been to correlate sequence data input with the occurrence of contacts between residues as output data. The input data of residue types are represented as binary numbers and the output as integer numbers of e.g. residue contacts correlated to others. In each instance of training a vector of input values of residue types, the size of the vector (window) representing the correlation among the residues, is to be related to a vector of output values of a given desired property corresponding to a specific residue (e.g. the middle of the window) in the input vector. The network study was carried out on several types of network architectures, one being for example 17×20 (17 is the window size) input elements, 10 hidden neurons and 1 output neuron.

It is important when utilizing neural networks to have a few basic facts of common knowledge about the architecture of the network in relation to the training. First of all the network should be dimensioned according to the training set, i.e. the number of adjustable parameters (the synaptic weights and thresholds) should not exceed the number of training examples. There is a heuristic rule that the number of training examples should be around 1 to 5 times larger than the number of synaptic weights. Basically the ability to learn and recall learned data increases with the size of the hidden layer while the ability to generalize decreases with an increasing number of hidden neurons above a certain limit. This fact can clearly be understood when one considers the network as essentially a curve fitter between points depicting relations between input and output data in the training set. Therefore it is also easy to understand that a network can be overtrained when the training process reaches the point where the spurious data points are memorized. The training process and the construction of the training set is of greatest importance because the networks predictive power is depending on how clearly the training set is defined and how many patterns that are exposed.

Most success in the present application was obtained with a training and construction procedure for feedforward networks called Cascade-Correlation [13]. This algorithm optimizes the weights in a feedforward network *and* the number of hidden units by adding units during the training process. The initial network contains only input and output units and is first trained using the normal delta-rule which is the special case of the backpropagation algorithm without hidden units. Thus the first phase of the training leads to the same solution that would be obtained by a perceptron and maps only those input patterns that may be separated linearly onto different output patterns. This linear part of the mapping may cover already a lot of input / output pattern pairs in the training set. To further reduce the error one hidden unit that is initially not connected to the output layer is added. The weights leading into this unit are adapted by maximizing the correlation between the activity of this unit and the residual error occurring at each output unit. After this adaptation, all weights into this unit are frozen and the new hidden unit is connected to the output layer with all new weights set to 0. All weights connected to the output units are trained again to minimize the error function. The process of adding new hidden units that maximize the correlation between their activity and the error remaining at the output layer is repeated until the mapping has the desired accuracy. Since each new hidden unit is also connected to all existing hidden units, the network contains as many hidden layers as hidden units.

In order to evaluate the performance of the network various statistical measures have been proposed. In the case of a dual valued output we shall be using the so-called Mathews coefficient[14, 15]. If we denote the two possible output values by 1 and 2 (signifying binding or no-binding) and if p is the number of correctly predicted examples of 1s, $\bar{p}$ the number of correctly predicted examples of 0s, q the number of examples of 1s incorrectly predicted and $\bar{q}$ is the number of examples of 0s incorrectly predicted then we define the coefficient C_M as:

$$C_M = \frac{p\bar{p} - q\bar{q}}{\sqrt{(p+q)(p+\bar{q})(\bar{p}+q)(\bar{p}+\bar{q})}} \tag{3}$$

For complete coincidence with the correct decisions (ideal performance) the measure is 1 and for complete anti-coincidence C_M is -1. A poor net will give $C = 0$ indicating that it does not capture any correlation in the training set in spite the fact that it might be able to predict several correct values.

3   Implementation

The actual neural networks for predicting fold classes are constructed from the SNNS (Stuttgart Neural Network Simulator) environment[19] and are mostly of the feedforward type. The networks are trained on a selection of proteins from each of 42 fold classes containing domain segments of proteins or often whole proteins. The input representation for each protein domain is a 20×20 matrix containing integer numbers corresponding to the absolute frequencies of dipeptides occurring in neighbouring positions in the primary sequence of the domain. All protein domains are transformed this way into one input pattern of fixed size. Insertions and deletions from the protein sequence cause only small changes in the dipeptide frequencies. The same holds true for rearrangements of larger elements in the sequence. There are many cases where members of the same fold class differ mostly by permutations of sequence elements. Such permutations of the primary sequence lead to very similar dipeptide matrices which supports similar classification results. Each fold class is represented by one output unit which should have an activation close to 1.0 if the domain coded in the input layer is a member of that fold class. In all other cases the activity should be close to 0. When an unknown sequence is classified, the fold class corresponding to the largest activation at the output unit is assigned to the sequence. This is the usual winner-takes-all evaluation of the output of a classifier.

In order to facilitate the interpretation of misclassifications we grouped the fold classes in larger super classes that have a natural one-dimensional order inferred from physical properties of the folds. The super class prediction and the fine grained classification should then assign classes that are close in this order.

As mentioned in the introduction a general prediction of the 3-dimensional structure of a novel protein on the basis of its sequence of amino acids is only likely to be successful by computer technics, and especially neural networks, if the fold-class to which the protein belongs to can be determined first. If that is the case a subsequent determination of the 3-dimensional structure of the protein can be obtained through a prediction of the distance matrix that represent the 3-dimensional backbone structure. The distance matrix prediction can be carried out by a neural network trained on the protein folds from the same fold-class.

The neural networks for distance matrix prediction will be described in details elsewhere but basically the most successful network were recurrent. Sequences are read into a symmetrical window of the size of 30 residues and the correlated output would be a real valued distance information between the middle residue and the 30 residues to the left. The C_α atomic position was used as the coordinate for each residue. Thus having an input window scanning over the sequence of a protein domain makes the network produce a band of distance values along the diagonal of a distance matrix. The bandwidth is determined by the "horizon" of correlation of outputs. It is possible to have a larger output "window" than an input window in a recurrent network due to the pipeline and feedback processing. These distance matrix predicting networks are trained on known proteins or protein domains from the fold class that had been determined from the previous network. The training set corresponding to each fold class will on average contain 10 proteins or protein segments each around 100 residue long.

4   Fold–classification from packing analysis

Up to now we have been using protein fold classifications from the literature. We also made another more novel attempt to classify protein structures from an analysis of the

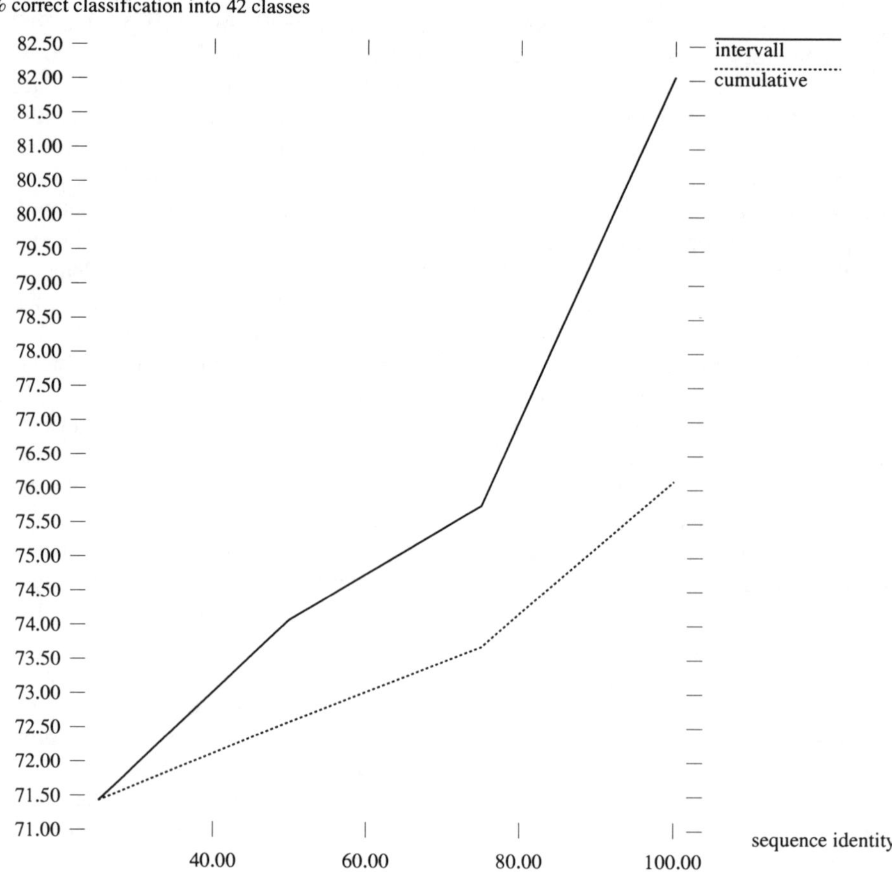

Figure 2: The figure shows test prediction accuracy vs. homology with the training set.

degree of packing. More precisely this was done by making distribution of pair correlations within each protein domain. The frequency of each distance between residues (C_α atomic positions) within a domain or a whole protein was plotted with the measure of distances in Ångstrom along the horizontal axis and the normalized frequency (occurrence) along the vertical axis. Thereby one gets a characteristic distribution of each structure of protein domains. Some structures are represented by a very broad distribution while others have a sharp delta-like distribution. Especially the hight of the distribution is telling about the underlying lattice structure in the domain. For example a typical protease structure, like a zig-zag lattice, will have a distinct peak in the pair correlation distribution at the lattice spacing length. The position, τ, of the peak in the distribution was taken as a simple measure of the domain structure and all the domain structures were hence classified into distinct groups of folds. Folds with the smallest values of peak positions, τ, turned out to be small peptides while intermediate ranges of τ usually could represent globular proteins. Large values of τ represented immunoglobulins and ac-proteases. Small values of τ thus signified little regularity and large values big regularity in the underlying lattice frame. In the following chapter results are given for how well neural networks could predict membership

true foldclass		predicted foldclass																																									
		01	02	03	04	05	06	07	08	09	10	11	12	13	14	15	16	17	18	19	20	21	22	23	24	25	26	27	28	29	30	31	32	33	34	35	36	37	38	39	40	41	42
01	cytc	2	0	0	0	0	0	0	0	0	0	0	0	0	0	0	0	0	0	2	1	0	0	0	0	0	0	0	0	0	0	0	0	0	0	0	0	0	0	0	0	0	0
02	hmr	0	1	0	0	0	0	0	0	0	0	0	0	0	0	0	0	0	0	1	0	0	0	0	0	0	0	0	0	0	0	0	0	0	0	0	0	0	0	0	0	0	0
03	wrp	0	0	0	0	0	0	0	0	0	0	0	0	0	0	0	0	0	0	1	0	0	0	0	0	0	0	0	0	0	0	0	0	0	0	0	0	0	0	0	0	0	0
04	crn	0	0	0	1	0	0	0	0	0	0	0	0	0	0	0	0	0	0	0	0	0	0	0	0	0	0	0	0	0	0	0	0	0	0	0	0	0	0	0	0	0	0
05	globin	0	0	0	0	8	0	1	0	0	0	0	0	0	0	0	0	0	0	0	0	0	0	0	0	0	0	0	0	0	0	0	0	0	0	0	0	0	0	0	0	0	0
06	lsm	0	0	0	0	0	1	0	0	0	0	0	0	0	0	0	0	0	0	1	0	0	0	0	0	0	0	0	0	0	0	0	0	0	0	0	0	0	0	0	0	0	0
07	cpp	0	0	0	0	0	0	1	0	0	0	0	0	0	0	0	0	0	0	0	0	0	0	0	0	0	0	0	0	0	0	0	0	0	0	0	0	0	0	0	0	0	0
08	cyp	0	0	0	0	0	0	0	1	0	0	0	0	0	0	0	0	0	0	1	0	0	0	0	0	0	0	0	0	0	0	0	0	0	0	0	0	1	0	0	0	0	0
09	inhibit	0	0	0	0	0	0	0	0	2	0	0	0	0	0	0	0	0	0	0	0	0	0	0	0	0	0	0	0	0	0	0	0	0	0	0	0	0	0	0	0	0	0
10	wga	0	0	0	0	0	0	0	0	0	4	0	0	0	0	0	0	0	0	0	0	0	0	0	0	0	0	0	0	0	0	0	0	0	0	0	0	0	0	0	0	0	0
11	ca_bind	0	0	0	0	0	0	0	0	0	0	4	0	0	0	0	0	0	0	0	0	0	0	0	0	0	0	0	0	0	0	0	0	0	0	0	0	0	0	0	0	0	0
12	cpa	0	0	0	0	0	0	0	0	0	0	0	1	0	0	0	0	0	0	0	0	0	0	0	0	0	0	0	0	0	0	0	0	0	0	0	0	0	0	0	0	0	0
13	sns	0	0	0	0	0	0	0	0	0	0	0	0	1	0	0	0	0	0	0	0	0	0	0	0	0	0	0	0	0	0	0	0	0	0	0	0	0	0	0	0	0	0
14	ferredox	0	0	0	0	0	0	0	0	0	0	0	0	0	2	0	0	0	0	0	0	0	0	0	0	0	0	0	0	0	0	0	0	0	0	0	0	0	0	0	0	0	0
15	plipase	0	0	0	0	0	0	0	0	0	0	0	0	0	2	0	0	0	0	0	0	0	0	0	0	0	0	0	0	0	0	0	0	0	0	0	0	0	0	0	0	0	0
16	xia	0	0	0	0	0	0	0	0	0	0	0	0	0	0	0	1	0	0	1	0	0	0	0	0	0	0	0	0	0	0	0	0	1	0	0	0	0	0	0	0	0	0
17	fxc	0	0	0	0	0	0	0	0	0	0	0	0	0	0	0	0	1	0	0	0	0	0	0	0	0	0	0	0	0	0	0	0	0	0	0	0	0	0	0	0	0	0
18	kinase	0	0	0	0	0	0	0	0	0	0	0	0	0	0	0	0	0	0	1	0	0	0	0	0	0	0	0	0	0	0	0	0	0	0	0	1	0	0	0	0	0	0
19	nbd	0	0	0	0	0	0	0	0	0	0	0	0	0	0	0	0	0	0	9	0	0	0	0	0	0	0	0	0	0	0	0	0	0	0	0	0	0	0	0	0	0	0
20	binding	0	0	0	0	0	0	0	0	0	0	0	0	0	0	0	0	0	0	1	2	0	0	0	0	0	0	0	0	0	0	0	0	0	0	0	0	0	0	0	0	0	0
21	tln	0	0	0	0	0	0	0	0	0	0	0	0	0	0	0	0	0	0	1	0	0	0	0	0	0	0	0	0	0	0	0	0	0	0	0	0	0	0	0	0	0	0
22	barrel	0	0	0	0	0	0	0	0	0	0	0	0	0	0	0	0	0	0	0	0	0	2	0	0	0	0	0	0	1	0	0	0	0	0	0	0	0	0	0	0	0	0
23	eglin	0	0	0	0	0	0	0	0	0	0	0	0	0	0	0	0	0	0	1	0	0	0	0	0	0	0	0	0	0	0	0	0	0	0	0	0	0	0	0	0	0	0
24	pti	0	0	0	0	0	0	0	0	0	0	0	0	0	0	0	0	0	0	0	0	0	0	0	1	0	0	0	0	0	0	0	0	0	0	0	0	0	0	1	0	0	0
25	gap	0	0	0	0	0	0	0	0	0	0	0	0	0	0	0	0	0	0	0	0	0	0	0	0	0	0	0	0	1	0	0	0	0	0	0	0	0	0	0	0	0	0
26	cts	0	0	0	0	0	0	0	0	0	0	0	0	0	0	0	0	0	0	0	0	0	0	0	0	0	1	0	0	0	0	0	0	0	0	0	0	0	0	0	0	0	0
27	rdx	0	0	0	0	0	0	0	0	0	0	0	0	0	0	0	0	0	0	0	0	0	0	0	0	0	0	3	0	0	0	0	0	0	0	0	0	0	0	0	0	0	0
28	pgk	0	0	0	0	0	0	0	0	0	0	0	0	0	0	0	0	0	0	1	0	0	0	0	0	0	0	0	0	0	0	0	0	0	0	0	0	0	0	0	0	0	0
29	virus	0	0	0	0	0	0	0	0	0	0	0	0	0	0	0	0	0	0	1	0	0	0	0	0	0	0	0	0	7	0	0	0	0	0	0	0	0	0	0	0	0	0
30	virus_prot	0	0	0	0	0	0	0	0	0	0	0	0	0	0	0	0	0	0	0	0	0	0	0	0	0	0	0	0	0	3	0	0	0	0	0	0	0	0	0	0	0	0
31	dfr	0	0	0	0	0	0	0	0	0	0	0	0	0	0	0	0	0	0	1	0	0	0	0	0	0	0	0	0	0	0	1	0	0	0	0	0	0	0	0	0	0	0
32	gcr	0	0	0	0	0	0	0	0	0	0	0	0	0	0	0	0	0	0	0	0	0	0	0	0	0	0	0	0	0	0	0	3	0	1	0	0	0	0	0	0	0	0
33	igb	0	0	0	0	0	0	0	0	0	0	0	0	0	0	0	0	0	0	0	0	0	0	0	0	0	0	0	0	2	0	0	0	15	0	0	0	0	0	0	0	0	1
34	il	0	0	0	0	0	0	0	0	0	0	0	0	0	0	0	0	0	0	0	0	0	0	0	0	0	0	0	0	0	0	0	0	0	2	0	0	0	0	1	0	0	0
35	pap	0	0	0	0	0	0	0	0	0	0	0	0	0	0	0	0	0	0	0	0	0	0	0	0	0	0	0	0	0	0	0	0	0	0	0	1	0	0	0	0	0	0
36	sbt	0	0	0	0	0	0	0	0	0	0	0	0	0	0	0	0	0	0	0	0	0	0	0	0	0	0	0	0	0	0	0	0	1	0	0	1	0	0	0	0	0	0
37	ac_prot	0	0	0	0	0	0	0	0	0	0	0	0	0	0	0	0	0	0	1	0	0	0	0	0	0	0	0	0	0	0	0	0	0	0	0	0	5	0	0	0	0	0
38	tox	0	0	0	0	0	0	0	0	0	0	0	0	0	0	0	0	0	0	0	0	0	0	0	0	0	0	0	0	0	0	0	0	0	0	0	0	4	0	0	0	0	0
39	s_prot	0	0	0	0	0	0	0	0	0	0	0	0	0	0	0	0	0	0	1	0	0	0	0	0	0	0	0	0	0	0	0	0	0	0	0	0	0	0	6	0	0	0
40	plasto	0	0	0	0	0	0	0	0	0	0	0	0	0	0	0	0	0	0	0	0	0	0	0	0	0	0	0	0	0	0	0	0	0	0	0	0	0	0	0	2	0	0
41	ltn	0	0	0	0	0	0	0	0	0	0	0	0	0	0	0	0	0	0	2	0	0	0	0	0	0	0	0	0	0	0	0	0	0	0	0	0	0	0	0	0	0	0
42	hoe	0	0	0	0	0	0	0	0	0	0	0	0	0	0	0	0	0	0	0	0	0	0	0	0	0	0	0	0	0	0	0	0	0	0	0	0	0	0	0	0	0	1

Figure 3: The figure shows the commutation matrix for the prediction of test cases among the 42 fold-classes.

of fold-classes based on the τ measure.

5   Results

The main results in this paper are concerning the prediction of fold classes from sequence alone since that is the most novel element and distance matrix prediction from a homologous training set is well-known and described elsewhere[9].

The fold class predictions are performed in 3 different levels of detail. The first classification uses 26 fold classes each containing 3 members or more. By using the τ measure we define a set of 13 superclasses that are used for prediction of the coarse grained fold class. Finally a more extended classification of 42 fold classes is described. The set of 26 fold classes is a subset of the larger set. All fold classes are listed in fig.3.

The networks were trained to an accuracy of 100 % on the training set and with a test score of 71 % in predicting a fold class correct on the basis of the sequence. In the case of the 42 fold-classes we could obtain a score of correctness of around 73 % and with a Mathews coefficient of around 0.60. In table 2 true and predicted fold classes for the 26 and 42 classes are opposed in a 2-dimensional array.

We took two cases of proteins, the protease 2TRM and Rubredoxin 3RXN, predicted by a network to belong to the correct fold class and then trained networks on these fold classes, S_PROT and RDX, respectively, to produce the corresponding distance matrices. Finally these predicted distance matrices were used in a minimization procedure[17] to construct the 3-dimensional backbone structure for the two proteins with their side-chains positioned from data produced by a rotamer predicting feed forward neural network that, on the basis

true foldclass		predicted foldclass												
		1	2	3	4	5	6	7	8	9	10	11	12	13
1	gp1 mlt	0	0	0	0	0	0	0	0	0	0	0	0	1
2	crn	0	0	0	0	0	0	0	0	0	0	1	0	0
3	inhibit pti rdx tox	0	0	4	0	0	0	0	1	0	0	0	0	0
4	ctf eglin hoe sn3	0	0	0	0	0	0	0	1	0	0	1	0	0
5	b5c gn5 hip utg	0	0	0	0	0	0	0	2	0	0	0	0	0
6	cc5 rnt	0	0	0	0	0	0	1	0	0	0	0	0	0
7	256b acx cytc cytc3 ferredox fxc hmr il pab plasto plipase sns ssi tmv tnf virus_prot wrp	0	0	0	0	0	0	10	2	1	0	0	0	0
8	ca_bind cla dfr etu gap gcr globin hmg-2 igb lzm pap sod wga	0	0	0	0	0	0	3	27	3	0	2	1	0
9	blm carbonic cyp hmg-1 ltn pgm pyp rhd s_prot virus	0	0	0	0	0	0	1	2	6	0	0	1	0
10	binding cpa kinase sbt tln	0	0	0	0	0	0	0	0	0	2	2	0	1
11	aat cpp icd nbd pgk xia	0	0	0	0	0	0	0	0	0	1	6	1	0
12	ac_prot barrel cat cts gls	0	0	0	0	0	0	0	2	0	0	2	4	0
13	acn	0	0	0	0	0	0	0	1	0	0	0	0	0

Figure 4: The figure shows the permutation matrix for the prediction of the fold-classes derived from packing analysis.

of sequence data, can predict side-chain orientations on a fixed backbone.

The prediction of the 42 fold-classes is nicely illustrated in the permutation matrix shown in figure 3. It is seen that most mistakes are made between classes close to the diagonal which means among similar classes. The classes are arranged so the alpha-helical rich protein classes are at the upper end and the beta-strand rich protein classes are on the bottom. Figure 4 shows a similar matrix for predictions of the fold-classes based on the packing density.

This system of many networks, each trained on a specific set of proteins from each fold class, can be implemented on a parallel processing system in order to speed up the procedure in going from a protein sequence to a final structure. We chose to implement this system of many neural networks on a connection machine, CM5, such that each processor would run a network trained on each one of the fold classes. Such a system would run in an MIMD implementation and turned out to be efficient.

6 Discussion

An artificial neural network system has been constructed to classify 3-dimensional protein structures by predicting what fold class they belong to on the basis of their sequence. Earlier a simple perceptron network has been used to classify proteins according to 4 classes[20].

The present networks appear to train acceptably well (about 90% correct and a similar Matthews's coefficient) on the task of predicting fold classification and distance matrix geometry. Their predictive performance turned out to be rather successful with a score of around 73% for predicting fold classes (with a total of 42 classes) and with a score of 71% in the case of 26 fold-classes. In the case of predicting a correct super fold-class (out of a total of 13 classes) we obtained a correctness score of 65%.

Interestingly it seems that neural networks may not be able to achieve greater than about 70% accuracy in predicting the fold classes like (and probably due to the same reasons) when predicting the secondary structure of nonhomologous proteins[16]. One explanation for that may be due to the postulate that around 70% of the secondary structures found in the native structure are formed at an early stage (i.e. msec) of protein folding and thus without training the network on intermediate structures the performance will never surpass the 70%. The determination of the folds is similar to the determination of the topology of the protein backbone and that, on the other hand, depends on the fixation of secondary structures. Thus there is an argument to the correlation of secondary structures and fold-classes. Furthermore the new classification of folds that we proposed is quite depending on the content of secondary structures. Low values of the τ parameter represent alpha-rich fold-classes and high values of τ represent beta-rich fold-classes.

Acknowledgements

We should like to thank the staff and students of the physical chemistry department at the University of Illinois as well as the molecular bio-physics group at the DKFZ institute at Heidelberg for technical assistance and stimulating discussions. Especially we acknowledge Drs Richard Goldstein, Zan Schulten, Joseph Bryngelson and J. A. McCammon for help. We also acknowledge NCSA for providing us with computer resources especially at the connection machine CM5.

References

[1] S. Pascarella and P. Argos, *Protein Engineering*, **5**, 121-137 (1992).

[2] L. Holm and C. Sander, *J. Mol. Biol.*, **233**, 123-138 (1993).

[3] C. Chothia, *Nature*, **357**, 543-544 (1992).

[4] D. J. Jones, W. R. Taylor and J. M. Thornton, *Nature*, **358**, 86-89 (1992).

[5] W. Kabsch and C. Sander, *Biopolymers*, **22**, 2577-2637 (1983).

[6] N. Qian and T. J. Sejnowski, *J. Mol. Biol.*, **202**, 865-884 (1988).

[7] H. Bohr, J. Bohr, S. Brunak, R. M. J. Cotterill, B. Lautrup, L. Nørskov, O. H. Olsen and S. B. Petersen, *FEBS Lett.*, **241**, 223-228 (1988).

[8] L. H. Holley and M. Karplus, *Proc. Natl. Acad. Sci. USA*, **86**, 152-156 (1989).

[9] H. Bohr, J. Bohr, S. Brunak, R. M. J. Cotterill, H. Fredholm, B. Lautrup and S. B. Petersen, *FEBS Lett.*, **261**, 43-46 (1990).

[10] S. Brunak, J. Engelbrecht and S. Knudsen, *Nature*, **343**, 123 (1990).

[11] D. E. Rumelhart, J. L. McClelland et al., Parallel Distributed Processing, MIT Press, 1986.

[12] R. J. Williams and D. Zipser, *Neural Computation*, **1**, 270-280 (1989).

[13] S. E. Fahlman and C. Lebiere, in Advances in Neural Information Processing systems II, D.S. Touretzky, Ed. Los Altos, CA: Morgan Kaufmann, 1990, 524-532.

[14] B. W. Mathews, *Biochem. Biophys. Acta*, **405**, 442 (1975).

[15] P. Stolorz, A. Lapedes and Y. Xia, "Predicting Protein Secondary Structure Using Neural Net and Statistical Methods", Los Alamos Preprint LA-UR-91-15, 1991.

[16] H. Bohr, R. Goldstein and P. G. Wolynes, AMSE Periodicals, Modelling, measurement and control, *C*, **31**, 55 (1992).

[17] J. Bohr, H. Bohr, S. Brunak, R.M.J. Cotterill, H. Fredholm, B. Lautrup and S.B. Petersen, J. Mol. Biol., **231**, 861-869, (1993).

[18] R. A. Goldstein, Z. A. Luthey-Schulten and P. G. Wolynes, *PNAS USA*, **89**, 9029-9033 (1992).

[19] A. Zell, N. Mache, T. Sommer, and T. Korb, in Proc. Applications of Neural Networks Conf., SPIE, Aerospace Sensing Intl. Symposium, Orlando Florida, 1469, 708-719 (1991).

[20] I. Dubchak, S. Holbrook and S Kim, *Prot. Struc., Func. and Genetics*, **16**, 79-91 (1993).

Distance-based Approaches to Protein Structure-Function Analysis

Michael N. Liebman

Bioinformatics Program, Amoco Technology Company, Mail Code F-2, Naperville, IL
60563

Abstract

Our research focuses on the development of computational approaches for use
in rational protein engineering, biological pathway engineering and drug design and risk assessment. Algorithm development requires access to a wide
range of both public domain and proprietary databases, comprising information
about molecular structure and function, physico-chemical properties, biological organization and biochemical and clinical effects. These databases are
typically compiled and maintained by "domain experts", e.g. spectroscopists,
crystallographers, clinicians, molecular biologists, rather than integrated within a comprehensive database architecture. We attempt to integrate this diverse
data to enable modeling, simulation and analysis, by developing tools for data
abstraction, representation and evaluation which emphasize the "functional"
characteristics or object-oriented nature of the data. Several models for incorporation of the distributed data are under development including conformal
mapping tools, logic programming and Petri net analysis.

1   Introduction

We have been developing tools to evaluate the relationship between structure and function
in proteins with interest in using the acquired knowledge to directly manipulate proteins
and enzymes through rational protein engineering, as well as indirectly through rational
design of targeted modulators. Our analysis examines the natural state of these molecules
and their response to natural mutation, e.g. disease states, or modulation, e.g. inhibition.
An important component of our analysis includes discrimination between *in vitro* and *in
vivo* functionality for use in identifying appropriate targets to attain optimal response, i.e.
selectivity, specificity, reactivity. To achieve this we both develop as well as incorporate
computational and experimental approaches.

A typical goal might be states as the need to design a low molecular weight, active site
directed inhibitor for a specific enzyme target. Additional unstated but significant characteristics might include ease of synthesis, low cost to produce, high specificity and ability to
establish commercial value through patent protection. Our analysis would involve examination of natural substrates and inhibitors of the enzyme, both molecular and macromolecular,
evaluation of the suitability of the selected target and its potential for side–effects based on
its participation in a biological pathway, analysis of information on genetic variants of the
enzyme and related clinical effects, experimental evaluation of conformational response of

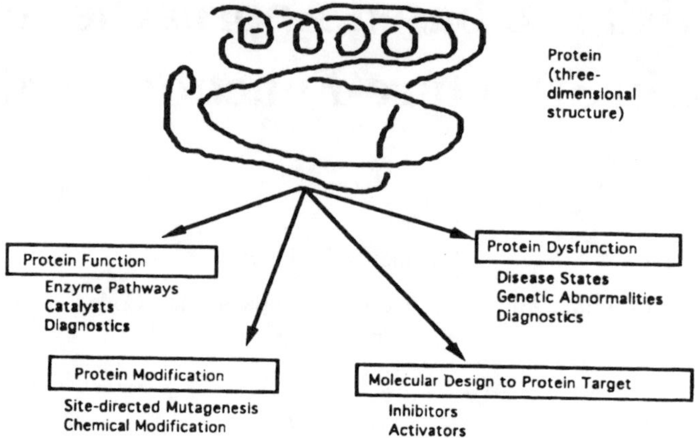

Figure 1:

the enzyme structure both within and outside of the active site, etc. We use the term "bioin-formatics" to describe the methodologies by which we integrate data as well as measurable computational and experimental input into the problem-solving process.

The majority of problems which confront research programs can derive benefit from solutions of an interdisciplinary nature. This is in contrast with the education and training of a typical research which tends to focus within a specific discipline, e.g. physical chemistry, chemical engineering, molecular biology, inorganic chemistry. Databases were traditionally developed and maintained on single computer systems and provided access to a restricted community of trained users. Advances in computer technology can readily provide access to computational methodologies and data from a wide-range of disciplines. The evolution of a "data highway" or information network enhances the efficiency of this access. The integration of interdisciplinary data stored in multiple databases on local, wide-area and international networks can be utilized to address many research problems can be utilized to address many research problems, provided a means for that data integration exists. The tools we are developing work to enable that integration.

Conventional data integration favors implementing a standard database architecture to enable ready exchange of data, but the imposition of standards could impede development of discipline-specific databases. The typical database focuses on needs of that discipline, e.g. the protein crystallographic database contains atomic coordinates, crystal symmetry, etc. and limited reference to spectroscopic, kinetic, genetic or clinical data. The key to enabling access to an interdisciplinary solution using such databases requires providing both cross-referencing and the potential for expanding the interdatabase linkages. Our bioinformatic methodology emphasizes functional linkages and develops these using conformal mapping to generalized forms of data representation. This approach is independent of the structure and location of component database and it lays the foundation for establishing the virtual reality of a "distributed database".

The strategic goal of our research is in developing technology to realize value from data such as the results of the Human Genome Project. As such, we do not address areas which focus on collecting and collating the genomic sequence, itself, but rather on its downstream processing and utilization, as gene products. The tools and methodologies which we are attempting to develop are not solely targeted at questions which might exist today, but also

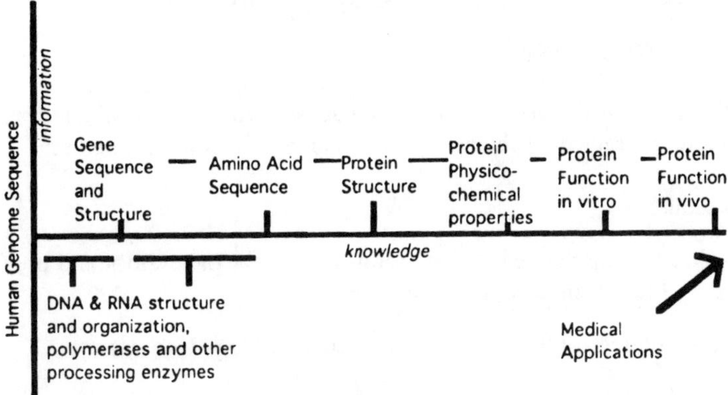

Model of the BioInformation Pathway for the Human Genome

Figure 2:

at those questions which cannot yet be asked. Although we describe the application of these methods to biological and biophysical data, we believe that the approach is potentially generalizable to other areas which are technology-rich as well and may serve as a model for use in transfer of this technology and approach.

2 Background

We have attempted to represent the relationship between information and knowledge in biology from the viewpoint that these orthogonal. Based on this, the plot shown below is a model for the organization of information starting from the genome, using information as the ordinate and knowledge as the abscissa. This plot suggests the hierarchical relationship between information database, which would be vertical lines, and the increase in value which correlates with progression along the knowledge axis from genome sequence towards medical application. We specifically note the difference between function *in vitro* and function *in vivo* because of its significance in utilizing information to analyze specific problems, i.e. the relevance of using a given assay to predict function in real–time, *in vivo* delivery systems. In an academic sense, this representation describes all of the biological and interdisciplinary interfaces listed above. This representation presents a framework for organizing scientific experimentation and results well beyond that attainable within oneþs lifetime. The value of the organization is the ability to plot where domain-specific databases exist and what upstream (or downstream) linkages may be of value for predicting and evaluating other information, e.g. could we compute a proteins structure from its amino acid sequence or its physical properties from its structure or sequence? The focus of our tool development is to solve problems by developing tools to bridge between such data and to evaluate which linkages are essential to a specific problem or potentially generalizable to larger classes of problems. We consider the construct of these linkages as developing a virtual database which enables the integration, at a distance, of these underlying databases.

3 Conventional Databases Comprising the Virtual Database

The conventional categorization of existing databases which are available for access and integration within biological problem-solving includes, but is not limited to:

Nucleic acid sequences:

1. GenBank(GB), maintained at Los Alamos National Laboratory and previously distributed by IntelliGenetics (Mountain View, CA);

2. EMBL, maintained at the European Molecular Biology Laboratory;

3. Backbone Database (NCBI) maintained by the National Center for Biotechnology and Informatics at the National Library of Medicine;

4. genome sequences for model organisms, e.g. C. elegans (Univ. Arizona); mouse (Whitehead Institute and MIT)

5. human chromosome databases. e.g. chromosome 16(CH16)Alamos National Laboratory)

Note: genomic and chromosome databases also contain information on genetic and physical maps.

Amino acid sequences:

1. Protein Identification Resource (PIR) maintained at Georgetown University National Biomedical Research Foundation;

2. SwissProt (Swiss) maintained by Amos Bairoch, Luzerne, Switzerland

3. X-ray crystallographically determined protein structures (PDB) maintained by Brookhaven National Laboratory

4. natural mutant hemoglobin sequences (HMB) maintained at the University of Michigan

5. Bio-engineered mutants (PERI) maintained by the K. Nishikawa, Protein Engineering Research Institute, Osaka, Japan

Structural databases:

1. Protein structures from x-ray crystallography (PDB) maintained as Brookhaven National Laboratory's Protein Data Bank

2. Protein structures from nuclear magnetic resonance (PDB and NMR) where NMR is maintained at the University of Wisconsin NMR Protein Resource

3. small molecule x-ray crystallographic structures (CCDB) maintained by the Cambridge Crystallographic Data Bank, Cambridge, UK

4. Protein secondary structure analysis: Greer-Levitt algorithm (G-L) published but not maintained; Kabsch-Sanders algorithm (K-S) published but not maintained.

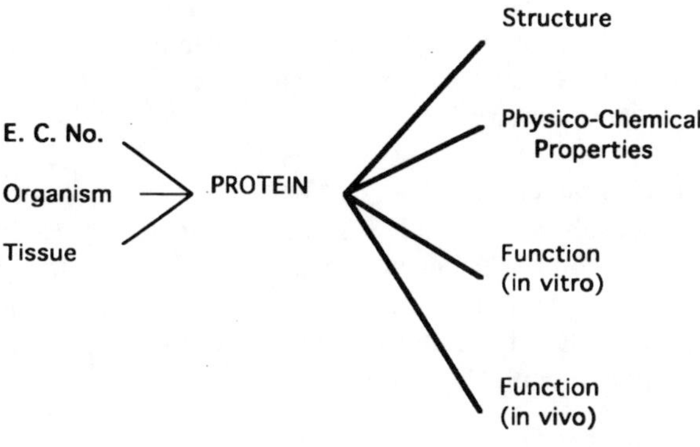

Figure 3:

Clinical databases:

1. Hemoglobinopathies (HMB) maintained at the University of Michigan

2. Coagulation Factor IX (FIX) maintained by S. Sommer at Mayo Clinic

Proprietary databases:

1. All pharmaceutical and biotechnology industrial groups maintain proprietary databases of the classes listed above. These are typically in a format standard to that used in public domain databases to facilitate use of commercial access tools.

Several of these databases do contain some level of cross-referencing through comments and notations, e.g. PDB contains the Enzyme Commission number (E.C.) for each enzyme which relates to catalytic function and the constituent amino acid or nucleic acid sequence. These references are limited and not uniformly defined, e.g. secondary structure in PDB is not analytically assigned although the full atomic coordinates are included. A relational database model for PDB has been developed which links the existing record classes, but does not support ready extension to derivable data classes, e.g. molecular volumes, surface area, or to non-structural data, e.g. substrate/inhibitor identification, participation in biological pathway.

4　Basis for ATC Representations and Database Integration

The basis for developing the representations we have used to integrate the underlying databases involves developing the concepts of inheritable characteristics which a protein exhibits. These are shown below (see Fig. 3):

Within these concepts, a set of descriptors can be listed which reflect the individual characteristics which are measurable, calculable or definable. Examples of these descriptors, within these concepts, are presented below:

Structural parameters:
folding domains; sequence of secondary structure (substructure); substructure

composition; substructure hierarchy; cis-peptides; deviations from ideal configuration in distances or angles; structure determination; structural resolution;
structural refinement method; method of data collection; crystallization conditions, e.g. solvent, pH, temperature; presence of non-crystallographic intra-
and inter-molecular symmetry.

Physico-chemical parameters:
molecular volume, solvent accessible surface; rugosity (surface area to volume
ratio); ellipsoid of revolution parameters, e.g. axial ratio through center of mass,
orientation; radius of gyration; molecular dipole moment, e.g. backbone and
side-chains; packing density; bulk hydrophobicity; amphiphilicity; molecular
weight; link to spectroscopic data bases.

Function (*in vitro*):
substrate specificity; inhibitor specificity; binding pocket based on substrate/inhibitor
contacts; mechanism; binding constant/turnover rate; antigenic determinants.

Function (*in vivo*):
phylogenetic source of protein; organ site of activity; environmental conditions
at site of activity; substrate specificity; inhibitor specificity; cofactors; binding
constant/turnover rate; participation in metabolic pathway or cascade; link to
nucleic acid data base sequences, e.g. exon boundaries; Enzyme Commission
classification; membrane localization; post-translational modification, e.g. glycosylation, sites and type; isolatable form, e.g. zymogen, prepro-, proenzyme.

Focusing on these descriptors, the existing databases are observed to contribute data
to multiple characteristics. PDB contains data on specificity and conformational response
to *in vitro* and *in vivo* interactions, solvent perturbation, pH and temperature variation,
site-directed mutations (natural and bio-engineered), variation in organism, evolutionary
relationships, etc., and it can thus be used to address a variety of questions by appropriate
selection of the constituent data. In addition, flexibility is required to adapt to new descriptors
which might be developed and do not exist within the present database (see Fig. 4).

5 Computational Tools to Support Database Integration Through Data Representation

The key to integrating data/information from the various experimental and computational
areas described above is dependent upon the ability to: 1) represent the data in a form
which minimizes biases, or permits their quantitative evaluation ; and 2) optimize access
to structural, functional and physico-chemical data through the use of common algorithms.
We show below the example of bias which is indeterminate in this case. We assume that all
proteins which we study can be classified as soluble or membrane bound, and that they can
be used to represent the complete universe of proteins. The bias, or difference between what
is known and what may exist that is unknown, is indeterminate in size and in its potential
impact on our ability to generalize our results.

This type of bias (see Fig. 5) can also be true of methodologies developed for problem-
solving, and we attempt to evaluate and utilize the inherent bias of an approach by building in

Examples of concepts and their potential database sources are listed below:

```
Structure                                 Databases(examples)
    primary structure
            nucleic acid                  GB, EMBL
                  exon/introns            ATC
            amino acid                    PIR, Swiss
            mutations
                  naturally occurring     HMB, FIX, ATC
                  bio-engineered              PERI, ATC
            super-families                    PIR
    secondary structure
            composition                   PDB
            ordering                      PDB, G-L, K-S
            predicted or modeled
    substructures                         ATC
    domains                               ATC
    tertiary structure
            x-ray crystallography             PDB
            nuclear mag. resonance        NMR
            modeled by homology
    quaternary structure
            crystal packing                   ATC

Physico-chemical Properties
    intrinsic
            hydrophobicity                derived
            dipole moment                     ATC
            solvent accessibility         derived
            van der Waals surface             derived
    extrinsic
            molecular recognition             ATC
            molecular reactivity
            molecular dynamics            Traj

Function- in vitro
    structural change
            x-ray crystallography             PDB
            nuclear mag. resonance        NMR
            Fourier transform infrared    ATC
            Circular Dichroism            ATC-USDA
    molecular interactions
            recognition                   ATC
            specificity
    kinetics
    thermodynamics
    solvent interactions                  ATC
    metal/ion interactions                    ATC
    solution vs crystal environment       ATC

Function-in vivo
    biological pathways                   ATC
            molecular recognition
            molecular specificity
    clinical observations                 HMB, FIX, ATC
    chromosomal abnormalities             various, ATC
    post-translational modification
```

Note: ATC represents proprietary databases under development within the Bioinformatics Program, Diagnostics Sector, Amoco Technology Company.

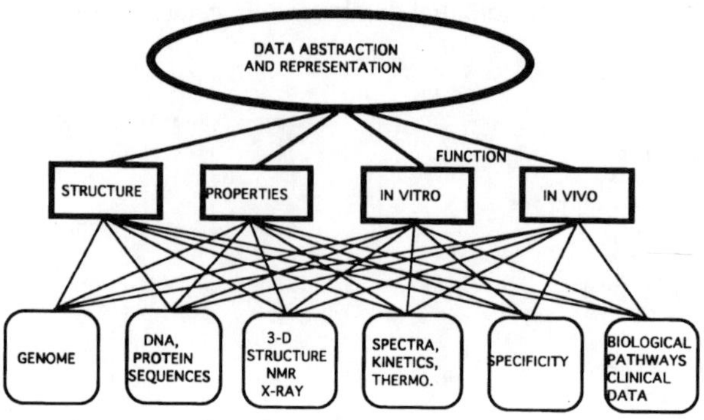

Figure 4:

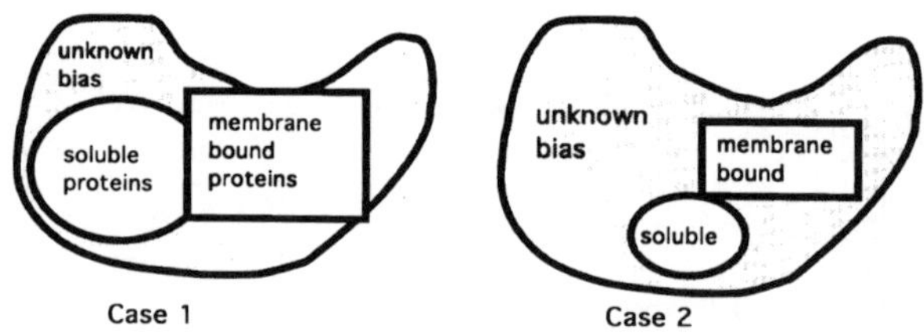

Figure 5:

consistency checks. We describe our methods for representation, analysis and comparison in terms of their dimensionality, i.e. one-dimensional representations relate descriptors to a relative position within the nucleic acid/amino acid sequence, using the sequence to derive the list or ordering; two–dimensional representations also use a sequence-based ordering to generate a matrix containing descriptors which relate pairs of sequence elements, i.e. residues, which may not be contiguous; three-dimensional representations relate descriptors to positions relative to a defined coordinate system, typically Cartesian or polar coordinate space; and, four-dimensional representations involve descriptors which incorporate time with lower dimension descriptors. Structural, functional and physico-chemical descriptors using each of these formats are described below, along with their respective advantages and limitations. An important benefit results from the regions of overlap or redundancy which exist amongst these methods of representation as it enhances the opportunity to make observations and to validate their consistency.

5.1

One-dimensional representations use the position within a sequence to represent a descriptor based on that specific sequence element, or evaluated within a window of sequence elements which bounds that specific sequence element. With this representation we use descriptors

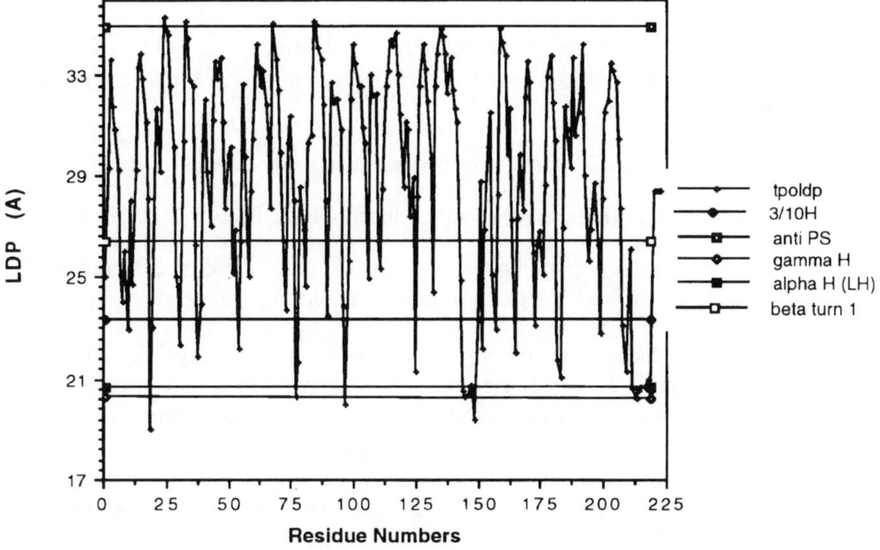

Figure 6:

for:

1. structure- Linear distance plot (LDP); bond dihedral angle plot(BDA)

2. physico-chemical properties- as derived from statistical analysis of observed protein/nucleic acid structures, and including hydropathy, flexibility, bulk, surface area, pK, etc.

3. functional properties- relative dipole orientation of sequence elements

Advantages which are realized with one-dimensional representations include their use in rapid pattern recognition and analysis of structure, function and properties, separately and in combination; invariance of a given mapping to rotation and translation of the reference frame of the descriptor, e.g. rotation or translation of a molecule does not alter its linear distance plot; extendability to other descriptors as developed, e.g. intron-exon boundaries, enzyme-inhibitor contact regions; and suitability for direct comparison of localized regions between closely related molecules, e.g. local conformational perturbation upon inhibitor binding into an enzyme. The main limitations involve the chiral ambiguity in structural representation of a protein with the LDP, i.e. distance information within a local structure lacks chirality; selection of "window" size to preserve information content when representing window-based descriptors, e.g. averaging hydropathy over a variable window size of residues and representing a single value for a specific sequence position. Analysis and comparison tools which have proved most useful with these representations include use of signal processing methods such as simple addition or subtraction to highlight local regions among representations of closely related molecules, and dynamic programming algorithms for pattern recognition which encompasses potential insertions and deletions and has been used to develop a protein substructure library.

5.2

Two-dimensional representations are based on the generation of an n-dimensional square matrix, where n is the number of elements within the sequence list. With this representation format we use descriptors for:

1. structure- distance matrix, partitioned distance matrix, both intra- and inter-molecular

2. physico-chemical properties- weighted hydrophobicity, base-pairing potential

3. function- electrostatic potential matrices for backbone, side-chains including dipole-dipole, charge-dipole and charge-charge interactions, dipole-interaction matrix

Advantages of this representation format similarly are for use in analysis by pattern recognition of structural organization and orientation and comparison of independent molecules or conformational states of a specific molecule. The natural organization of this transform yields sequence-adjacent features, i.e. secondary structure-related, near to the matrix diagonal, and more sequence-distant features more distant from the matrix diagonal. As with the one-dimensional representation, these representations are invariant to the orientation of the reference frame for these descriptors. The lack of chiral information in one-dimensional representation of individual segments is significantly negated as the occurrence of macro-molecules as mirror equivalents is much lower than that for small molecules. Comparison and analysis utilize pattern recognition and vision-analysis based tools, and are presently being evaluated within a neural network training project.

5.3

Three-dimensional representations involve the direct referencing of the structure/function/property descriptors to an external (cartesian) or internal coordinate (polar) reference frame. These representations comprise the more typical manner for viewing the three–dimensional structure with interactive, three–dimensional computer graphics displays. The descriptors which we use in this form include:

1. structure- atomic coordinates (orthogonal reference frame), fractional atomic coordinates (crystallographic reference frame), polar plot around designated atom/point centers, molecular graphics representation

2. physico-chemical properties- solvent accessible surface, van der Waals surface, molecular dipole orientation

3. function- hydrodynamic shape, molecular electrostatic potential surface, ion/solvent channels, internal cavities

The advantages of this form of representation are in its wide–spread acceptance due to the increased availability of interactive computer graphics displays. This advantage, which has evolved to make these representations readily accepted by the non-specialist scientific community, may prove to be a long–term limitation because of its interpretation as representing "what a molecule looks like". The most notable limitation of this form of representation is the dependence of a particular viewing perspective on the orientation within a coordinate reference frame. Thus it is necessary to rotate a molecule, i.e. apply a rotation matrix to the atomic coordinate list and re-display, to obtain a different view. This reveals the difficulty in applying analytical methods to the analysis of these representations. Comparison and analysis are performed by effecting the direct superposition of one molecule onto another by generation of a rotation-translation matrix, orientation sampling within the display reference frame, e.g. rotating the molecule in 5-degree intervals about a specific viewing axis, and visual comparison.

5.4

Four-dimensional representations involve those descriptors which contain an element of time in conjunction with other relevant descriptors. Thus any of the descriptors described above could be utilized in a four–dimensional representation due to fluctuation of structure, function or physico-chemical properties with time. Those descriptors which we identify to have specific temporal components are:

1. structure- dynamical nature of a molecule as viewed by computation of a molecular dynamics trajectory; structural disorder observed in crystallographically determined structures; structural dynamics as viewed by NMR

2. physico-chemical properties- measurable properties which change with time, principally due to the occurrence of functionally-derived processes

3. function- enzyme cascades, metabolic pathways, allosteric processes as mapped through sparse matrix methods

The advantage of this form of representation is the ability to incorporate processes which are the true physico-chemical and functional states of the molecules under study. The dynamic nature of macromolecules may be closely linked with functions of reactivity and specificity. Analysis and comparison of pathways which contain molecules of common evolutionary origin may reveal the extent to which evolutionary influence is exerted at the physiological level and lead to the understanding of structure-function at a higher level of development, and a much fuller appreciation of the information actually encoded at the genetic level.

5.5   Petri Net Analysis

The Petri net is presented as a model for a discrete–event system. It is utilized in our representation and analysis of (bio)chemical pathways to enable quantitative pathway comparison, discrete–event simulation and analysis of complex pathway behavior. Petri Net methodology supports application of this approach in systems where incomplete data may exist, e.g. kinetic parameters or even pathway linkages, to allow for semi–quantitative modeling.

A Petri net (PN) is a graph formed by two kinds of nodes, called places (pj) and transitions (tj). Directed edges, called arcs, connect places to transitions, and transitions to places. A non-negative integer number of tokens is assigned to each place, and it can vary based on the state of the Petri net. Each arc has a weight, a positive non-zero integer, assigned to it. Pictorially, places are represented by circles, transitions by bars or boxes, arcs by lines ending in an arrow, and tokens as black dots placed in the circles. Generally if there is no arc–weight specified on the graph we assume it to be equal to one.

Each transition (event) is associated with a finite number of input places (pre-conditions) and output places (post-conditions). A transition is enabled when the number of tokens in its input places is greater than or equal to the weights on the arcs connecting the places to the transition. A transition with no input places, called a source transition, is always enabled.

An enabled transition can fire, depositing tokens in its output places, again their number determined by the arc-weights. A transition with no output places, called a sink transition, can fire when enabled, consuming the tokens from its input places. The state of a PN, which is the number of tokens present in the individual places, is denoted by M and called the marking of the net. The initial marking of a PN is denoted by M0. Thus, the firing of a sequence of transitions may change the marking of the net. A marking M is reachable from M0 if there exists some firing sequence that accomplishes this change.

> Mathematically a Petri net is represented as PN= (P, T, E, W, M0);
> where P = { p1, p2, p3, ... pm} is a finite set of places
> T = { t1, t2, t3, ... tn} is a finite set of transitions is a set of arcs
> W: E { 1, 2, 3, ...} is a weight function
> M0: is the initial marking
> P and T being disjoint sets

Extendibility: Petri nets are fundamentally extendible. A transition represents an event which requires certain pre-conditions and results in some post-conditions if the event actually occurs. If this event is a combination of other events, it can be visualized as a transition that emulates a Petri sub-net. Any modification to this sub-net is reflected in the behavior of the original transition. This can be well understood in the context of an Object Oriented Programming framework, where one object inherits attributes of other objects that it contains.

Liveness: A Petri net is said to be live, if from any marking reachable from M0 it is possible to fire any transition in the net through some further firing sequence.

Reachability: A necessary condition for a marking Mn to be reachable from the initial marking M0 is the existence of a non-negative solution. This condition becomes sufficient with the following requirement. All markings in the firing sequence from M0 to Mn must be coverable, i.e. there must exist a minimum marking such that the transitions in the firing sequence from M0 to Mn are enabled.

Reversibility: A Petri net is said to be reversible if the initial marking M0 is reachable from all other possible markings in the set of markings reachable from M0. A marking M' is a home state if it is reachable from all other markings in the set of possible markings of a Petri net.

Fairness: There are different definitions to fairness, but the one that is of relevance is Unconditional (or Global) fairness. If in a firing sequence s, every transition occurs infinitely often, then s is globally fair. A Petri net is globally fair if every s in all possible markings from M0 is globally fair. A pathway that has the property of global fairness, suggests the existence of a state of continuous operation starting from an initial state, without outside intervention.

Structural reduction: Large Petri nets can be reduced by the substitution of certain combinations of places and transitions with smaller units without sacrificing the original properties of the net

5.6   Logic Programming

Prolog has been utilized to examine the use of logic programming in developing a database structure to permit complex data representation and analysis. In particular, biochemical pathways have been represented as a database with Prolog, e.g. intermediary metabolism, where the characteristics of inheritance and recursive application have been utilized. Recognizing the variable nature of the information contained within our knowledge of biological pathways, the use of Prolog has enable development of tools which can be applied recursively, at the appropriate level of granularity merited by the quality of the pathway data being analyzed. These tools are unique in their recursive nature, but accomplish this with a loss of speed in application.

6   Capitalizing on Bias Using the Object-Oriented Approach

We cannot expect to completely eliminate the bias which may exist in data or in methodologies as noted earlier. At best we can implement self–consistency tests where appropriate. The object-oriented methods we use to describe proteins and enzymes provide a means to benefit from the existing bias by inverting our approach from bottom–up to top–down. Thus our interest in and inability to accomplish developing general rules for protein structure prediction from amino acid sequence may indicate that critical information is both missing and unassessable from the existing data. By contrast, we can construct, using object definitions, a series of structurally and functionally related proteins. We should expect that analysis of this set would be able to predict the next member of this class, and failure would indicate the inability to generalize to other classes. The characteristics actually learned may be class-associates, although some subset of these characteristics may be generally applicable. The difficulty and the value comes in recognizing and separating these characteristics. We accomplish some of this through parallel analysis of multiple classes (see Figure 6) and integration of data within a class to address problems of a specific application as described in example of Factor IX, below.

7   Application of the Distributed Database

The virtual database, described above, exists as a series of domain-specific databases, residing on a variety of computers, at a series of international sites and linked through a set of tools which focus on data representation and abstraction. The utility of this construct can best be seen by examining an application of its wide-ranging capabilities. The problem we present addresses the analysis of Factor IX, an enzyme in the blood coagulation cascade for which certain chromosomal abnormalities, i.e. natural single site mutations, yield a coagulation deficiency termed hemophilia Bm.

Information which exists, initially, is that Factor IX contains a protease domain which appears to be a serine protease and that the gene that physically maps to this condition, and therefore is responsible for the DNA coding for this enzyme is located on the X chromosome. We describe the linkages, accomplished with the representations and tools described above, to a variety of specific databases and the composite information which has been derived from the virtual database.

Structural information at the level of primary structure enabled the mapping of the protease domain of Factor IX (FIX) to the family of trypsin–like enzymes. Three–dimensional structural information existed for several members of this family, and the identification of

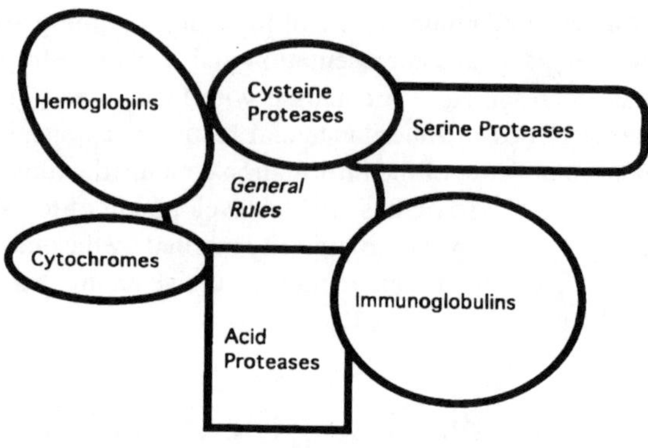

Figure 7:

a structural core of conserved three–dimensional structure and conserved primary structure could be observed and mapped to FIX. Identification of structurally variable regions of these enzymes, and their mapping to observed *in vitro* function (e.g. binding of calcium ions), suggested a potential site for modification of FIX which might relate to physiological function. Analysis of trypsin–like enzymes indicate a correlation with physiological function and the ability to modulate inhibitor/substrate interactions through binding to this site. Representation and searching of the CCDB revealed presence of candidates for binding to this site which could modulate *in vitro* and potentially *in vivo* function of these enzymes. Conformational response of the enzymes, in contrast to their binding activity, specificity and reactivity, could be evaluated against both *in vitro* and *in vivo* inhibitors. Spectroscopic measurements, FT-IR and CD, both revealed common patterns of structural response which differ between *in vitro* and *in vivo* enzyme response. Correlation of spectroscopy and x-ray crystallographic analysis was made by evaluation of the solution conditions used for crystallization and evaluation of the environments influence on the enzyme's physical state. Molecular dynamics analysis indicated the occurrence of conformational perturbation outside of the classically defined active site in this enzyme family. Experimental observations revealed that these regions were localized, and correlated specifically with interaction with *in vivo* and not *in vitro* inhibitors. Mapping of the chromosomal changes which result in clinical observations of hemophilia Bm indicate that these same non-active site regions of the enzyme are identifiable from the sequence-clinical data as well. Evidence was compiled concerning the mechanism of action of the disease, potential means for modulating genetic defects and rationally designing site-directed mutations or drug/modulator design or selection from large databases through the integrated use of the distributed database, relying on the tools for data abstraction and representation to integrate the data. In addition, representation and analysis of the biological pathway in which FIX participates, blood coagulation, and correlation with physical mapping within the region on the chromosome for the FIX gene, suggest alternative sites for modulating and controlling the impact of genetic disease and the potential for secondary or tertiary effects which result from the genetic event itself as well as the proposed means for controlling it. Experiments to verify these hypotheses and extend the existing knowledge base have been planned and the results will further extend the underlying data and concepts used in this application.

8   Conclusions

We have described the ability to integrate a wide–range of data and databases, extending over many computer systems, sites, disciplines and areas of concentration, through the use of data abstraction and representation techniques. The ability to define higher order concepts which can unify the underlying data, e.g. structure, properties, function (*in vitro* and *in vivo*) serve as a key to developing this approach. The distributed database approach outlined here has proven to be flexible and extendible in that we can readily incorporate new data into the process without interruption of existing linkages. The complexity of the problems we have applied this approach to, and the ability to identify linkages in information which are not readily apparent even to specialists in a particular application, suggest the potential for generalization of this approach. We are currently extending our research into database mining using neural network methodologies to use this virtual database concept.

Acknowledgements

We would like to acknowledge the contributions to both development and application of the methodologies described in this report by Sal Amato, Steve Prestrelski, Ron Buono, Nilofer Jiwani, Joe Jesson, Steve Kleinman, Rina Dukor, Gary Klein, Ann Brugge, Toni Kazic, V. Reddy and Michael Mavrovouniotis.

References

[1] F. C. Bernstein, T.F. Koetzle, G.J.B. Williams, E.F. Meyer, Jr., M.D. Brice, J.R.Rodgers, O.Kennard, T. Shimanouchi and M. Tasumi, *J. Mol. Biol.*, **112**, 535-542 (1977).

[2] G. L. Wilcox, M. Poliac and M. N. Liebman, *Tetra. Comp. Let.*, 191-220 (1990).

[3] M. N. Liebman and A. L. Brugge, Santa Fe Institute Studies in the Sciences of Complexity, Volume VII, eds. G. Bell and T. Marr, Addison-Wesley Longman Publishing Group, 183-202, 1989.

[4] M. N. Liebman, *J. Comp.-Aided Molecular Design*, **1**, 323-341 (1987).

[5] M. N. Liebman, *J. Indus. Microbiology*, **3**, 127-137 (1988).

[6] M. N. Liebman, *Enzyme*, **36**, 115-140 (1986).

[7] S. J. Prestrelski, A. L. Williams and M. N. Liebman, *Proteins*, **14**, 430-439 (1992).

[8] G. C. Klein and M. N. Liebman, unpublished results

[9] S. J. Prestrelski, D. M. Byler and M. N. Liebman, *Proteins*, **14**, 440-450 (1992).

[10] M. N. Liebman, C. A. Venanzi, H. Weinstein, *Biopolymers*, **24**, 1721-1758 (1985).

[11] J. Jesson and M. N. Liebman, unpublished results

[12] V. N. Reddy, M. L. Mavrovouniotis, and M. N. Liebman, in press.

[13] N. Jiwani and M. N. Liebman, in press.

Quantification of secondary structure prediction improvement using distantly related proteins

Jonathan M. Levin, Steffano Pascarella*, Patrick Argos* and Jean Garnier

Unite d'ingenierie des proteines, INRA, 78352 Jouy-en-Josas, France
*EMBL, P. 102209, D600, Heidelberg, Germany

Abstract

The limitations of various secondary structure prediction algorithms are discussed in terms of local sequence and tertiary structure effects. Evidence is presented which suggests that when one single polypeptide chain is predicted the limitation of prediction accuracy is 63–65% due to consideration of only local sequence effect. Improvements in accuracy brought about by using evolutionary related proteins is analyzed. Seven different protein families (82 sequences of known structure) were considered to provide a range of alignment and prediction conditions. The results of the alignments obtained by spatial superposition of main chain atoms in known tertiary protein structures showed an average of 88.8% (SD of 3.8) agreement between the observed secondary structures. The range of 89%±7.6% should represent the maximum range of accuracy that one can expect from a secondary structure prediction. This maximum value is achieved with the program SIMPA for homologous proteins when one of them has its structure experimentally determined and is used for the prediction of the others. Using the structural alignments without adding the knowledge of a homologous structure allowed an average of 8% points improvement in secondary structure prediction accuracy, when compared to those obtained from the individual sequences. Substitution of structural alignments by those determined directly from an automated sequence alignment algorithm showed variations in the prediction accuracy that correlated with the quality of the multiple alignments and evolutionary distances of the primary sequences. Secondary structure predictions can be reliably improved using alignments from an automatic alignment procedure with a mean increase of 6.8%, if there is a minimum 25% sequence identity between all sequences in a family. One can then expect when no homologous structure is known that the use of related sequences can bring the overall prediction accuracy of secondary structure prediction to the 68.5%–72.3% range depending upon the method of prediction used.

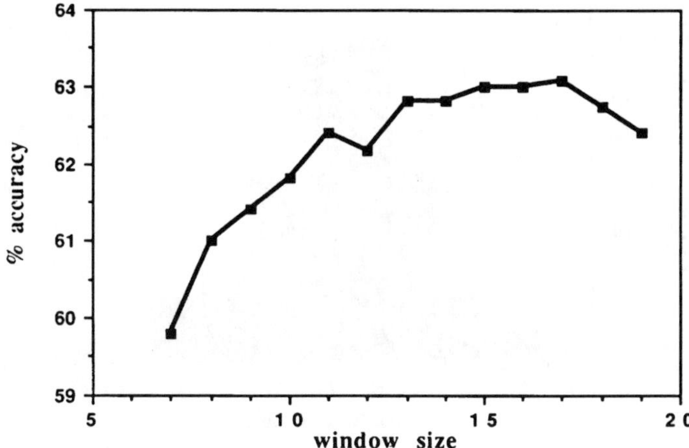

Figure 1: Effect of the window size in number of amino acids residues on the accuracy of secondary structure prediction by the program SIMPA (Levin and Garnier, [7]).

1   Introduction

Relating the amino acid sequence of a protein to its 3 dimensional or tertiary structure is presently an important issue in biology; it is sometimes referred to as the second genetic code. It is well documented that the tertiary structure of proteins governs their biochemical function, for example catalysis, protein–protein or protein–nucleic acid recognition and is the source of specificity which is a fundamental principle in the complex organized systems found in biology.

A limited but important stage towards the understanding of the second genetic code is the prediction of secondary structures from the amino acid sequence. Secondary structure predictions using single protein amino acid sequence are presently performing not better than 63-65% of correctly predicted residues in three states, helix, extended and other (see recent reviews in Garnier and Levin [1], Rost and Sander [2]). As this accuracy was obtained from algorithms with different bases such as information theory (Garnier and Robson, [3]), neural network [4, 5, 6] or similarity analysis [7] one has to raise the question of the origin of such a limited and yet common accuracy. Following the pioneering work of Zvelebil et al [8] who found that the GOR method could be substantially improved by using multiple sequences of proteins belonging to the same evolutionary family, several attempts have been made to use this information to improve secondary structure predictions. However good benchmark studies are lacking. Before entering into the quantification of the possible improvement in prediction and its limits brought about by the use of distantly related proteins and their alignments, we want to analyze first some features of the prediction algorithms which allow us to understand the limitation in accuracy when only the information from one single amino acid sequence is used for the prediction. This should be of help in developing new predictive algorithms and interestingly the conclusions that one can draw are in agreement with recent views and experiments concerning the mechanism of protein folding. This mechanism

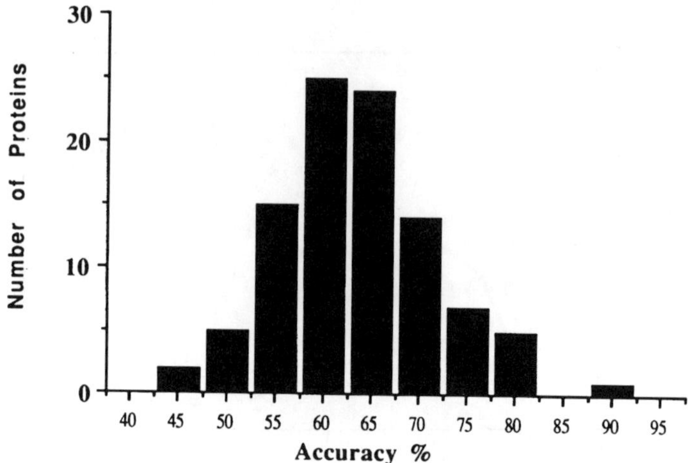

Figure 2: Variation of secondary structure prediction accuracy of the GOR III method with the number of proteins. 98 proteins were examined. (Reproduced from Robson and Garnier, [10]).

would include first a rapid formation of secondary structures from the local sequence and then a variable readjustment of these secondary structures during the formation of the molten globule and possibly also in the final stage of acquisition of the tertiary structure. Secondary structure predictors have addressed their algorithms to the first stage of the folding process, unfortunately with parameters evaluated from experimental data which include the last stages or tertiary structure effect. Consequently these parameters contain some background noise due to the tertiary structure effect, this noise can be eliminated in some conditions, (see below section 2.3) allowing the estimation of the relative contribution of these stages to the final structure.

2 Local sequence information and secondary structure prediction accuracy

2.1 Data base size

This issue is permanently raised but several instances of increasing the size of the data base [3] or simulating an increase in size [4] did not seem to be very effective and was of the order of 1 or less percent point improvement after cross validation when doubling or even tripling the data base size. This improvement is about one standard deviation on the average accuracy. However other considerations (see below) have to be made before making a decision about the present data base size.

2.2 Window size effect

All the secondary structure prediction methods considering only a single protein sequence for the prediction have in common the use of the amino acid sequence of a short segment of the polypeptide chain in their prediction algorithms, usually a sequence length of 17 amino acids [9]. Attempts [4, 5, 7, 9] to increase the length of that window not only did not improved the accuracy but decreased it (see fig.1). The origin of this failure to take into account the whole sequence has to be searched in the tertiary structure of proteins : the six to eight residues in contact with each single residue that probably control its secondary structure are unevenly scattered along the sequence, and no good statistics can yet be derived from the present data base to relate pair residue interactions and position along the whole sequence. It is precisely for that reason that the present size of the data base is too limited.

2.3   Variability of accuracy between proteins

Another feature worth considering is the large variability of prediction accuracy within proteins (Robson and Garnier, [10]). The standard deviations run currently from 6 to 9 % depending on the method of prediction. An example of such variability is presented in fig. 2 where the standard deviation is 6.4%. One way to understand this variability is to consider that for each non-homologous protein, a specific environmental effect resulting from its tertiary structure would affect the secondary structure formation. This specific effect will add to the local sequence effect. It is only by pooling enough proteins together, that one can expect this effect to cancel, that is to consider all the amino acids of the data base irrespective of the proteins of which they come from. One can then expect to have a good estimate of the effect of the local sequence. In that condition, it is interesting to note that the calculated probability of a residue to be in a given conformation considering only the local sequence matches relatively well with the observed probability to be in that conformation in the pooled data base (see fig 3). Thus the data base is large enough to evaluate correctly the local sequence effect and any improvement of prediction accuracy from 63–65% to the estimated maximum of 89% (see below) has to be gleaned from the tertiary structure effect. A similar conclusion has been recently reached by Rao et al. [11]. If for the tertiary structure effect the data base of known structures is too limited, the use of homologous proteins has proved a beginning to take into account this effect.

3   Use of homologous proteins to improve secondary structure prediction

An obvious and now classical way is to use modeling by homology.

3.1   Use of sequence homology when one member of the family has a known structure

The basic principle is that homologous proteins (evolutionary related proteins) of the same family share the same tertiary folding including the same secondary structure. This has been well documented by results from X-ray data [12]. Modeling by homology consists in aligning the homologous sequences and to affect the coordinates of the known structure to the aligned residues then locating the secondary structures in the unknown. Considering the difficulties in performing correct sequence alignment for low identify and/or different length sequences, an alternative consists of using the homology with the known structure to predict the secondary structure without using an amino acid alignment. Most secondary structure prediction methods are sensitive to the presence of homologous proteins in the

Table 1: Secondary structure prediction of homologous proteins with the SIMPA method*.

File Name	Identity %	Basic Prediction %	Homologous Prediction %
2 ACT	48	53.7	83.5
2 ALP	39	50.0	77.8
2 ABX	42	81.1	75.7
5 CHA (1)	34-43	65.7	89.3
5 CHA (2)	41-45	60.8	87.6
3 C2C	40-44	64.3	87.5
2 EST	30-43	62.1	86.7
2 EBX	42	71.0	96.7
1 FDX	43	70.4	75.9
2 FDI	43	64.2	70.8
2 HHB (1)	35-45	61.7	92.9
2 HHB (2)	45	56.9	89.7
2 LHB	35	73.2	88.6
1 MCP (2)	44	67.6	88.7
1 FB4 (2)	44	75.6	87.3
2 PKA (1)	34-42	63.8	96.3
2 PKA (2)	30-41	60.5	91.5
1 MBN	25	80.4	88.9
1 PPD	48	63.7	92.0
2 SGA	39	54.1	76.8
3 RP2	34-38	56.7	84.8
1 TPO	36-45	61.4	89.7
Global: 22	**25-48**	**64.5 (SD 8%)**	**86.3 (SD 6.7%)**

*Selected from Levin and Garnier [7]. The basic prediction was obtained when all homologous proteins of more than 22% identity were removed from the data base. The homologous prediction is explicitly using the homologies in the data base.

data base which artificially improves the accuracy of the prediction of those proteins. Some methods are more sensitive than others specially neural network and methods based on homology. This property has been used to improve secondary structure prediction with the program SIMPA [7] and an example of such an improvement is presented in table 1. Here a close to the maximum of the possible accuracy is achieved, 86.3 % accuracy (SD of 6.7) per chain for a three state prediction and for proteins of relatively low identity (25 to 48%). This accuracy will be higher for more homologous proteins. In several instances it has been observed to be useful in checking the quality of the multiple alignment by controlling if the predicted secondary structures are correctly aligned to the known structure [13] or to define more precisely, for example, the hypervariable loops in antibody modeling (data not shown).

3.2   Use of multiple sequence alignments when the structure of an homologue is not known

The most frequent case encountered is that there is no known homologous structure but several sequences of homologous proteins are known. Multiple alignment algorithms for amino acid sequences of these proteins aim to match residues conserved during evolution of the amino acid sequences and to detect positions of deletions or insertions in the sequence. The results of these alignments often correspond to those obtained by superposition of the

```
   Res    Proteins
          12345
   1      EEEET     HHHHH     HHHHH     HECHH     HHHHH
   2      GSSSS     CHCHH     HHHHH     HECHH     HHHHH
   3      LNNGG     CCCCC     CCCCC     HHCCH     HHHHH
   4      -FFYY     -CCCC     -CCCC     -HCCC     -CCCC


          SEQ       OBS       MAX       PRED      +C
                              89%       58%       63%
```

tertiary structures. The consequence is that aligned residues are those having the same secondary structures and the same contact residues in the tertiary structure. By averaging the predicted secondary structures at that position on all homologous sequences one should improve the prediction as when one measures a property by doing several experiments and taking the mean of the experimental values. The more distantly related are the sequences, the greater is the corresponding number of "experiments" providing the alignment is correct, therefore we consider that the sequence of an homologous protein is an "experimental" result of various mutations having maintained the same secondary and tertiary structures. The mean of the predictions will have a value with less noise than each single prediction and then will be more accurate.

3.2.1 Method used to include predictions of homologous sequences

The following constitutes an example of the prediction from 5 different homologous proteins at 4 residue positions and for three secondary structures, helix (H). extended (E) and others (C). SEQ refers to the protein sequence in single letter code. OBS corresponds to observed conformations and MAX is the best possible prediction assuming that for each residue position only one conformation can be chosen, which here is the most frequently observed. PRED is the individual predictions of secondary structure either from GOR III [14] or SIMPA [7]. +C is the consensus prediction obtained by choosing the most frequently predicted conformation at each residue position as the unique conformation for that position and for all the proteins. This consensus prediction has been done with the GOR III, SIMPA or both predictions (COM). An attempt to use the relative probabilities of the predicted conformations gave similar results [15] however we found it useful in the few cases when an equal number of two different conformations were predicted at the same residue position. Prediction accuracy is given as a percentage, i.e. the number of residues correctly predicted divided by the total number of predicted residues in order to compare with previous published data in the most objective way. For the consensus predictions the results were calculated as follows : at each residue position a correct conformation is counted when the consensus prediction agrees with the observed conformation for each residue at that position. Then the reported accuracy concerns all the residues of all the sequences in a given family including the residues which could not be meaningfully superimposed from the known 3 D structures.

Seven families were chosen to represent a wide range of alignment environments and sequences, all having their structures solved : nucleotide binding domains (NBD) mainly dehydrogenases[10], flavodoxin with 15 proteins; serine proteases (S_PR), mammalians and fungus with 12 proteins; aspartic proteases (AC_PR), pepsins and chymosin with 8 proteins; cytochrome cs (CYTC) with 6 proteins; globins (GLOB), hemoglobins, myoglobins and

Table 2: Accuracy of prediction (%) using structurally aligned sequences.

PROTEIN METHOD	NBD	S_PR	AC_PR	CYTC	GLOB	IGB	LZM	AVE	SD	DIF
SIMPA	61.2	60.7	60.4	61.1	66.7	62.9	67.4	62.9	1.1	0
GOR	58.8	56.4	58.8	59.5	63.8	60.0	63.8	60.2	1.0	0
SIMPA+C	67.4	62.4	65.4	62.8	75.6	70.5	70.9	67.9	1.8	5.0
GOR+C	63.9	62.9	66.3	62.6	72.1	69.6	66.5	66.3	1.3	6.1
COM+C	69.9	67.3	69.2	62.0	76.3	70.6	71.7	69.6	1.5	<u>8.05</u>
MAX	88.2	90.1	87.9	89.6	91.9	81.2	92.7	88.8	1.4	
HOMOLOGY	18	29	24	40	30	23	35			

Adapted from[15]. **AVE** mean score over all families. **SD** is the standard deviation of the means, i.e. the standard deviation between families divided by $(7-1)^{1/2}$. **DIF** refers to the difference in the prediction of the mean SIMPA/GOR III single prediction and the prediction COM +C based on the multiple alignment. **HOMOLOGY** is the average of the percent of identical residues for all the proteins in each family aligned pairwise.

leghemoglobin with 10 proteins; immunoglobulins (IGB), heavy and light chains with 26 proteins and lysozymes (LZM), lysozymes and a lactalbumin with 5 proteins. There were 82 proteins in total. For further details, see Levin et al. [15].

Table 3: Accuracy of prediction (%) using CLUSTAL sequence alignments.

PROTEIN METHOD	NBD	S_PR	AC_PR	CYTC	GLOB	IGB	LZM	AVE	SD	DIF
SIMPA	61.2	60.7	60.4	61.1	66.7	62.9	67.4	62.9	1.1	0
GOR	58.8	56.4	58.8	59.5	63.8	60.0	63.8	60.2	1.0	0
SIMPA+C	57.0	63.4	60.6	62.2	75.7	66.3	59.9	63.6	2.1	0.7
GOR+C	54.0	61.1	58.6	60.6	72.8	68.1	58.3	61.9	2.2	1.7
COM+C	56.8	63.4	63.8	62.6	77.1	68.7	61.9	64.9	2.2	<u>3.4</u>
MAX	67.0	86.8	75.3	89.5	92.2	77.8	83.0	81.7	3.0	-7.1
HOMOLOGY	17	30	22	41	32	23	38			
HOMOLOGY > 25%, COM+C								68.5		<u>6.8</u>

Adapted from [15]. See table 2 for definitions.

Table 4: Percentage residue identities and prediction differences for the family S_PR.*.

1	2	3	4	5	6	7	8	9	10	11	12	
6	13	10	10	6	10	23	22	7	-9	-16	-3	DIF
100	54	44	42	35	30	33	34	29	18	16	16	
	100	43	44	35	33	33	35	28	20	18	22	
		100	74	44	39	36	35	35	18	17	19	
			100	45	39	36	37	31	19	18	17	
				100	41	34	35	34	21	18	16	
					100	44	34	36	16	21	21	
						100	35	29	19	21	18	
							100	28	17	18	21	
								100	21	17	20	
									100	65	34	
										100	37	
											100	
									3	-1	10	DIF

*From [15]. The percentage of identity is calculated from CLUSTAL alignment for each sequence of the family. DIF as in table 2 and given for each protein. The values of DIF in the last row of the table are calculated from the predictions obtained using a CLUSTAL alignment of only the last three sequences (fungus proteases).

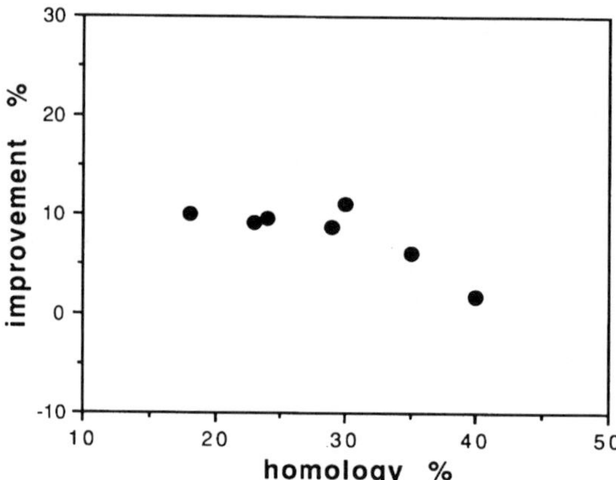

Figure 3: Effect of homology on the prediction improvement of secondary structures with structurally aligned residues. The homology is the average homology (percent of identical residues) for all the proteins in each family aligned pairwise. The prediction is the combination of GOR III and SIMPA and the difference is made with the average of the GOR III and SIMPA accuracy's from individual sequence predictions.

3.2.2   Results of the structural alignment

Results after cross-validation are represented in table 2 with the accuracy of prediction given for each family. The highest accuracy score of 69.6 % with a standard deviation (SD) of the mean of 1.5 percent points was achieved for the combined prediction of GOR III and SIMPA. This corresponds to an improvement of 8.05 percent points compared with the averaged accuracy's of GOR III and SIMPA over individual sequences. Noticeable are the values of MAX which score at an average of 88.8 % (SD between families of 3.8%) and not up to 100%. We obtain here a more restricted and defined range than the 70% to 100% range proposed by Russel and Barton [16]. According to crystallization conditions and crystal forms, the resolution and refinement of the tertiary structure and the algorithms used for superposition, small conformational differences occurs within the same protein and in our examples also between homologous proteins. This results in an upper limit for the accuracy of prediction of secondary structures which cannot be better than the experimental data from where they are derived or compared with. It is remarkable that the improvement increases as the homology decreases (see fig. 4). This might well have been expected since the more different are sequences, the better are the consensus predictions.

3.2.3   Results of automated alignments

Several automated alignments were used and yielded similar results. An example of the use of CLUSTAL[17] is given in table 3. In this case the average improvement drops

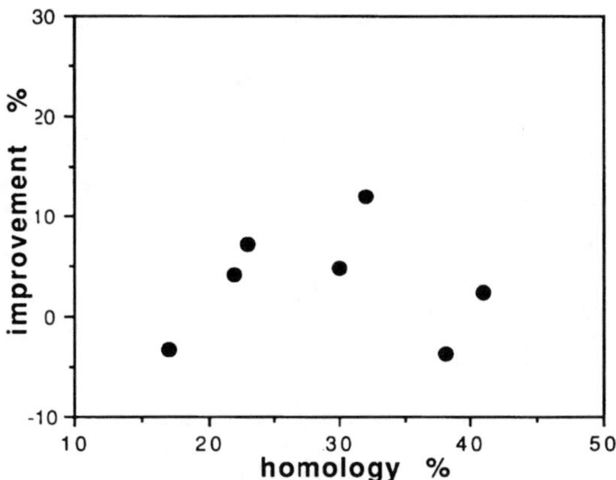

Figure 4: Effect of homology on the prediction improvement of secondary structures with CLUSTAL alignment. See comments in fig. 4 legend.

significantly from 8.05% down to 3.4% with a greater standard deviation of 2.2 percent points. Still the combination of the two methods of prediction is better than each one taken separately. Also the average MAX value has been reduced down to 81.7% indicative of a defect in the alignments. The average homologies between each of the sequences within a family (table 2) are relatively low, 18 to 40% of identity. Thus misalignments must occur and perturb the prediction. This would explain the drop in accuracy for distant homologies with an optimum around 30-35% of identity and a new drop in accuracy for higher homologies as found for structural alignments (see fig. 5). An example is presented in table 4 for the serine protease family. Clearly the fungus proteases, rows 10, 11 and 12 (table 4) which display a low identity with the other mammalian protease sequences, 22% and lower, lose prediction accuracy through the use of the automated alignment (see also fig. 6). If these three fungus proteases are aligned only among themselves, see bottom of the table, a net increase in accuracy of 4% is achieved. This phenomenon was also observed with the other families. As the recognition of homology by sequence alignment for sequences greater than 80 residues requires at least 25% of identity [18], we reduced the data base to those which have at least 25% identity. In this case the average improvement for the combined prediction GOR III and SIMPA is 6.8% yielding after cross–validation a total accuracy of 68.5% for the seven families (table 3). This improvement is very close to the one observed from the structural alignment of the homologous proteins of 25% or more identity, i.e. 6.9%. From figs. 4 and 5 the improvement seems to become marginal above 40% identity.

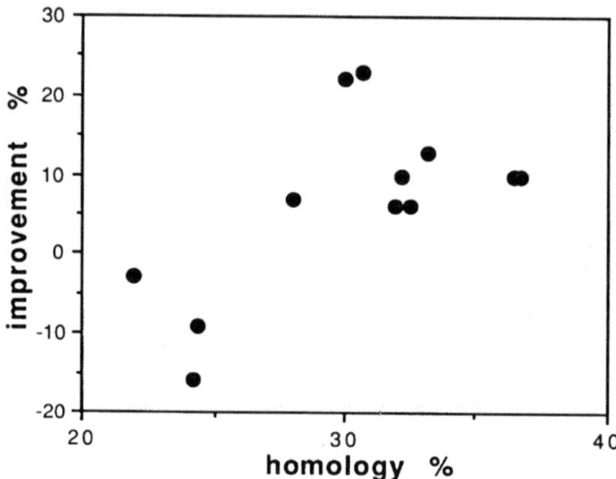

Figure 5: Effect of homology in the serine protease family on the prediction improvement of secondary structures with CLUSTAL alignment. Same comments as in fig. 4. The homology is the average of each polypeptide chain homology with the eleven other chains.

4   Conclusions

The range of maximum accuracy of 89% ± 7.6% that one can expect from a secondary structure prediction is interesting although not surprising, since it is the result of experimental accuracy in structure determination. The limited accuracy of the three state predictions to 63–65% by using only local sequence information allows us to estimate the tertiary effect in the formation of secondary structures in proteins at about 25%. One should keep in mind that these are average numbers and we should expect variations of the tertiary effect between proteins which explain that some proteins can be well predicted with local information and others not. The lesson that use of homologues can improve secondary structure prediction is not really new [7, 8] but its more recent exploitation [6, 15, 19] is effective. From benchmarks discussed here a simple consensus algorithm can extract an average improvement of near 7% bringing the accuracy close to or over 70 % with a standard deviation on this average value of about 1 percent point (Rost and Sander,[6] and Levin et al, [15]). But we expect that these attempts can be made more refined in the future and then bring more improvements if one can determine the cause of the variability in prediction improvement observed between individual sequences and effectively use it. Thus secondary structure predictions should become an essential tool in predicting the tertiary fold of a protein.

References

[1] J. Garnier and J.M. Levin, The protein structure code : what is its present status?, *CABIOS*, 7, 133-142 (1991).

[2] B. Rost and C. Sander, Progress in protein structure prediction?, *TIBS*, **18**, 120-123 (1993).

[3] J. Garnier and B. Robson, The GOR method for predicting secondary structures in proteins. In Prediction of Protein Structure and the Principles of Protein Conformation (G.D. Fasman, ed.) Plenum Press, New York, 10, 417-465, (1989).

[4] N. Qian and T.J. Sejnowski, Predicting the secondary structure of globular proteins using neural network models, *J. Mol. Biol.*, **202**, 865-884 (1988).

[5] H.W. Holley and M. Karplus, Protein secondary structure prediction with a neural network, *Proc. Natl. Acad. Sci. USA*, **86**, 152-156 (1989).

[6] B. Rost and C. Sander, Prediction of protein secondary structure at better than 70% accuracy. J. Mol. Biol., 232, 584-599 (1993)

[7] J.M. Levin and J. Garnier, Improvements in a secondary structure prediction method based on a search for local sequence homologies and its use as a model building tool, *Biochim. Biophys. Acta*, **955**, 283-295 (1988).

[8] M.J. Zvelebil, G.J. Barton, W.R. Taylor and M.J.E. Stenberg, Prediction of protein secondary structure and active sites using alignment of homologous sequences, *J. Mol. Biol.*, **195**, 957-961 (1987).

[9] J. Garnier, D.J. Osguthorp and B. Robson, Analysis of the accuracy and implications of simple method for predicting the secondary structure of globular proteins, *J. Mol. Biol.*, **120**, 97-120 (1978).

[10] B. Robson and J. Garnier, Protein structure prediction, *Nature*, **361**, 506 (1993).

[11] S. Rao, Q-L. Zhu, S. Vajda and T. Smith, The local information content of the protein structural database, *FEBS Letters*, **322**, 143-146 (1993).

[12] C. Chothia and A.M. Lesk, The relation between the divergence of sequence and structure in proteins, *Embo J.*, **5**, 823-826 (1986).

[13] J. Garnier, Levin, J.M., J.F. Gibrat and V. Biou, Secondary structure prediction and protein design in : Protein structure prediction and design. Biochem. Soc. Symp., 57, J. Kay, G.G. Lunt and D.J. Osguthorpe, Ed. Portland Press Ltd. London, 11-24 (1990).

[14] J.F. Gibrat, J. Garnier and B. Robson, Further development of protein secondary structure prediction using information theory. New parameters and consideration of residue pairs, *J. Mol. Biol.*, **198**, 425-443 (1987).

[15] J.M. Levin, S. Pascarella, P. Argos and J. Garnier, Quantification of secondary structure prediction improvement using multiple alignments, *Protein Eng.*, **6**, 849-854 (1993).

[16] R.B. Russel and G.J. Barton, The limits of protein secondary structure prediction accuracy from multiple sequence alignment, *J. Mol. Biol.*, **234**, 951-95 (1993).

[17] D.G. Higgins and P.M. Sharp, Clustal : a package for performing multiple sequence alignment on a microcomputer, *Gene*, **vol. 73**, 237-244 (1988).

[18] C. Sander and R. Schneider, Database of homology-derived structures and the structural meaning of sequence alignment, *Proteins : Struct. Funct. Genet.*, **9**, 56-68 (1991).

[19] S.A. Benner and D. Gerloff, Patterns of divergence in homologous proteins as indicators of secondary and tertiary structure : a prediction of the structure of the catalytic domain of protein kinases, *Adv. Enz. Regul.*, **31**, 121-181 (1990).

Delineating the mainchain topology of four-helix bundle proteins using the genetic algorithm and knowledge based on the amino acid sequence alone

Patrick Argos and Thomas Dandekar

European Molecular Biology Laboratory, Meyerhofstrasse 1, Postfach 10 22 09, 69012 Heidelberg, Germany

Abstract

The genetic algorithm was used to search mainchain conformational space in folding four-helix bundle proteins from knowledge derived from the amino acid sequence alone. The folding simulations were grid-free and relied on a selection of representative dihedral angular configurations for the backbone atoms. Successful fitness criteria and weights, optimized for an idealized helix bundle, included hydrophobic interactions, structural compactness without atomic clashes, and secondary structural nucleation regions predicted from the primary sequence. The proper mainchain fold for the all-helical proteins cytochrome b_{562}, hemerythrin, and cytochrome c' was achieved. The relative simplicity of the fitness function suggested that only a few basic energetic principles need be considered to achieve a molten-globule folding intermediate characterized by a lack of complete and specific sidechain interactions.

1   Introduction

Nature has produced (and continues to do so) surviving species in the Earth's planetary setting. Large phylogenetic spaces are scanned by introducing various combinational genetic species into the environment with the fittest, or at least those most facilely surviving, providing the chromosomal information for succeeding and more optimal generations. The natural techniques used to achieve a high quality solution from the near infinite genetic possibilities include selection of optimal species, mutation within a given individual, and recombination (cross-over) amongst chromosomes. John Holland [1] originally introduced these natural adaptation methods under the guise of the "genetic algorithm" for the solution of technical problems requiring searches over large combinatorial character spaces such as gas pipeline flow and artificial intelligence [2].

The prediction of a protein's fold and structure from only a knowledge of its amino acid sequence is also a prime target for the genetic algorithm, especially given the near limitless possibilities theoretically available for the atomic coordinates generally ranging in the thousands; for example, there are about 10 atoms on average composing each residue and around 350 residues in an average protein. Nature has found and finds unique folds for unique and surviving amino acid sequences. In the work described here the genetic algorithm, based on a grid-free and three-dimensional system relying on standard and representative dihedral angles applied to the backbone atoms and taken from known tertiary protein structures, is applied to the amino acid sequence of three four-helix bundle proteins. Only simple and a few fitness criteria are utilized: hydrophobic interactions, nucleating secondary structural segments predicted by techniques reliant upon the primary structure alone, and structural compactness without atomic clashes. Selection of criteria and appropriate weights for the fitness terms were determined from an idealized four-helix bundle topology. Details of this effort can be found in Dandekar & Argos [3]. Previous efforts in this area include that of the present authors [4] who used the genetic algorithm for various sequence design tasks and grid-bound and idealized folding applications, Unger & Moult [5] who demonstrated that genetic algorithms are superior to Monte Carlo simulations in searching conformations for two-dimensional protein models, and Sun [6] who relies on full energy terms in the fitness function to elicit tertiary structure for very small proteins such as mellitin.

2   Materials and Methods

2.1   Models and structures

The chromosome was represented as a successive string of binary digits encoding for the various dihedral angles at the C_α atom of the mainchain from the N- to C-terminus. Several rotation angles (ϕ, Ψ) that characterize the conformational space utilized by known protein structures [7] were each assigned a specific binary code. The configurations included those of the α-helix (-65, -40); 3_{10} helix (-89, -1); β-strand (-117, 142); extended structure (-69, 140); glycine ((78, 20) and (103, -176)); and cis-proline often observed in turns (-82, 133). Standard peptide bond angles and distances were maintained amongst the mainchain atoms: C_α-C', 1.53Å; C'-O, 1.24Å; C'-N, 1.32Å; and N-C_α, 1.47Å. Sidechain atoms were not included though the entire backbone group was represented (C', O, C_α, N); nonetheless, sidechain characteristics such as hydrophobicity were assigned to each mainchain C_α atom for use in the fitness function.

Fitness criteria and weights were tested on an idealized four-helix bundle [8] with topology $\alpha_{10}L_5\alpha_9L_5\alpha_9L_5\alpha_{10}$ where α_N designates an α-helix of residue length N and similarly for L_N representing loop segments. Residues were assigned as hydrophobic (A) or non-hydrophobic (a) according to an amphipathic wheel [9] with pattern AaaAaaaAaa. Figure 1 illustrates the bundle fold and distribution helical C_α atom positions projected onto a circular wheel.

Simulation trials based on the use of the genetic algorithm and known primary structures were applied to three proteins with experimentally determined atomic coordinates; namely, hemerythrin from sipunculid worm [10]; cytochrome b_{562} from *E.coli* [11]; and cytochrome c' from *rhodospirillum molischianum* [12]. The respective codes given to the proteins in the Brookhaven data-bank of tertiary structures [13] are 1HMD, 256B, and 1CRN. The observed and predicted C_α positions were superposed by the method of McLachlan [14]. The overall root-mean-square distance deviation (RMSD) between structurally equivalent observed and simulated C_α atoms was used as a criterion for the accuracy of the predicted topology.

Predictions of secondary structural regions (α-helix, β-strand, coil) based on the amino acid sequence alone were taken from the method of Ptitsyn & Finkelstein [15]. This

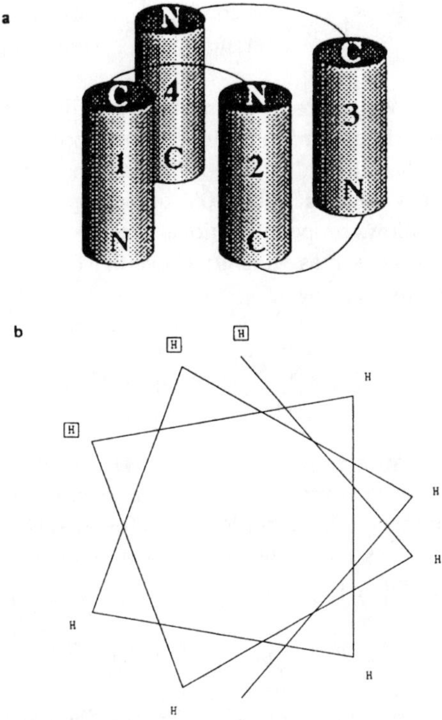

Figure 1: a) Schematic topology of the idealized four-helix bundle where helices are shown as cylinders with respective N- and C-termini along the primary sequence. Connecting loops are shown as thin, arched lines. b) Illustration of a helical 10-residue amphipathic wheel which is a projection of C_α atom positions at line intersections. The peak hydrophobic residues are boxed.

technique did not rely on sequence statistics that could have included the proteins tested here, thereby biasing the simulation results. Instead, their method was based upon stereochemical theory and hydrophobic templates.

2.2   Genetic algorithm

In the genetic algorithm procedures two parents were selected from the population with a probability of 0.2 and allowed one random crossover site. The number of generations allowed was set high to assure adequate convergence in the simulation runs (632 individuals for 632 generations); the mutation rate was set as one mutated bit per individual per generation.

Proper fitness criteria and associated weights are critical for successful optimization by the genetic algorithm. The fitness weights can be positive (reward) or negative (punishment) in sign. An idealized four-helix bundle as previously described was used to optimize the criteria and weights. The fitness function (ff) summed four terms:

$$ff = C + \text{clash} + \frac{\text{secondary}}{\text{structure}} + \frac{\text{tertiary}}{\text{structure}} , \tag{1}$$

The term C is defined by (W_C*lchrom) where W_C is a weight factor, asterisk (*) refers to multiplication, and lchrom is the chromosomal length corresponding to the number of binary slots required to represent the dihedral angles of the protein investigated. The constant C normalizes the overall fitness value such that 10% of the population has no positive fitness in the first generation; this allows proper evolutionary progress in converging to an optimal (fittest) population. W_C was chosen as +350 after several trials on the idealized bundle.

The clash term factors are given by

$$W_{CI} * \sum_{i=1}^{N} \sum_{j=i}^{N} \text{overlap}\,(i,j), \tag{2}$$

where overlap = 1 if the distance between the respective ith and jth C$_\alpha$ atoms is less than or equal to 3.8 Å, mimicking real proteins; otherwise, overlap = 0. N is the number of mainchain atoms. The optimal clash weight W_{CI} was found to be -500.

The secondary structure (SS) term which involved cooperativity (coop) and dihedral angular preference (pref) is defined by

$$W_{ss}(\text{coop} + \text{pref} - 38), \tag{3}$$

where two successive and initiating residues in the same dihedral state were each awarded +10 under coop, followed by successive additions of +1 for each adjacent extension of the same conformation in either sequential direction. If residues were in the same secondary structural state as that predicted, their pref value was assigned a value of +20; extensions of this state under the aforementioned coop conditions was also allowed beyond the predicted regions up to 3 residues within another nucleation region or extension site. The subtraction of 38 normalized according to the number of helical residues in the idealized bundle. The preferred weight W_{ss} for the secondary structure component was +14.

The tertiary structure term consisted of the addition of two compactness measures:

$$\left(W_{gs} * \frac{\text{global}}{\text{scatter}} \right) + \left(W_{hs} * \frac{\text{hydrophobic}}{\text{scatter}} \right), \tag{4}$$

where scatter is simply the sum of the distances of each C$_\alpha$ atom from the center-of-mass of the current fold. The global scatter took into account all C$_\alpha$ atoms while the hydrophobic term involved only those associated with strongly hydrophobic residues (Met, Ile, Leu, Val, Tyr, Cys, Phe) (16). Under optimal conditions, $W_{gs} = -24$ and $W_{hs} = -19$.

Fitness terms were also used to mimic the formation of hydrogen bonds between a mainchain carbonyl oxygen and peptide nitrogen amongst any two residues in the fold or between residues following a helical pattern (residue i to residue $i + 4$). The H-bond effects are not discussed as such fitness terms did not foster the fold of the idealized bundle, either by their exclusion from the fitness function built with the other criteria or from their inclusion with one or more of the remaining criteria deleted. Dandekar & Argos [3] provide details.

Hemerythrin

```
  1-50        ....,....1....,....2....,....3....,....4....,....5
  SEQ         GFPIPDPYCWDISFRTFYTIVDDEHKTLFNGILLLSQADNADHLNELRRC
                            AAA      AAAAAAAAAAAAAAAAAAAA   AAAAAAAAAA
              U U  tEEEeEeE  EEeEEthhhhhHHhhHHHH tttttthHhhHhhH

 51-100       ....,....1....,....2....,....3....,....4....,....5
  SEQ         TGKHFLNEQQLMQASQYAGYAEHKKAHDDFIHKLDTWDGDVTYAKNWLVN
              AAAAAAAAAAAAAA       AAAAAAAAAAAAAAA      AAAAAAAAAA
                   UU     UU     U   U           hHHhhH  Ttt HhHhhhHHHh

101-113       ....,....1...
  SEQ         HIKTIDFKYRGKI
              AA    AAA
              hH  EeEeE   U
```

Cytochrome b$_{562}$

```
  1-50        ....,....1....,....2....,....3....,....4....,....5
  SEQ         ADLEDNMETLNDNLKVIEKADNAAQVKDALTKMRAAALDAQKATPPKLED
               AAAAAAAAAAAAAAAAAA    AAAAAAAAAAAAAAAAAAA     AAA
               HhhhHhhHhhhHhHHh         HhhhHhhHh   U       tttU

 51-100       ....,....1....,....2....,....3....,....4....,....5
  SEQ         KSPDSPEMKDFRHGFDILVGQIDDALKLANEGKVKEAQAAAEQLKTTRNA
                 AAAAAAAAAAAAAAAAAAAAAAAAAAA    AAAAAAAAAAAAAAAAAA
               tt  ttHhhHhhhHhHHHHhhHhhhHHhHh tt U        hhhhHhh

101-106       .....,.
  SEQ         YHQKYR
              AAAAA
              U   U
```

Figure 2: Secondary structure predictions for hemerythrin and cytochrome b$_{562}$ according to the method of Ptitsyn and Finkelstein [15] are either β-strand (e or E), α-helix (h or H), turn (t or T) or coil (blank or U). It must be emphasized that only helical and strand predictions were used as nucleation regions with fixed dihedral angles; the turn predictions shown are only informative. Observed secondary helix structures are also indicated by an A. Peak hydrophobic residues according to the scale of Manavalan and Ponnuswamy [16] are marked by capital letters in the predicted regions (H, E, or T) or U if no prediction was given by Ptitsyn and Finkelstein (1983). SEQ refers to the amino acid sequence given in single letter code.

3 Results

3.1 Idealized bundle

The accomplishment of an idealized four-helix fold without clashes as depicted in Figure 1 was used to judge the significance of the several folding principles represented in the fitness function. Tests from various combinations of criteria and weights, including exclusion of particular terms and variable helical (8 to 16) and loop residue (3 to 6) lengths, showed that compactness of all residues with emphasis on the strongly hydrophobic ones was essential

Table 1: Comparison of C_α RMSD (Å) fits amongst observed and optimal simulated folds including loop atoms.

| | Hemerythrin | | Cytochrome b_{562} |
	simulated	observed	observed
Cytochrome b_{562} simulated	11.9	11.2	6.1
Cytochrome b_{562} observed	12.9	12.7	
Hemerythrin observed	6.7		

along with secondary structural nucleation sites (as small as three residues) for each helix of the bundle and subsequent cooperative extension. Nucleation sites at the helix termini were particularly effective.

3.2   Real all-α structures

The criteria and weights determined effective in the idealized case were then applied in genetic algorithm simulations involving folding of sequences from four-helix proteins with experimentally determined tertiary structures; namely, hemerythrin, cytochrome c' and cytochrome b_{562}. The Ptitsyn/ Finkelstein secondary structure prediction method was applied to each of the amino acid sequences; those segments predicted as helix and occasionally strand were assigned as preferred regions (Eq. (3)). Typical predictions are shown for hemerythrin and cytochrome b_{562} in Figure 2. In ten simulation trials for each of the proteins with different randomly chosen starting conformations, one-half generally achieved the proper four-helix topology, handedness and orientation; i.e., each helical axis was 20° within that observed, antiparallelity in successive helices along the sequence was maintained, and secondary structural length was within 1 or 2 turns of that observed for individual helices. In each case, the fold selected was that with the largest fitness value in the various populations. Folds with lower values gave higher C_α RMSD values relative to the observed structure and did not display proper topology.

Table 1 lists the RMSD values for all structurally equiva- lent simulated and observed C_α atom positions for each of the three protein structures tested and over all atoms excluding loops and only over each helix observed. It is clear that the residue phasing of all helices is proper except for helix I in hemerythrin which is rotated by one-residue. These results are achieved despite cases where only one-third of an observed helix sequence span was predicted, single helical predictions were interrupted by prediction of a turn span, or N- or C-terminal regions (or both) were not predicted or mispredicted. The hemerythrin four-helix fold was also accomplished despite a long symmetry breaking N-terminal extended region. It is clear that the genetic algorithm achieved near proper orientation of the helices despite many other possible outcomes. For example, largely skewed helices could well minimize scatter about the center-of-mass. The simultaneous action of all the criteria in the fitness function is critical.

Table 2: Comparison of structurally equivalent simulated and observed C_α atom positions using RMSD values given in Å.

Protein	with loops	without loops	individual helices (from N–terminus)			
(fittest fold)			I	II	III	IV
Cytochrome b_{562}	6.1	5.8	0.4	0.6	0.8	1.2
Hemerythrin	6.7	5.6	2.7	1.7	0.4	0.7
Cytochrome c'	6.1	3.9	1.4	1.2	0.4	0.9

Table 3: Comparison of absolute and component values in the fitness functions. Notes: The absolute values for the fitness function components as well as the total values obtained in the simulation for cytochrome c' are given. Values are also given for the observed folds where predicted and observed secondary structures are taken as the same. Further, fitness values averaged over the failed simulation trials are listed.

	secondary structure (minus clashes	tertiary structure gs	tertiary structure hs	total fitness (+ positive constant C)
Cytochrome c'	52604	-25782	-6353	62819
observed fold	53480	-26958	-5371	63501
failed simulations	42010	-24544	-5706	54110

The genetic algorithm responds to the specific sequence under test. Table 2 lists the RMSD values for equivalent and superposed C_α atoms in cytochrome b_{562} and hemerythrin over all possible cross comparisons of simulated and observed topologies. The two proteins as simulated possessed the same number of residues allowing direct equivalencing. The observed and simulated folds for a given sequence are much closer ($\sim$6Å RMSD) than are those, simulated or observed, for different sequences ($\sim$12Å). Exemplary values for the various criteria in an optimized cytochrome c' fitness function resulting from a successful simulation are listed in Table III. These values are compared to those obtained from observed and average failed structures. The observed fold produces the best (maximum) fitness value with the simulated close by while the failed structures are clearly lower. Hydrophobic interactions and compactness contribute about 40% of the optimal value while secondary structure nucleation and subsequent cooperativity are responsible for the remaining 60%.

Figure 3 shows a stereo mainchain trace of cytochrome b_{562} where the observed structure and optimal simulation can be compared. Figure 4 illustrates a failed hemerythrin simulation

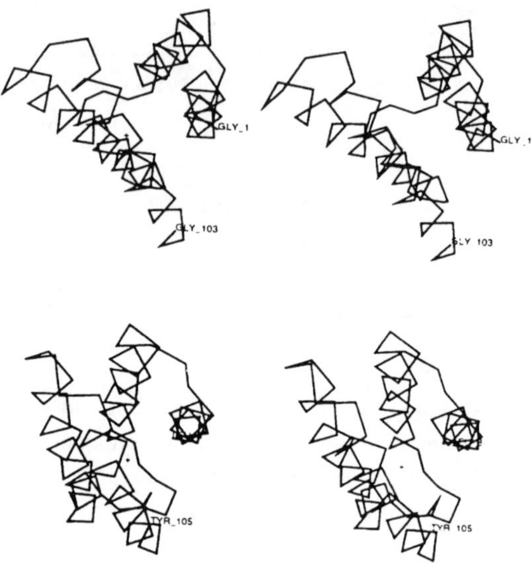

Figure 3: Fittest and observed folds (stereo view) of cytochrome b_{562}. The secondary structure predictions of Ptitsyn and Finkelstein [15] were used as nucleation sites (Figure 2). The top view illustrates the final fold in the simulation while the bottom is that of the experimentally determined fold (corresponding amino acids given). Virtual bonds connecting successive C_α atoms are shown as connecting lines.

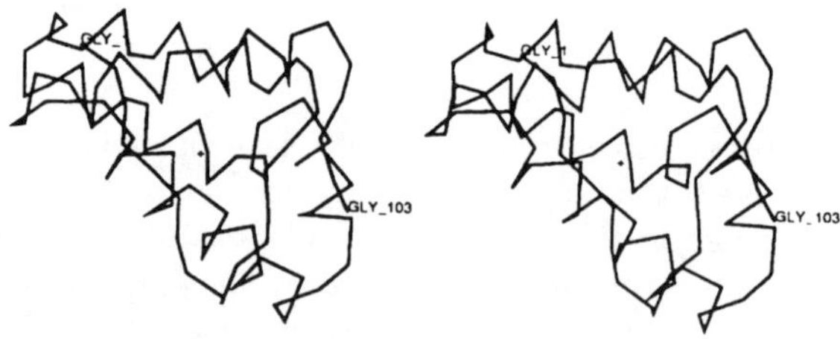

Figure 4: Example of a failed simulation trial for hemerythrin. Virtual bonds connecting sequential C_α atoms are shown as connecting lines in the stereo view.

where a bent region in one helix disturbs proper helix aggregation and topology for the bundle. Figure 5 shows exemplary and successive folding steps from early to late generations in the simulation of a small idealized helix bundle.

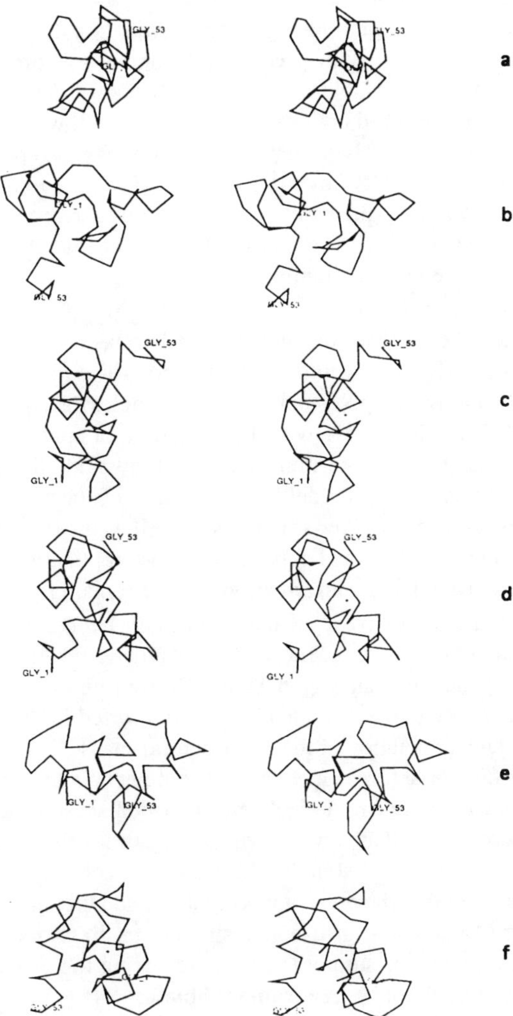

Figure 5: *Ab initio* folding simulation of a small idealized four-helix bundle as described in Material and Methods. A representative individual (a) is shown from the random start population, as are the fittest individuals after (b) 1, (c) 10, (d) 30, (e) 100 generations, and (f) the optimal fold at the simulation end. Virtual bonds connecting successive C_α atoms are given as connecting lines. Double images of each case are shown for a stereo perspective. Terminal residues are indicated as Gly.

4 Discussion

The results presented here show that the mainchain topology of small all-helical proteins can be identified by the genetic algorithm when used as a conformational search tool and coupled with surprisingly simple folding (fitness) rules and with only a knowledge of the protein's amino acid sequence and information derivable from it (hydrophobic residue distribution and secondary structural prediction). It would also seem possible to achieve such optimal folds for all β-strand and mixed strand/helix proteins given the often amphipathic character of extended structure and the ability of prediction techniques to provide nucleation sites of such conformations with reasonable accuracy. Inclusion of further fitness terms such as residue size, shape, and preferred contacts will hopefully allow better orientation of the substructural units to be achieved. Hydrogen bond criteria may also be necessary to induce formation of sheet structures. Further fitness terms gleaned from available experimental data could involve disulphide bridging, proximity of amino acids derived from crosslinking and spatially close catalytic residues and cofactor binding sidechains or hydrophobic residues well conserved in multiple sequence alignments.

The genetic algorithm could be the first step in folding proteins *ab initio* from primary structural information alone. If the mainchain trace has been sufficiently delineated by the genetic algorithm procedures, it would be possible to utilize homology modelling techniques which place sidechains onto given mainchain topologies through energy calculations. Eisenmenger et al. [17] provide a review and an example of such techniques. No doubt further development of the genetic algorithm and homology modelling methods is required for such applications. In homology modelling, relaxation of the assumed but partly erroneous backbone fold is required. Larger structures as well as those with various secondary structural types must enter the domain of genetic algorithm application. The calculationally problemsome increases in available conformational space remain formidable.

The illustrative folding pathway of an idealized helix bundle given in Figure 5 corresponds to suggestions made from recent nuclear magnetic resonance experiments (see Baldwin and Roder [18] and Chelvanayagam et al. [19] for references) on folding intermediates where some secondary structural nucleation is purported in the early folding phase, followed by extension and association of the substructural spans, first locally and then globally. The effectiveness of simple forces in eliciting the mainchain topology from the genetic algorithm is also consistent with the molten globule intermediate state [20], characterized by secondary structure formation and association yet lacking in specific sidechain interactions. Observation of such states implies that the backbone fold is achieved by relatively nonspecific interactions as also found in the present work. Chan and Dill [21, 22] and Skolnick and Kolinski [23] who use Monte Carlo grid-bound simulations in two- and three-dimensions also emphasize the importance of simple forces in achieving the proper fold though their particular grid topologies may not be free from fold biases [24].

Complex physico-chemical forces between sidechains, secondary structural propensity and the overall cooperativity of protein folding have been translated in this work into abstract and simple rules which rely only on knowledge of the protein amino acid sequence. For the foreseeable future, until the exact physicochemical forces are known and can be modelled and calculated with sufficient detail and speed, the genetic algorithm approach provides a way to bridge the gap between secondary structure predictions and tertiary fold.

Acknowledgement

The authors are grateful for a postdoctoral fellowship to Thomas Dandekar from the Deutsche Forschungsgemeinschaft without whose financial support this work could never have been accomplished.

References

[1] J. Holland, Adaptation in Natural and Artificial Systems. The University of Michigan Press, Ann Arbor, 1975.

[2] D.E Goldberg, Genetic algorithms in search, optimization and machine learning. Addison Wesley Publ., Reading, Massachu- setts, 1989.

[3] T. Dandekar and P. Argos, Folding the mainchain of small proteins with the genetic algorithm, *J. Mol. Biol.*, in press, (1994).

[4] T. Dandekar and P. Argos, Potential of genetic algorithms in protein folding and protein engineering simulations, *Protein Eng.*, **5**, 637-645 (1992).

[5] R. Unger and J. Moult, Genetic algorithms for protein folding simulations, *J. Mol. Biol.*, **231**, 75-81 (1993).

[6] S. Sun, Reduced representation model of protein structure prediction: Statistical potential and genetic algorithms, *Protein Sci.*, **2**, 762-785 (1993).

[7] M. J. Rooman, J.-P. A. Kocher and S. J. Wodak, Prediction of protein backbone conformation based on seven structural assignments, *J. Mol. Biol.*, **221**, 961-979 (1991).

[8] P. Argos, M. G. Rossmann and J. E. Johnson, A four-helical supersecondary structure, *Biochem. Biophys. Res. Commun.*, **75**, 83-86 (1977).

[9] M. Schiffer and A. B. Edmundson, Use of helical wheels to represent the structures of proteins and to identify segments with helical potential, *Biophys. J.*, **7**, 121-135 (1967).

[10] R. E. Stenkamp, L. C. Sieker and L. H. Jensen, Restrained least-squares refinement of *themiste dyscritum* methydroxohemerythrin at 2.0Å resolution, *Acta Cryst.*, **B38**, 784-792 (1982).

[11] F. Lederer, A. Glatigny, P. H. Bethge, H. D. Bellamy and F. S. Mathews, Improvement of the 2.5Å resolution model of cytochrome b_{562} by redetermining the primary structure and using molecular graphics, *J. Mol. Biol.*, **148**, 427-448 (1981).

[12] B. C. Finzel, P. C. Weber, K. D. Hardman and F. R. Salemme, Structure of ferricytochrome c' from rhodospirillum molischianum at 1.67Å resolution, *J. Mol. Biol.*, **186**, 627-643 (1985).

[13] F. C. Bernstein, T. F. Koetzle, G. J. B. Williams, E. F. jr. Meyer, M. C. Brice, J. R. Rodgers, O. Kennard, T. Shimanouchi and M. Tasumi, The protein data bank: A computer-based archival file for macromolecular structures, *J. Mol. Biol.*, **112**, 535-542 (1977).

[14] A. D. McLachlan, Gene duplications in the structural evolution of chymotrypsin, *J. Mol. Biol.*, **128**, 49-79 (1979).

[15] O. B. Ptitsyn and A. V. Finkelstein, Theory of protein secondary structure and algorithm of its prediction, *Biopolymers*, **22**, 15-25 (1983).

[16] P. Manavalan and P. K. Ponnuswamy, Hydrophobic character of amino acid residues in globular proteins, *Nature*, **275**, 673-674 (1978).

[17] F. Eisenmenger, P. Argos, and R. Abagyan, A method to configure protein sidechains from the mainchain trace in homology modelling, *J. Mol. Biol.*, **231**, 849-860 (1993).

[18] R. L. Baldwin and H. Roder, Characterizing protein folding intermediates, *Current Biology*, **1**, 218-220 (1991).

[19] G. Chelvanayagam, Z. Reich, R. Bringas, and P. Argos, Prediction of protein folding pathways, *J. Mol. Biol.*, **227**, 901-916 (1992).

[20] K. Kuwajima, The molten globule state as a clue for understanding the folding and cooperativity of globular-protein structure, *Proteins 6*, **87-103** (1989).

[21] H. S. Chan and K. A. Dill, Origins of structure in globular proteins, *Proc. Natl. Acad. Sci. USA*, **87**, 6388-6392 (1990).

[22] H. S. Chan and K. A. Dill, The protein folding problem, *Physics Today*, **46**, 24-32 (1993).

[23] J. Skolnik and A. Kolinski, Simulations of the folding of a globular protein, *Science*, **250**, 1121-1125 (1990).

[24] L. M. Gregoret and F. E. Cohen, Protein folding. Effect of packaging density on chain conformation, *J. Mol. Biol.*, **219**, 109-122 (1991).

Correlation between protein secondary structure and the mRNA nucleotide sequence

Søren Brunak, Jacob Engelbrecht and Can Kesmir

Center for Biological Sequence Analysis
Department of Physical Chemistry, Building 206,
The Technical University of Denmark, DK-2800 Lyngby, Denmark

Abstract

A complete search through GenBank was performed in order to locate mRNA sequences coding for structural entries included in the Brookhaven Protein Data Bank. When compared to the amino acid sequence, the mRNA sequence has an additional potential for carrying structural information due to the degeneracy of the genetic code. The search produced 319 protein chains with matching amino acid sequence, mRNA sequence and secondary structure assignment. Analysis of the data revealed strong signals in mRNA sequence regions preceding helices and sheets. These signals originate from common features in codons coding for very abundant amino acid residues at the N-termini outside of these structures. No correlation between the positioning of rare codons and the location of structural units of secondary structure was found.

1   Introduction

The problem of predicting the 3-dimensional functional state of a protein based on the information encoded in its primary structure is often referred to as "the second half of the genetic code". The focus of this work however has been the importance of the genetic code itself: is the final conformation of the polypeptide chain correlated with the specific coding sequence of nucleotides in the mRNA? Correlations may be present locally at the single codon level and also on a more global contextual scale covering many consequtive nucleotides.

Earlier work on cotranslational folding argued that protein secondary structures are separated by rare codons in order to guarantee high accuracy of translation in protein parts critical for their three dimensional structure[1, 2], and that pausing of ribosomes on mRNA's close to the junction between domains — at rare codons — might influence folding pathways[3, 4]. The protein folding pattern and the mRNA sequence may be correlated due to differentiation in the translational speed caused by the choice of codons[5, 6, 7, 8, 9]. Pausing during translation may give the unfinished polypeptide chain time to explore more deeply a limited conformational space and thus reach the correct final conformation more rapidly and reliably.

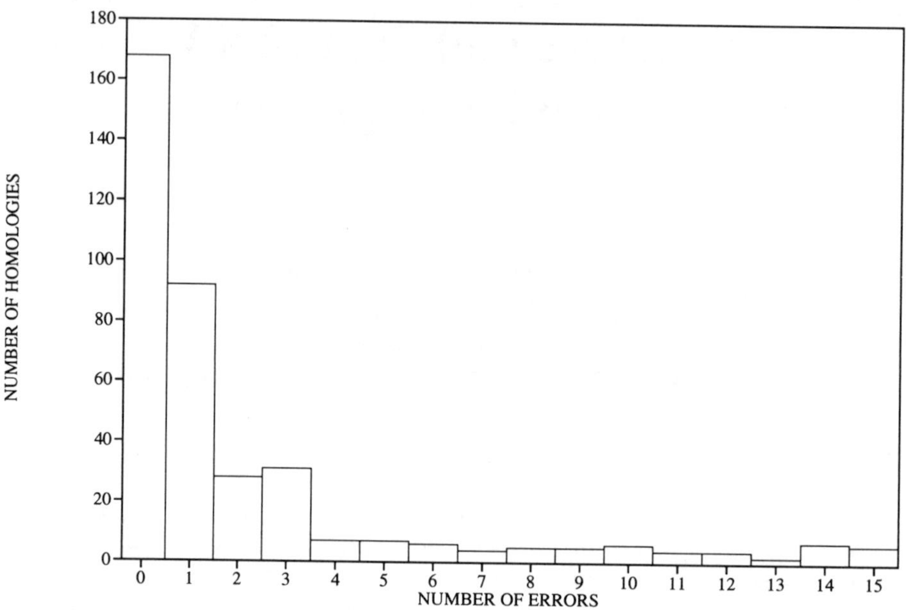

Figure 1: Histogram showing the number of strong PDB chain homologies found in GenBank and GenPept. The largest fraction of the sequences had mRNA sequences coding perfectly for the PDB sequence. A minor fraction had between one and three (amino acid) errors stemming primarily from posttranslational modification.

In contrast to this view, we find strong correlations between three features of protein formation: abundant amino acids, their non-rare codons and secondary structure.

2   Materials and Methods

Data sources and search methods

To construct the database, GenBank was searched for sequences consistently coding for polypeptide entries in the Brookhaven Protein Data Bank (PDB). Surprisingly, only a little over one half of the PDB sequences had GenBank counterparts with but a few errors in the coding sequence. Figure 1 shows the distribution of homologies found with fifteen or less (amino acid) errors. Errors in the range $1 - 3$ primarily stem from posttranslational modification, either of the newly synthesized protein, or later during protein purification. Many mismatches occur between glutamic acid and glutamine, and aspartic acid and asparagine. Some are caused by virus strain differences. Only chains with three or less errors (319) were considered for further analysis.

Only unique protein sequences in PDB were included in the search. Prior to the search the PDB sequences were assigned secondary structure by the Kabsch and Sander program DSSP[10]. Entries with incomplete coordinates not producing secondary structure output

Figure 2: See Color Plate, page xxii.

were discarded together with PDB model structures. The largest fraction of the sequences found had mRNA sequences coding perfectly for the PDB sequence. A minor fraction had between one and three (amino acid) errors stemming primarily from posttranslational modification.

The homologies were found by searching GenBank (rel. 76, April 93) and Brookhaven Protein Data Bank (ver. 61, July 1992) by the two programs tfasta and fasta[11]. The tfasta program compares a protein sequence with a DNA sequence library translated in all six reading frames, while fasta compares a protein sequence with a protein sequence library. The GenPept database (a translation of all annotated GenBank reading frames) was used for the fasta runs. The two search programs have different advantages and they both failed in some cases to find perfect homologies present in GenBank or GenPept. The database construction required a substantial amount of manual work. The DSSP program makes chain breaks not present in the PDB files due to the absence of some side–chain coordinates. A correction for this problem was implemented prior to the secondary structure assignment by comparing the original PDB sequences with the sequences produced by DSSP.

3   Results

As codon usage differs between organisms[12], the two largest homogeneous groups of data were extracted for separate analysis: enterobacteria (7378 residues) and mammals (14618 residues). Figure 2 shows how the 61 codons correlate with four specific conformational categories of secondary structure — helix, sheet, turn and random coil. Only unique sequences were included in the calculation. The categories are color coded: helix (red), sheet (green), turn (yellow), and random coil (blue). The histogram in black indicates the codon usage. The left pair of indicators represent data for enterobacteria, the right pair data for mammals; note the well known fact that the codon usage in enterobacteria is more skew than in mammals. (Percentages of amino acid content in the two groups were: A (9.9/6.9), C (1.0/3.0), D (6.5/5.7), E (6.5/6.0), F (3.6/4.2), G (7.8/7.9), H (2.2/2.5), I (6.5/4.8), K (6.7/6.7), L (8.8/8.7), M (2.3/2.0), N (4.6/4.5), P (4.1/4.6), Q (4.2/4.1), R (4.3/4.2), S (4.8/7.1), T (5.0/5.7), V (6.8/6.8), W (1.1/1.3) and Y (3.3/3.4). One stop codon (TGA) was found coding for the non-standard amino acid selenocysteine in the gene coding for glutathione peroxidase (1GP1_A/BOVGPX1)).

The data for the two groups show no indication of a preference for rarely used codons in the turn or coil conformation. The distribution of distances between rare codons in the two divisions was also calculated. When taking into account that genes for proteins expressed at different levels have variable contents of rare codons[13], the data provides no evidence for a deviation from the geometrical distribution $p^{n-1}(1-p)$. This means that no additional clustering of rare codons has been identified.

In order to explore to what extent the more information–rich mRNA sequences provided a better basis for the prediction of protein secondary structure[14, 15, 16, 17] a low–similarity subset of the data was extracted from the complete set[18]. The representative low–similarity subset of the data[18] had with a 30% similarity cutoff 167 chains; increasing the cutoff to 35% left 178 chains in the subset. Each set was subsequently divided in five different ways into a training set (2/3) and a testing set (1/3) and used as input data for the

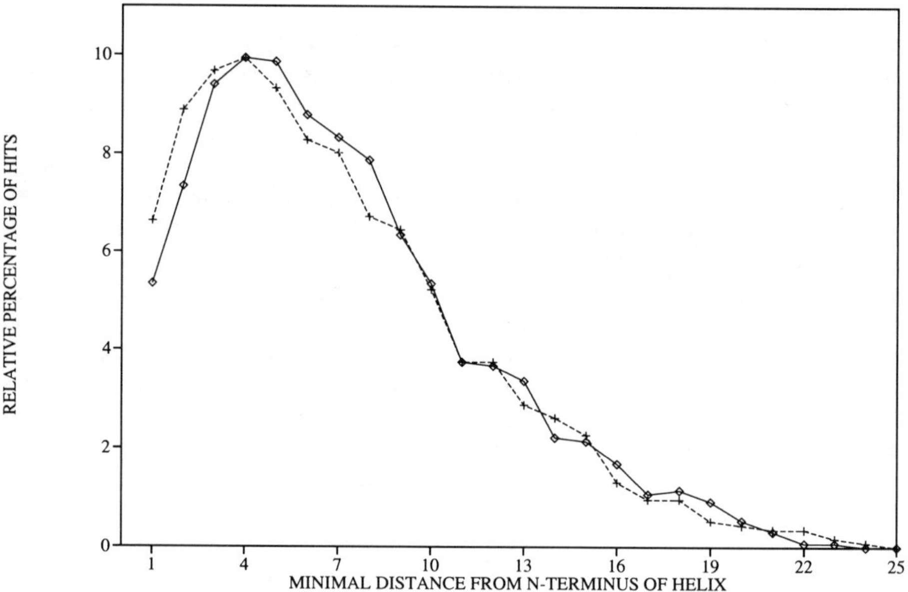

Figure 3: Predictive performance of network algorithms in the N–terminal part of helices given for the amino acid (——) and mRNA (- - - -) representations. In the initial parts of these secondary structures (where the statistics is best), the mRNA network clearly has a better relative performance compared to the conventional amino acid representation.

networks.

The representative subset served as input data for feed–forward neural network algorithms[19]. The architecture has been reviewed in many recent papers[15, 20, 21] where details may be found. Networks were trained on two versions of the data, one with matching amino acid sequence and secondary structure assignments, another with the amino acid representation replaced with the corresponding mRNA nucleotides.

Input windows in the two network types had conventional sparse coding[15, 19] and, in addition to the input, two layers of non–linear processing neurons. Input windows in the network architecture were varied from $7, 9, \cdots, 21$ amino acids, $(21, 27, \cdots, 63$ nucleotides respectively), and numbers of hidden units were varied in steps of one up to 50, testing 400 networks of each type. The networks were trained by presenting the window segments in the training set in random order. For the mRNA networks, best performance was obtained by keeping windows with non–truncated codons only. This reduced the total number of mRNA training examples by a factor of three making the training set size exactly the same as for the amino acid representation.

The average predictive performance (three–category prediction, helix, sheet, and coil) was best for a 13 amino acid window reaching correlation coefficients[22] of $C_\alpha = 0.46$, $C_\beta = 0.36$, and, $C_c = 0.43$. No improvement in the predictive performance of the mRNA networks could be obtained beyond a window of 33 nucleotides with $C_\alpha = 0.34, C_\beta = 0.28$,

Figure 4: See Color Plate, page xxiii and xxiv.

and, $C_c = 0.35$. When grouping nucleotides in the input representation as either purines (A/G) or pyrimidines (C/T) the performance of the mRNA networks did drop, but was still relatively good, while a coding based on strong (C/G) versus weak (A/T) hydrogen bonding was found to be useless. Altogether it means that in general it is much harder to predict protein secondary structure directly from the mRNA sequence than by a conventional approach based on the amino acid sequence. Apart from the missing correlation between specific codon usage and secondary structure two factors contribute to this situation. Firstly, the nucleotide space is larger than the amino acid space ($4^{3 \cdot 13} \gg 20^{13}$), and thus harder to cover by examples. Secondly, the genetic code fails to be linearly separable[19, 23] due to the scattering of codons for one amino acid, the six codons for serine.

This means that the mRNA networks in addition to learning the mapping between the amino acid representation and the secondary structure assignments must learn the genetic code mapping itself. This task is, non-linear, and requires at least two hidden units, as shown in ref. [23]. The fact that a genetic code network (nucleotide triplets as input and one of twenty amino acids as output) needs at least two hidden units to learn the code means that the code is non-linearly separable only. This holds true for the (otherwise sensible) sparse encoding of the nucleotides used here and also by most other workers. Computerised analysis of DNA or pre-mRNA striving to relate patterns in the nucleotides to amino acids[21] will therefore be a non-linear problem regardless of the algorithm applied.

4 Discussion

The performance of the overall superior amino acid networks is known[15, 24, 25] to be low at the N– and C– termini of helices and sheets. We investigated whether this is also the case for the mRNA representation and found the surprising result that the relative number of hits in the N–termini of helices and sheets is higher for this representation. The effect is stronger for helices, leading to the conclusion that the contrast in the coil–to–helix transition region is larger at the mRNA level than at the amino acid level. Figure 3 shows the percentage of hits as function of the distance (in amino acids) from the N-terminus of the helix. For the first, second and third positions in the helix the mRNA has a better performance.

This is also confirmed by showing the coil–to–helix transition region as a sequence logo[26]. Figure 4 shows the Shannon information content in the coil–to–helix region for enterobacteria *(a)* and mammals *(b)*. The largest sequence specificity of all the portions of the polypeptide and mRNA sequences was found here, with the position preceding the helix having the most specific population of amino acids and nucleotides. In figure 4 the coil region has negative positions and the helix region positive positions. The upper and lower panels in each section correspond to the mRNA and amino acid representations respectively. The height of the letters indicates their frequency at that position; the vertical bar gives the scale, 0.25 bits (mRNA) and 0.50 bits (amino acids). The height of each column is computed by the formula $I = \log_2(L_A) + \sum_{\alpha=1}^{L_A} p_\alpha log_2(p_\alpha)$, where L_A is the length of the alphabet (20 or 4). Using the probabilities p_α, the measure I quantifies the similarity of all amino acids in a column. The color scheme in the logos of proteins is based on the amino acid divisions: green (neutral and polar); blue (basic); red (acidic); black (neutral and hydrophobic). Analysis of the coil–to–sheet transition regions gave a similar result: strong amino acid and nucleotide specificity prior to the secondary structure. At the

C–termini of helices and sheets, the specificity for certain amino acids is not reflected in a strong signal at the mRNA level due to the variability of the codons involved (data not shown).

This finding agrees with the N–terminal part of the 'helix boundary hypothesis'[27]. Non–rare codons completely dominate the codons coding for the amino acids flanking the N–terminal of helices. The similarity of these codons gives a very strong signal at the mRNA level, a signal exploited by the mRNA network to obtain high performance in this region.

The known[28, 29] periodical preference for G and non-G nucleotides in the first and second codon positions (the framing code $(G\text{-}nonG\text{-}N)_n$) is clearly seen in Figure 4; a previously unnoticed strong preference for C and T in the third position in the last codon before the helix start is also clearly visible. Three complementary $(NNC)_n$ binding sites in the 16S ribosomal RNA are known[30, 31, 29] to be in contact with the mRNA during translation and are possibly responsible for maintaining the correct reading frame. Interestingly, the first anti-codon position in the complementary pattern of the mRNA binding sites in 16S rRNA is occupied almost solely by A's and G's. The (non-rare) codons preceding helices have therefore an additional potential for increasing the stability of the transient contacts with the rRNA, thus slowing down the sequential translocation of the ribosome.

Acknowledgements

We thank Drs. John Ipsen, Carsten Petersen and Henrik Nielsen for helpful discussions. This work was supported by the Danish Natural Science Research Council and the Danish National Research Foundation.

References

[1] I.A. Krasheninnikov, A.A. Komar, and I.A. Adzhubei. The role of clusters of rare codons in determining the boundaries of portions ofthe polypeptide chain with a monotypic secondary structure in the process of co-translational folding of the protein. *Dokl. Akad, Nauk. SSSR*, **303**, 995–999, 1988.

[2] I.A. Krasheninnikov, A.A. Komar, and I.A. Adzhubei. Nonuniform size distribution of nascent globin peptides, evidence for pause localization sites, and a cotranslational folding model. *J. Prot. Chem*, **10**, 445–453, 1991.

[3] I.J. Purvis and *et al*. The efficiency of folding of some proteins is increased by controlled tates of translation *in vivo* — a hypothesis. *J. Mol. Biol.*, **193**, 413–417, 1987.

[4] T. Crombie, J. Swaffield, and A.J.P. Brown. Protein folding within the cell is influenced by controlled rates of polypeptide elongation. *J. Mol. Biol.*, **228**, 7–12, 1992.

[5] P.M. Sharp, T.M.F. Tuohy, and K.R. Mosurski. *Nuc. Acids Res.*, **14**, 5125–5143, 1986.

[6] J. Kypr and J. Mràzek. Occurrence of nucleotide triplets in genes and secondary structure of the coded proteins. *Int. J. Biol. Macromol.*, **9**, 49–53, 1987.

[7] H. Liljenström and G. von Heijne. Translation rate modification by preferential codon usage: Intragenic position effects. *J. Theor. Biol.*, **124**, 43–55, 1987.

[8] G.C. Candelas, A. Ortiz, and N. Ortiz. *Biochem. Cell. Biol.*, **67**, 173–176, 1989.

[9] M.A. Sørensen, C.G. Kurland, and S. Pedersen. Codon usage determines translation rate in *escheria coli. J. Mol. Biol.*, **207**, 365–377, 1989.

[10] W. Kabsch and C. Sander. Dictionary of protein secondary structure: Pattern recognition of hydrogen–bonded and geometrical feature. *Biopolymers*, **22**, 2577–2637, 1983.

[11] R. Pearson. Rapid and sensitive sequence comparison with fastp and fasta. In R.F. Doolittle, editor, *Meth. Enz.* **183**, pages 63–98, New York, 1990. Academic Press, Inc.

[12] K. Wada, Y. Wada, H. Doi, F. Ishibashi, T. Gojobori, and T. Ikemura. Codon usage tabulated from the genbank genetic data. *Nuc. Acids Res.*, **19**, 1981–1986, 1991.

[13] P.M. Sharp and W.H. Li. Codon usage in regulatory genes in. *Nuc. Acids Res.*, **10**, 7737–7749, 1986.

[14] J. Garnier and J.M. Levin. The protein structure code: what is its present status? *CABIOS*, **7**, 133–142, 1991.

[15] N. Qian and T. J. Sejnowski. Predicting the secondary structure of globular proteins using neural network models. *J. Mol. Biol.*, **202**, 865–884, 1988.

[16] H. Bohr, J. Bohr, S. Brunak, R.M.J. Cotterill, B. Lautrup, L. Nørskov, O.H. Olsen, and S.B. Petersen. Protein secondary structure and homology by neural networks: The α-helices in rhodopsin. *FEBS Letters*, **241**, 223–228, 1988.

[17] B. Rost and C. Sander. Jury returns on structure prediction. *Nature*, **360**, 540, 1992.

[18] U. Hobohm, M. Scharf, R. Schneider, and C. Sander. Selection of representative protein data sets. *Protein Science*, **1**, 409–417, 1992.

[19] J. Hertz, A. Krogh, and R.G. Palmer. *Introduction to the theory of neural computation.* Addison–Wesley, Santa Fe Institute, Studies in the Sciences of Complexity, 1991.

[20] M.J. MacGregor, T.P. Flores, and M.J.E. Sternberg, M. J. E. Prediction of beta-turns in proteins using neural networks. *Protein Engineering,* **2**, 521–526, 1989.

[21] Prediction of human mRNA donor and acceptor sites from the DNA sequence. Brunak, S., Engelbrecth, J., & Knudsen, S. (1991). *J. Mol. Biol.*, **220**, 49–65.

[22] B.W. Mathews. Comparison of the predicted and observed secondary structure of t4 phage lysozyme. *Biochim. Biophys. Acta*, **405**, 442–451, 1975.

[23] N. Tolstrup, J. Toftgård, J. Engelbrecht and S. Brunak. Neural network model of the genetic code is strongly correlated to the GES scale of amino acid transfer free energies. *CBS-preprint*, 1994.

[24] S. Brunak. Doing sequence analysis by inspecting the order in which neural networks learn. In D.M. Soumpasis and T.M. Jovin, editors, *Computation of Biomolecular Structures — Achievements, Problems and Perspectives*, pages 43–54, Berlin, 1993. Springer–Verlag.

[25] S. Hayward and J.F. Collins. Limits on α-helix prediction with neural network models. *PROTEINS*, **14**, 372–381, 1992.

[26] T. D. Schneider and R. M. Stephens. Sequence logos: A new way to display consensus sequences. *Nucl. Acids Res.*, **18**, 6097–6100, 1990.

[27] L.G. Presta and G.D. Rose. Helix signals in proteins. *Science*, **240**, 1632–1641, 1988.

[28] E.N. Trifonov. Translation framing code and frame–monitoring mechanism as suggested by the analysis of mRNA and 16S rRNA nucleotide sequences. *J. Mol. Biol.*, **194**, 643–652, 1987.

[29] E.N. Trifonov. Recognition of correct reading frame by the ribosome. *Biochimie*, **74**, 357–362, 1992.

[30] M.I. Oakes and J.A. Lake. Dna–hybridization electron microscopy. localization of five regions of 16s rRNA on the surface of 30S ribosomal subunits. *J. Mol. Biol.*, **221**, 897–906, 1990.

[31] P. Wollenzien, A. Expert-Bezancon, and A. Fauvre. Sites of contact with 16S rRNA and 23S rRNA in the *escherichia coli* ribosome. *Biochem.*, **30**, 1788–1795, 1991.

Author Index